VOLUME SIXTY

THE ENZYMES

The Matrix Metalloproteinases

Series Editor

CLAUDIU T. SUPURAN

Neurofarba Department, Pharmaceutical and Nutraceutical Section, University of Florence, Sesto Fiorentino, Florence; Neurofarba Department, Sezione di Chimica Farmaceutica e Nutraceutica, Universita degli Studi di Firenze, Firenze, Italy

VOLUME SIXTY

THE ENZYMES

The Matrix Metalloproteinases

Edited by

CLAUDIU T. SUPURAN

Neurofarba Department, Pharmaceutical and Nutraceutical Section, University of Florence, Sesto Fiorentino, Florence; Neurofarba Department, Sezione di Chimica Farmaceutica e Nutraceutica, Universita degli Studi di Firenze, Firenze, Italy

ANTOINE DUFOUR

Department of Physiology and Pharmacology, Snyder Institute for Chronic Diseases, McCaig Institute for Bone and Joint Health, and Hotchkiss Brain Institute, University of Calgary, Calgary, Canada

ELSEVIER

ACADEMIC PRESS
An imprint of Elsevier
elsevier.com/books-and-journals

Academic Press is an imprint of Elsevier
125 London Wall, London EC2Y 5AS, United Kingdom
50 Hampshire Street, 5th Floor, Cambridge, MA 02139, United States

ISBN: 978-0-443-47088-2
ISSN: 1874-6047

For Information on all Academic Press publications visit our website at
https://www.elsevier.com/books-and-journals

Publisher: Zoe Kruze
Acquisitions Editor: Leticia Lima
Editorial Project Manager: Palash Sharma
Production Project Manager: Abdulla Sait
Cover Designer: Muthukumaran Thangaraj

Typeset by Aptara, New Delhi, India

Contents

3. Classical MMP inhibitors

Andrea Trabocchi

4. Matrix metalloproteinase-13 selective inhibitors

Simone Giovannuzzi and Claudiu T. Supuran

5. Tissue inhibitors of metalloproteinases (TIMPs)

Niccoló Paoletti, Antoine Dufour and Claudiu T. Supuran

Contributors

Pirjo Aström
Research Unit of Biomedicine and Internal Medicine, Faculty of Medicine, University of Oulu, Oulu, Finland; Biocenter Oulu, University of Oulu, Oulu, Finland

Kévin Baranger
La Rochelle Université, UMR 7266 CNRS LIENSs, La Rochelle, France

Cyrille Barthélémy
La Rochelle Université, UMR 7266 CNRS LIENSs, La Rochelle, France

Laurent Devel
CEA, INRAE, Médicaments et Technologies pour la Santé (MTS), SIMoS, Université Paris-Saclay, Gif-sur-Yvette, France

Antoine Dufour
Department of Physiology and Pharmacology, Snyder Institute for Chronic Diseases, McCaig Institute for Bone and Joint Health, and Hotchkiss Brain Institute, University of Calgary, Calgary, Canada

Marielle Frank
La Rochelle Université, UMR 7266 CNRS LIENSs, La Rochelle, France

Dimitris Georgiadis
Department of Chemistry, Laboratory of Organic Chemistry, National and Kapodistrian University of Athens, Panepistimiopolis, Zografou, Athens, Greece

Simone Giovannuzzi
NEUROFARBA Department, Pharmaceutical and Nutraceutical Section, University of Florence, Firenze, Italy

Krista Juurikka
Research Unit of Biomedicine and Internal Medicine, Faculty of Medicine, University of Oulu, Oulu, Finland; Faculty of Biochemistry and Molecular Medicine, University of Oulu, Oulu, Finland

Niccolò Paoletti
Neurofarba Department, Pharmaceutical and Nutraceutical Section, University of Florence, Sesto Fiorentino (FI), Italy

Konstantinos Patiniotis
Department of Chemistry, Laboratory of Organic Chemistry, National and Kapodistrian University of Athens, Panepistimiopolis, Zografou, Athens, Greece

Santiago Rivera
Aix-Marseille Univ, CNRS, INP, Inst Neurophysiopathol, Marseille, France

Tuula Salo
Oral Pathology, Research Unit of Population Health, Faculty of Medicine, Medical Research Center, University of Oulu and Oulu University Hospital, Oulu, Finland; Department of Oral and Maxillofacial Diseases, Translational Immunology Program (TRIMM), iCAN Digital Precision Cancer Medicine Flagship, University of Helsinki, Helsinki, Finland; Department of Pathology, Helsinki University Hospital, Helsinki, Finland

Mary Sharon Stack
Harper Cancer Research Institute, University of Notre Dame, South Bend, IN, USA

Brighton C. Strasiotto
Harper Cancer Research Institute, University of Notre Dame, South Bend, IN, USA

Claudiu T. Supuran
Neurofarba Department, Pharmaceutical and Nutraceutical Section, University of Florence, Sesto Fiorentino, FI, Italy

Andrea Trabocchi
Department of Chemistry "Ugo Schiff", University of Florence, Sesto Fiorentino, Florence, Italy

Jennifer Vandooren
Department of Microbiology, Immunology and Transplantation, KU Leuven, Kortrijk, Belgium

Preface

Volume 60 of *The Enzymes* presents the matrix metalloproteinases (MMPs). MMPs are a family of zinc-dependent endopeptidases first identified over 60 years ago through their roles and ability to degrade interstitial collagen during tadpole tail resorption. Since those early observations, the MMP family has expanded to 23 members in humans, broadly classified as collagenases, gelatinases, stromelysins, matrilysins, and membrane-type MMPs (MT-MMPs), together with a growing number of atypical members that fall outside these historical groupings. Although initially studied almost exclusively for their capacity to remodel the extracellular matrix (ECM), MMPs are now recognized as multifunctional proteases that process a broad range of non-ECM substrates, including cytokines, chemokines, growth factors, and cell-surface receptors, thereby regulating cell signaling, immune function, tissue repair, and development, well beyond simple ECM turnover. Dysregulated MMP activity is implicated in a wide range of diseases, not only in humans, among which cancer invasion and metastasis, arthritis, fibrotic disease, cardiovascular disease, neurodegeneration, oral disease, making this family of proteases a longstanding and still highly active target for drug discovery. Despite decades of effort, the clinical translation of MMP inhibitors has been hampered by challenges in achieving isoform selectivity and by the complex, at times opposing, roles that individual MMPs play across different tissues and disease stages. This volume brings together the current mechanistic, structural, and translational knowledge on MMPs, intending to clarify these challenges and highlight emerging strategies to overcome them.

This volume is divided into 10 chapters which address various topics related to MMP biology, detection, inhibition, and disease.

Chapter 1, by Dufour and Supuran, provides a general overview of the MMP family, introducing their classification, domain architecture, catalytic mechanism, and regulation, and setting the stage for the more focused chapters that follow.

Chapter 2, by Devel, examines activity-based probes designed to detect MMP activity, discussing how these chemical tools allow direct visualization and quantification of active, rather than merely present, MMPs in complex biological samples, an important distinction given that MMPs are extensively regulated at the post-translational level via zymogen activation and endogenous inhibition.

Chapter 3, by Trabocchi, reviews classical and most-studied MMP inhibitors, tracing the evolution of early broad-spectrum hydroxamate and non-hydroxamate zinc-binding inhibitors, the lessons learned from their limited clinical success, and the structural insights that paved the way for more selective drug design strategies.

Chapter 4, by Giovannuzzi and Supuran, focuses specifically on MMP-13-selective inhibitors, an MMP of particular interest in osteoarthritis and cancer, reviewing the structural determinants of selectivity and the most promising chemical scaffolds developed to date.

Chapter 5 by Paoletti et al. discusses the tissue inhibitors of MMPs (TIMPs).

Chapter 6, by Juurikka and Åström, reviews the role of MMPs in oral diseases, including periodontitis and oral cancers, describing how MMP-mediated remodeling of oral connective tissue and epithelium contributes to disease progression and how this informs diagnostic and therapeutic approaches.

Chapter 7, by Frank et al., examines the involvement of MMPs (with a focus on MT1-MMP and MT5-MMP) in neurodegenerative diseases, discussing their roles in blood-brain barrier integrity, neuroinflammation, and amyloid processing, and the therapeutic opportunities and challenges associated with modulating MMP activity in the central nervous system.

Chapter 8, by Vandooren, presents the role of MMPs in neutrophil biology, highlighting how neutrophil-derived MMPs, stored in specific granules and rapidly released upon activation, contribute to innate immune responses, tissue remodeling and inflammatory disease.

Chapter 9, by Strasiotto and Stack, discusses the role of MT1-MMP in ovarian cancer, describing how this membrane-anchored MMP promotes tumor cell invasion and dissemination within the microenvironment, and its potential as a therapeutic target in cancer.

Chapter 10, by Dufour and Supuran, closes the volume with a discussion of the challenges and opportunities in the drug design and pharmacological applications of MMP inhibitors, integrating the mechanistic and translational themes raised throughout the book and outlining the strategies most likely to finally deliver clinically viable, isoform-selective MMP-targeted therapeutics.

We are grateful to all authors who contributed to this book, providing insightful and quite detailed chapters. We also wish to thank Leticia Lima, the desk editor, Palash Sharma and the production editor, Abdulla Sait, for the constant help throughout the project.

Antoine Dufour and Claudiu T. Supuran
Calgary and Florence, 2026

CHAPTER ONE

Overview of matrix metalloproteinases (MMPs) in health and disease

Antoine Dufour[a,*], **Claudiu T. Supuran**[b]
[a] Department of Physiology and Pharmacology, Snyder Institute for Chronic Diseases, McCaig Institute for Bone and Joint Health, and Hotchkiss Brain Institute, University of Calgary, Calgary, Canada
[b] Neurofarba Department, Pharmaceutical and Nutraceutical Section, University of Florence, Sesto Fiorentino, FI, Italy
*Corresponding author. e-mail address: antoine.dufour@ucalgary.ca.

Contents

Abstract

Matrix metalloproteinases (MMPs) are a family of structurally related, zinc-dependent endopeptidases with broad substrate specificity, extending far beyond fibrillar collagens to include chemokines, cytokines, growth factors, cell-surface receptors and diverse extracellular matrix (ECM) components. In addition to their well-established proteolytic functions, MMPs can exert non-proteolytic effects through scaffolding roles, interactions with other membrane proteins, or modulation of cell–cell and cell–matrix contacts. Through these catalytic and non-catalytic mechanisms, MMPs influence cell signaling, immunity, barrier function, ECM remodeling, and tissue homeostasis, positioning them as central regulators in both normal physiology and a wide spectrum of acute and chronic diseases. Their dual capacity to either promote tissue repair and resolution

Fungal and Protozoan Carbonic Anhydrases, Volume 60
ISSN 1874-6047
https://doi.org/10.1016/bs.enz.2026.03.004

or drive inflammation, fibrosis and degeneration, together with the mixed history of broad-spectrum and selective MMP inhibitors in the clinic, has reframed MMPs from simple "matrix-degrading enzymes" to context-dependent, temporally regulated modulators that remain compelling yet challenging therapeutic targets. Importantly, many MMP family members remain poorly characterized in specific tissue, cellular or disease contexts, and for several MMPs, only a limited set of physiological substrates is known or confidently validated *in vivo*. This book will present contemporary aspects of MMP biology, their context-dependent roles in health and disease, current generations of MMP inhibitors, and both the conceptual and practical challenges, as well as emerging opportunities, in the rational design and development of MMP-targeted therapeutics.

Keywords: Cancer, Fibrosis, Inflammation, Matrix metalloproteinases, N-terminomics, Protease, Proteomics, Tissue inhibitors of metalloproteases

1. Introduction

Matrix metalloproteinases (MMPs) are a family of structurally related, zinc-dependent endopeptidases with broad substrate specificity, extending far beyond fibrillar collagens to include chemokines, cytokines, growth factors, cell-surface receptors and diverse extracellular matrix (ECM) components. In addition to their well-established proteolytic functions, MMPs can exert non-proteolytic effects through scaffolding roles, interactions with other membrane proteins, or modulation of cell–cell and cell–matrix contacts. Through these catalytic and non-catalytic mechanisms, MMPs influence cell signaling, immunity, barrier function, ECM remodeling, and tissue homeostasis, positioning them as central regulators in both normal physiology and a wide spectrum of acute and chronic diseases. Their dual capacity to either promote tissue repair and resolution or drive inflammation, fibrosis and degeneration, together with the mixed history of broad-spectrum and selective MMP inhibitors in the clinic, has reframed MMPs from simple "matrix-degrading enzymes" to context-dependent, temporally regulated modulators that remain compelling yet challenging therapeutic targets. Importantly, many MMP family members remain poorly characterized in specific tissue, cellular or disease contexts, and for several MMPs, only a limited set of physiological substrates is known or confidently validated *in vivo*. This book will present contemporary aspects of MMP biology, their context-dependent roles in health and disease, current generations of MMP inhibitors, and both the conceptual and practical challenges, as well as emerging opportunities, in the rational design and development of MMP-targeted therapeutics.

2. Overview of the history of matrix metalloproteinases (MMPs)

Work on collagenolytic activity from the 1960s laid the foundation for the MMP field when Woessner reported, in 1962, that a uterine protease catalyzed collagen breakdown during postpartum involution in mammals [1]. Shortly thereafter, Gross and Lapière demonstrated hormonally driven collagen turnover in tadpole metamorphosis, documenting extensive collagen degradation in the skin, gills, and gut and thereby providing the first description of what would later be recognized as a matrix metalloproteinase [2]. In 1966, a collagen-degrading enzyme isolated from tadpole tail fin and dorsal skin was characterized as "tadpole collagenase" at the time, and this protease is now recognized as MMP1 [3]. Over the next decade, related vertebrate collagenases were described and later renamed as MMP2 and MMP3, marking the initial expansion of the MMP family [4,5]. It was in 1988 that MMP2 was first cloned, sequenced and named 72-kDa type IV collagenase/gelatinase A to emphasize its activity toward denatured collagen and gelatin; it was subsequently purified from human rheumatoid synovial fibroblasts [6]. MMP3 was first reported as a 54-kDa "proteoglycanase" as it was shown to cleave proteoglycans and casein [7]. A few years later, MMP3 was purified from rabbit synovial fibroblasts and was named stromelysin [8].

The idea that MMPs form an interconnected protease network was introduced in 1987, when stromelysin (now called MMP3) was shown to convert latent or proMMP1 into its active form [9]. Proteases initially labelled MMP4, MMP5, and MMP6 were ultimately recognized as reidentifications of MMP1, MMP2, or MMP3 and were excluded from the standardized nomenclature. As additional enzymes were discovered and individual proteases accumulated multiple historical names, the term "matrix metalloproteinase" or MMP was proposed in the late 1980s and subsequently adopted by the International Union of Biochemistry and Molecular Biology. By 1991, seven human MMPs (now designated as MMP1, -2, -3, -7, -8, -9, and -10) had been identified, and completion of the human genome project established a total of 23 human MMPs with distinct domain organizations (Fig. 1).

3. MMP domains and structures

MMPs belong to the metzincin superfamily and share a conserved catalytic HEXXHXXGXXH motif (where H is histidine, E is glutamic

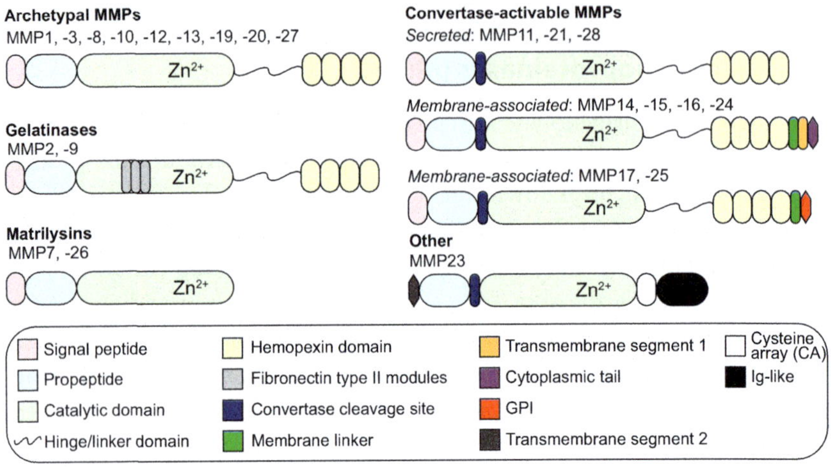

Fig. 1 All known domains of each of the 23 human MMPs.

acid, G is glycine, and X is any amino acid) that coordinates a catalytic zinc (Zn^{2+}) ion, yet individual family members differ markedly in domain composition, subcellular localization, and substrate range [5,10–12]. Humans express 23 MMPs that can be grouped into archetypal secreted enzymes, gelatinases, matrilysins, and membrane-type MMPs, each adapted to specific microenvironments through combinations of prodomain, catalytic, hemopexin, fibronectin-like, and membrane-anchoring modules (Fig. 1). Six MMPs are located at the cell surface and called membrane-type MMPs, or MT-MMPs (MT1-MMP or MMP14, MT2-MMP or MMP15, MT3-MMP or MMP16, MT4-MMP or MMP17, MT5- MMP or MMP24, and MT6-MMP or MMP25). There are two types of membrane-anchored domains: (i) via a transmembrane segment or (ii) via a glycosylphosphatidylinositol (GPI) anchor domain (Fig. 1).

The high structural homology of MMPs is seen in their catalytic domains. There is increased diversity of sequence homology in other domains, including the hemopexin domain and cytoplasmic tails. Therefore, the catalytic domain of MMPs regulates their proteolytic activity and the protein-protein interactions of MMPs are regulated by these other domains, which are more variable across the 23 MMPs. There is still little information known about some MMP domains, such as the Ig-like and CA domains of MMP23 and the GPI anchor of MMP17 and -25. It also remains unknown why MMP7 and -26 entirely lack a hemopexin domain. In MMPs, signal peptides, located at the N-terminus, vary in length and serve to direct newly synthesized proteins toward the secretory pathway. Following this, a prodomain of ~80 amino acids maintains the protease in an inactive state until it is cleaved by another protease or auto-activates to generate its active form [5,10]. The catalytic domain is typically ~170 amino acids long, contains the zinc-binding site and consists of five β-sheets and three α-helices [13,14]. Connecting this region to the hemopexin-like domain is a flexible hinge or linker segment, which ranges from 14 to 69 amino acids in length. The hemopexin domain itself comprises around 210 amino acids and adopts a four-bladed propeller configuration [13,15,16]. These differences in domain structures are likely tied to evolution, tissue specificity of certain MMPs and protein-protein interactors, which remain undercharacterized at the moment.

4. MMP proteolytic activity

MMPs are secreted as zymogens or latent enzymes and require a protease for activation. They contain a prodomain made of three α-helices, a connecting loop and a cysteine switch that blocks access to the catalytic domain [5]. The cysteine switch consists of a conserved sequence PRCGXPD that typically starts at amino acid 90 and ends at amino acid 99, where the processing occurs for removal [17,18]. The cysteine residue located at amino acid 92 interacts with the Zinc (Zn^{2+}) ion located in catalytic pocket of the active site. Upon MMP activation, this Cys-Zn^{2+} is lost and opens space for a water molecule to interact with the $Zn^{2+,}$ resulting in a fully active MMP.

To maintain tissue homeostasis and prevent unwanted proteolysis, there are several endogenous MMP inhibitors, including the four tissue inhibitors of metalloproteinases (TIMPs). TIMP1 was first identified in 1972 as a regulator of erythropoiesis and not as an MMP inhibitor *per se* [19]. It is in 1990 that TIMP2 was first identified and cloned [20]. TIMP3 was cloned in 1994 [21–25] and TIMP4 in 1996 [26]. The four TIMPs display varying degrees of glycosylation and have molecular weights ranging from 22 to 28 kDa [27]. Despite minor variations, mammalian TIMPs share a conserved structural organization, folding into a wedge-shaped configuration composed of two major domains [28]. The N-terminal domain contains 125 amino acids, while the C-terminal domain consists of 65 amino acids [29]. These domains are linked via six conserved disulfide bridges, which stabilize the structure and support distinct functional roles within the protein [30]. MMP inhibition occurs via the N-terminal domain when it binds to the catalytic cleft, coordinating with the zinc ion at the active site in a 1:1 inhibitor-to-enzyme ratio [28,31]. The C-terminal domain typically facilitates protein–protein interactions—most notably with the hemopexin domain of certain proMMPs [30]. Importantly, TIMPs act as dual regulators via MMP-dependent and -independent functions. They have been demonstrated to drive angiogenic functions, cell death, cell differentiation and cell movement [30]. Additionally, not all four TIMPs inhibit MMPs at the same rate. For example, TIMP1 is largely ineffective at inhibiting MMP14, MMP16 and MMP24 [30]. In contrast to TIMP1, TIMP3 can inhibit most MMPs in addition to other metalloproteases, such as a disintegrin and metalloproteinase (ADAM) and a disintegrin and metalloproteinase with thrombospondin motifs (ADAMTSs) families at varying degrees [30,32]. TIMP3 typically interacts with ECM proteins, whereas TIMP1, -2 and -4 are most often detected as soluble proteins. Interestingly, there are additional proteins that inhibit MMP functions, including a reversion-inducing cysteine-rich protein with Kazal motifs (RECK) and alpha-2 macroglobulin [33,34]. Likely, other interactors that may interfere with MMP functions have not been characterized yet. Therefore, MMPs' proteolytic activity is tightly controlled at several levels: stimulus- and cell type–specific transcription (dominated but not exclusively governed by AP–1 and other inflammatory transcription factors), zymogen activation via disruption of the cysteine switch or furin cleavage, and inhibition by TIMPs and other endogenous inhibitors [5]. This multilayered regulation allows MMPs to be rapidly induced during development, wound repair, or inflammation while limiting off-target proteolysis under homeostatic conditions.

5. Biological roles of MMPs and their substrates

MMPs were first discovered as enzymes that cleave collagen [10]. They were further defined by their ability to cleave fibrillar, basement-membrane collagens and other ECM proteins. With the rise of unbiased approaches such as proteomics and N-terminomics [35–38], numerous MMP substrates have been identified in the past two decades that are not related to ECM proteins. These substrates include chemokines [12,39–46], cytokines [47–51], growth factors [52–55], cell surface receptors [56–58], protease inhibitors [59], and intracellular proteins [60–62]. A full list of MMP substrates has been described previously in [5]. In fact, systematic proteomic mapping has shown that only about one-third of validated MMP substrates are classical ECM components, whereas the majority are soluble or membrane proteins that link proteolysis directly to cell communication and immune regulation [5,12,37,55]. Therefore, MMPs should be recognized as proteases able to process a wide diversity of both ECM and non-ECM substrates.

Tight control over MMP activity via TIMP inhibition, TIMP interaction via the hemopexin domain or other protein-protein interactions is therefore essential for determining which substrates MMP cleave, as their processing patterns vary depending on the cell type, tissue context, and pathological state. Shifts in the balance between MMPs and TIMPs may profoundly influence tissue homeostasis, the progression of disease and the resolution of inflammatory or fibrotic responses. For example, MMPs cleavage of chemokines or cytokines acts as molecular "on" or "off" switches, altering downstream signaling through receptors [55,63,64]. Through these proteolytic cleavages, MMPs act as upstream regulators of the recruitment of leukocytes to inflamed or infected sites. MMP8 and MMP9 derived from neutrophils were shown to cleave and activate interleukin-8 (IL-8/CXCL8) to generate a feed-forward loop that amplifies additional neutrophil recruitment [46,64]. Other MMPs, like MMP12, can terminate inflammatory signaling by cleavage of chemokines [43], interferon alpha (IFNα) [49] or interferon gamma (IFNγ) [48]. As a protease family, MMPs can both activate and deactivate chemokines, including CCL15 and CCL23, resulting in an enhancement of calcium (Ca^{2+}) signaling and increased chemotactic activity [39]. Collectively, MMPs exhibit broad substrate specificity, processing both extracellular matrix and non-ECM proteins, like chemokines, to fine-tune signaling networks and tissue remodeling. To understand their functions within a tissue or a disease context, all these factors need to be included in the assessment of their functions or the utility of using a broad-spectrum or selective MMP inhibitor.

6. Intracellular roles of MMPs

Although MMPs were initially identified as secreted enzymes, some studies have shown that their functional repertoire extends beyond this traditional view. There seem to be two mechanisms by which MMPs are located inside the cells. The first one is via alternative splicing, as shown in a 2002 study, which first demonstrated the presence of an intracellular form of MMP2 within cardiac myocytes [61]. In this specific context, due to the inefficient targeting of MMP2's signal peptide to the secretory pathway, approximately half of cellular MMP2 can remain within the cytosol, giving rise to three distinct intracellular isoforms produced via alternative splicing [62,65,66]. One of these MMP2 cytosolic variants lacks 50 amino acids at its N-terminus and is referred to as MMP2NTT50 and another splice variant, MMP2NTT76, misses 76 amino acids from its N-terminal region and was shown to be induced under oxidative stress conditions resulting to the activation of nuclear factor kappa-light-chain-enhancer of activated B cells (NF-κB) and nuclear factor of activated T-cells (NFAT) mitochondrial-to-nuclear stress signaling pathways [66]. Another example of a spliced isoform is MMP11, which was also identified in rhabdomyosarcoma cells [67].

The second way is via receptor endocytosis, such as clathrin-dependent or -independent mechanisms [60]. Examples of cargo molecules include synaptosomal-associated protein of 25 kDa (SNAP25) [68], low-density lipoprotein-related protein 1 (LRP1) [69–71] and nucleolin [49,72]. Although the mechanisms are not fully described yet, many MMPs can reach the cytosol or nucleus. For example, MMP3 displays an atypical intracellular localization. It is via a proline residue with MMP3's signal peptide that alters its conformation, leading to its presence in the nucleus as identified in human chondrosarcoma-derived chondrocytic HCS-2/8 cells [73]. Nuclear MMP3 has also been detected in HepG2 liver cancer cells and liver myofibroblasts, where it was shown to trigger cell death [74]. A similar role for MMP3 in cell death induction was reported in neurons [75]. In neurons, MMP10 was found to be proteolytically active and capable of cleaving huntingtin, resulting in cell toxicity and death [76]. During coxsackie virus infection, MMP12 was shown to increase NF-κB activity and IFNα production via a nuclear interaction [49]. The source of MMP12 was shown to be coming from macrophages, which also contain nuclear MMP14 that acts via a different signaling cascade: phosphoinositide 3-kinase delta (PI3Kd)/protein kinase B (Akt)/glycogen synthase kinase 3 beta (GSK3b) [77].

There are many reported MMP substrates, both outside and inside the cell, but only a handful have been fully validated *in vivo*, as most studies demonstrated these results using cell culture. It is likely that *in vivo* tissue architectures significantly impact what MMPs can have access to for cleavage, since factors such as ECM density, cellular organization, vascularization, and local mechanical forces can restrict or facilitate protease diffusion and substrate engagement. Depending on the type of organs and tissues, substrate diversity also impacts what MMPs can cleave, because each tissue has a distinct complement of ECM components, cell-surface receptors, and pericellular scaffolding proteins that together shape the local degradome and determine which candidate substrates are physically present, appropriately oriented, and biochemically permissive for proteolysis. In addition, cell-type–specific expression of cofactors, inhibitors such as TIMPs, and other proteases can modulate MMP activity and substrate processing, further emphasizing that substrate profiles defined in simplified systems may not accurately reflect the complexity of intact tissues. MMP substrates also tend to be different when comparing diseases versus healthy conditions, as pathological states such as chronic inflammation, fibrosis, cancer, and degenerative diseases remodel the extracellular and intracellular environment, alter gene expression programs, change protease and inhibitor levels, and expose neoepitopes that are not accessible under homeostatic conditions. These disease-associated changes can generate condition-specific substrate repertoires, shift cleavage site preferences, and lead to the production of bioactive fragments that drive feedback loops in tissue damage, repair, or signaling, thereby making it essential to define MMP substrates in the precise physiological or pathological context of interest. Therefore, additional work is needed to carefully validate all known (and still unidentified) MMP substrates, ideally by combining unbiased proteomic approaches, N-terminomics (degradomics), genetic models, substrate-trapping mutants, and *in vivo* labelling strategies with rigorous spatiotemporal analyses of proteolytic events, so that the field can move from catalogues of possible targets toward a mechanistically grounded, context-dependent understanding of physiologically relevant MMP–substrate relationships.

MMP expression and activity are shaped not just by enzyme abundance but also by where MMPs and their substrates are located in the body (Fig. 2). MMPs can reside inside cells, at the cell surface, within organelles, or in the extracellular space; therefore, only substrates that are co–localized are actually accessible for cleavage. Tissue architecture further constrains which substrates MMPs can reach; as shown for an epithelial surface, regional differences in MMP expression and localization along a tissue axis determine where cleavage can or cannot occur (Fig. 2). Different organs (such as the lung, liver, and joint) express distinct combinations of MMPs, inhibitors, and substrates, leading to tissue–specific proteolytic networks (Fig. 2). Between healthy and disease, the levels of MMPs, their inhibitors and substrates typically vary substantially and should be considered when evaluating MMPs as a drug target.

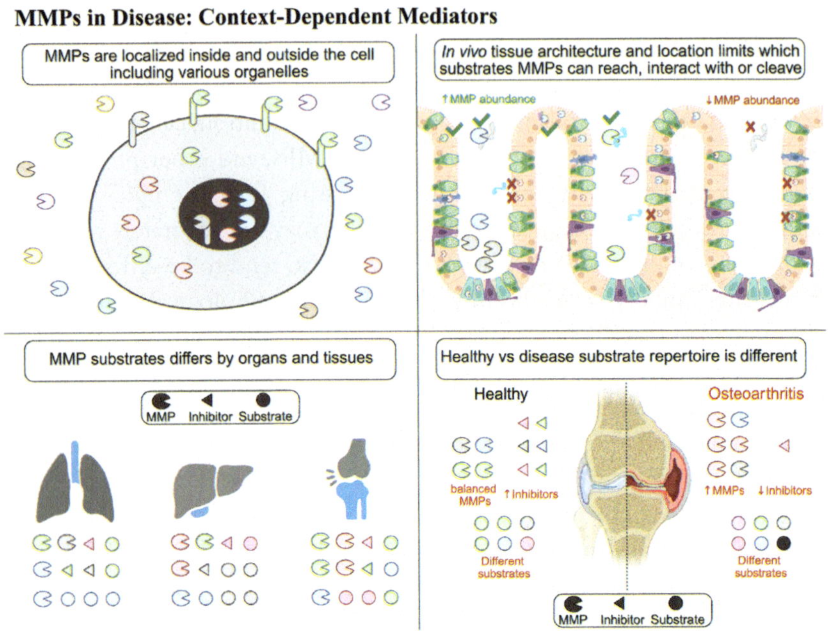

Fig. 2 MMPs in context of tissue homeostasis and human diseases.
Top left, MMPs have identified in multiple locations and organelles. *Top Right,* MMP abundance and their substrates may vary in tissue. This impacts what they can cleave as some substrates may not be present equally throughout the tissue. *Bottom left,* MMPs and their substrates are not equally expressed across organs and tissues. *Bottom Right,* The balance between MMPs and inhibitors is different between healthy and disease.

7. Cancer

Tumors are a dynamic milieu of interactors and often heterogeneous from patient to patient, comprising not only malignant cells but also stromal fibroblasts, endothelial cells, immune infiltrates, and ECM components that continually communicate through soluble factors, direct cell–cell contacts, and proteolytic remodeling [78,79]. This complex crosstalk generates highly variable biochemical and biophysical niches within and between tumors, so that two patients with the same histological cancer type can exhibit distinct molecular profiles, immune landscapes, and therapeutic vulnerabilities. Many proteases from both the cancer cells and immune cells promote tumor development and metastasis [78,80–83]. Interesting, not all proteases have detrimental roles and can be protective and should not be inhibited [80,84]. Importantly, MMPs have been demonstrated to be implicated, directly or indirectly, in virtually all major dysregulated pathways associated with cancer development and progression [5,85,86]. Over the past three decades, MMPs have been examined primarily for their roles in ECM degradation and activation of invasive programs that facilitate metastatic spread [55,80,81,86]. Although elevated MMP expression often correlates with advanced disease and poorer clinical outcomes, the initially dominant view of MMPs as uniformly deleterious in cancer has been substantially revised over the last twenty years. Early therapeutic strategies focused on broad inhibition of MMP proteolytic activity to curb ECM remodeling and thereby limit invasion and metastasis [81,87]. It is now evident that upregulation of an individual MMP does not automatically equate to enhanced tumor growth or dissemination, and at least ten MMP family members have been shown to exert context-dependent protective or tumor-suppressive effects [5,84]. Consequently, each MMP needs to be evaluated in detail, as its function is likely to depend on the specific cell type, tissue context, types of cancer and tumor setting.

Tumors have exceptionally high metabolic requirements and thus depend on a robust vascular network to deliver sufficient glucose and other nutrients [88]. To meet this need, they induce angiogenesis, the process by which new blood vessels sprout from existing ones, which represents a central hallmark of cancer progression [89,90]. Under hypoxic conditions, cancer cells upregulate and release proangiogenic growth factors such as the vascular endothelial growth factor alpha (VEGFA) to stimulate vessel growth and restore oxygen delivery [91]. MMPs are critical regulators of tumor-associated angiogenesis, and early attempts at broad MMP inhibition with compounds like batimastat (BB-94) were shown to block the angiogenic switch in premalignant tissues [92,93]. During sprouting angiogenesis, endothelial and infiltrating immune cells secrete MMPs and other proteases to remodel the surrounding ECM, enabling endothelial cell migration and new vessel formation [94]. For example, in pancreatic neuroendocrine tumors, increased MMP9 activity has been demonstrated to liberate VEGFA from the surrounding ECM, triggering a transition from a stable, non-sprouting vasculature to an actively angiogenic state [93]. Other MMPs (MMP1, -3, -7, -9, -14, -16, and -19) have been reported to modulate VEGF availability and thereby influence tumor vascularization [86]. Interestingly, proteolytic processing of ECM molecules, including collagen IV, collagen XVIII, and perlecan by MMP1, MMP2, MMP3, MMP9, or MMP13 can generate antiangiogenic fragments like tumstatin, endostatin, angiostatin, and endorepellin, which act to suppress new vessel growth [95]. Although many

MMPs participate in controlling angiogenesis, each individual enzyme tends to exhibit context-dependent roles that are specific to tissues, cell types, or types of tumors.

The ECM is an essential structural and signaling scaffold present in every tissue and organ, and when this balance is perturbed, such as in tumors with highly dynamic and dysregulated growth, ECM homeostasis is disturbed at biochemical, cellular, and structural levels [86,96]. Within tumors, ECM organization becomes profoundly altered and can exert both tumor-promoting and tumor-restrictive effects depending on the tumor types or the cancer stage. Elevated MMP8 expression illustrates this duality, correlating with favorable prognosis in oral squamous cell carcinoma and certain skin cancers, yet associating with worse clinical outcomes in ovarian, gastrointestinal, and hepatocellular malignancies [97–100]. The close link between MMPs and ECM remodeling impacts ECM degradation but also the liberation of membrane-bound cytokines, growth factors, and receptors, thereby modulating ECM integrity in an indirect manner as well [5,10,52,54,101,102]. Although many MMPs are capable of remodeling ECM proteins, strong evidence indicates that MMP14, MMP15 or MMP16 alone can be sufficient to support ECM transmigration and invasion [103–105]. Genetic studies in mice show that only *Mmp14*$^{-/-}$ knockout animals develop severe craniofacial abnormalities, arthritis, osteopenia, dwarfism, and soft tissue fibrosis linked to impaired type I collagen turnover and early postnatal death [106]. As per *Mmp15*$^{-/-}$ and *Mmp16*$^{-/-}$ mice remain viable with relatively mild phenotypes, supporting a dominant role for MMP14 as the main driver of collagen remodeling [104,107–109]. Careful examination of the role of each MMP is required to better understand their unique contribution to ECM remodeling in a tumor.

A unique feature of cancer cells is that they typically exhibit greater motility and migratory behavior than their normal (non-cancerous) counterparts [55,110]. This increase in migration ability is driven by MMPs in multiple ways. For example, in B-cell lymphocytic leukemia, the chemokine CXCL12 elevates MMP9 expression and promotes invasion and transendothelial migration through its interaction with the $\alpha4\beta1$ integrin [111]. Other chemokines, like CCL21, are similarly able to increase MMP9 levels to drive cell movement [112]. Paradoxically, when proMMP9 binds $\alpha4\beta1$ integrin, cell migration is suppressed based on MMP9 proteolytic activity, an effect not seen in healthy B cells [113]. When different MMPs are overexpressed in non-cancerous cells, they are shown to increase cell motility, even in the presence of broad-spectrum MMP inhibitors or added

TIMP1/TIMP2, indicating proteolysis–independent functions of MMPs to increase cell motility [114]. As different MMPs work via distinct mechanisms, some MMPs (MMP9 and -14) can enhance cell migration through interactions with the same proteins (e.g., CD44 and EGFR) but by binding to different regions of these proteins [115–117]. For MMP9, substitutions in amino acid residues within blades I and IV of its hemopexin domain reduce cell migration [115]. Drug-like peptides mimicking blade I or blade IV of the hemopexin domain were capable of inhibiting cancer cell migration [115]. For MMP14, these derived peptides contain an entirely different amino acid sequence within blade I and blade IV of its hemopexin domain, but were also able to significantly diminish cancer cell migration via similar mechanisms [116]. Therefore, the MMP-CD44-EGFR axis appears to be a key node in cancer cell migration and can be inhibited in various ways with different peptide inhibitors.

MMPs have been recognized for several decades as key contributors to virtually every major stage of tumor development, progression, and metastatic spread. Nonetheless, dissecting the precise function of each MMP remains difficult, including how they cooperate with one another and with other protease families via a protease web interactome, which substrates they process to activate or inactivate signaling pathways, and how these combined activities translate into overall harmful versus potentially protective effects within the tumor microenvironment [5,78,83]. As a result, despite intensive research into MMP inhibition as an anticancer strategy, neither broad-spectrum nor highly selective MMP–targeted drugs have yet gained regulatory approval for clinical use.

8. Acute inflammation

MMPs play a role in almost every phase of acute inflammation, from leukocyte recruitment and vascular leakage to cytokine activation and eventual resolution [5,118]. Importantly, their effects are context and time-dependent: the same MMP can sometimes amplify tissue injury or contribute to dampening inflammation and restoring homeostasis [43,46,55]. Acute inflammatory stimuli such as microbial products, cytokines, and hypoxia rapidly induce MMP gene expression and zymogen activation in resident and recruited cells [5,118]. Proinflammatory cytokines, including TNFα and IL–1β, upregulate MMP1, -3, -9, and often other MMPs in endothelial and stromal cells, creating a feed-forward loop [118,119]. Other proteases via proteolytic cascades involving plasmin, cathepsins, and other proteases can activate proMMPs. For example, plasmin activation of MMP9 is crucial for inflammatory macrophage migration in peritoneal inflammation [120]. Reactive oxygen species and mechanical stress can further modify MMP activity, while TIMPs and endogenous inhibitors (such as α2-macroglobulin) inhibit excessive proteolysis [118].

MMPs act early in acute inflammation to modulate endothelial and epithelial barrier integrity, thereby controlling exudate formation and leukocyte passage [121]. By cleaving junctional and ECM proteins, MMPs can increase vascular permeability and facilitate plasma protein leakage into tissues [122]. For example, MMP2 and -9 can cleave occludin, a tight junction, contributing to opening of the blood–brain barrier and enhancing leukocyte and fluid entry in CNS inflammation [47,123]. MMP7 can shed vascular endothelial cadherin and other adhesion molecules at endothelial junctions, weakening the adherens junctions that maintain barrier function [121].

MMPs have also been shown to enable neutrophils and monocyte–macrophages recruitment to traverse basement membranes and dense ECM, which is essential for rapid leukocyte influx in acute inflammation [5,43,46,120,121]. In many mouse models, peaks of activated MMP9 and -12 coincide with waves of neutrophil and then macrophage accumulation at inflamed sites [43,48,51,55,124]. Neutrophils store MMP8 and –9 in granules and release them upon activation, promoting their own migration through interstitial ECM matrices [43,51,58,124,125]. Beyond ECM degradation, MMPs precisely sculpt chemokine gradients that guide leukocytes within inflamed tissue [5,51,55,64,124]. Proteolytic processing can either enhance, reduce, or antagonize chemokine activity, thereby altering the magnitude and localization of leukocyte recruitment [12,39,41,47,126]. MMP-mediated cleavage of chemokines such as CXCL8 (IL–8) or CCLs can generate truncated forms with altered receptor specificity or potency, modulating neutrophil and monocyte chemotaxis [12,39,41,47,51]. Additionally, by shedding cell-surface heparan sulfate proteoglycans and accessory chemokine-binding proteins, MMPs influence how chemokines are retained or presented on endothelial and stromal surfaces [121–123]. Several MMPs directly process cytokines and cytokine precursors, thereby amplifying or reshaping acute inflammatory responses, including classic proinflammatory mediators such as TNFα and IL–1β [5,55]. MMP7 and –12 cleave membrane-bound proTNF–α to release soluble, active TNF–α from macrophages, augmenting local inflammation [127,128]. Multiple MMPs can activate or inactivate IL–1β and related cytokines, generating fragments with distinct biological activity and contributing to either escalation or attenuation of inflammation depending on context [5,55].

MMPs shape not only neutrophil entry but also their persistence and clearance at inflamed sites. In the lung, exogenous MMP12 can drive a rapid and severe neutrophil recruitment associated with transient spikes of inflammatory cytokines and chemokines in bronchoalveolar fluid [43,129]. MMP12 can regulate an apoptotic signal that terminates inflammation regulated by neutrophils [130]. In a collagen-induced arthritis in mice, MMP12 reduce neutrophils recruitments and the amount of neutrophil extracellular traps (NETS) [126]. Other examples include MMP–7-mediated cleavage of syndecan–1, which releases bound CXCL1, enhancing neutrophil recruitment to injured epithelium and intensifying acute responses [131].

Certain MMPs contribute to the resolution phase by promoting clearance of inflammatory cells, remodeling provisional ECM, and activating anti-inflammatory mediators [132]. For example, MMP9 can activate latent transforming growth factor–β (TGF–β), which suppresses inflammation and promotes tissue repair in later stages [133]. MMP8 has been implicated in both initiation and resolution of inflammation; *Mmp8*–deficient mice show altered neutrophil recruitment and delayed resolution in lung injury models [42,46,134]. Given their broad influence, MMPs are attractive but remain challenging drug targets for modulating acute inflammation. Early broad-spectrum MMP inhibitors showed dose-limiting toxicity and interference with physiological repair, emphasizing the need for selective strategies [5,121]. A key translational concept is to tailor MMP modulation to disease phase, aiming to dampen early barrier-disrupting and cytokine-amplifying actions while preserving or enhancing functions that promote resolution and tissue repair.

9. Chronic inflammation

Chronic inflammation arises when acute responses fail to resolve, leading to ongoing tissue injury, repair, and remodeling cycles in which MMPs are implicated. Persistent overexpression of MMPs has been implicated in most chronically inflamed tissues, including joints, gut, lung, liver, cardiovascular system, and chronically infected organs [5]. MMPs remodel the ECM, process cytokines and chemokines, and modulate cell–matrix interactions, thereby playing a key function in shaping the chronic inflammatory microenvironment [5,118,121]. MMPs help maintain continuous leukocyte recruitment by modifying chemokines, cytokines, and ECM–derived chemotactic fragments, and MMPs' activity ensures that inflammatory circuits do not switch off, even after the inciting trigger is controlled or removed [5,135]. For example, in inflammatory bowel disease (IBD), upregulated MMP8 and –9, together with neutrophil elastase, generate the matrikine N–acetyl–proline–glycine–proline (PGP), which drives persistent neutrophil infiltration and aggravates chronic colitis [135]. Additionally, MMP3 and –9, produced by colonic stromal and endothelial cells, are induced by cytokines and vasoactive mediators, integrating inflammatory signals into sustained leukocyte recruitment and microvascular activation [135,136]. In IBD, chronic relapsing gut inflammation with marked MMP overexpression has been demonstrated as many immune and stromal cells secrete MMPs in response to inflammatory cytokines and microbial products, producing diverse effects on barrier function and remodeling [5,137]. Individual MMPs often have distinct functions; therefore, they must be studied one MMP at a time to better understand their unique contribution. For example, MMP9 often exacerbates colitis by increasing epithelial permeability, driving neutrophil infiltration, and engaging Notch signaling that reduces goblet cell differentiation and mucin production [137]. By contrast, MMP2 appears protective for epithelial barrier integrity, whereas MMP10 and –19 have been implicated in promoting mucosal healing and balanced innate responses, indicating pro–resolution functions of specific MMPs [138].

In general, chronic overactivity of MMPs produces progressive structural damage that underlies many chronic inflammatory diseases: MMP1, –3, -9, -12 and –13 can mediate or directly degrade fibrillar collagens, proteoglycans, and basement membranes, eroding tissue integrity over years as in rheumatoid arthritis or osteoarthritis [5,139–142]. Additionally, MMPs participate in both promoting and modulating fibrotic remodeling via the regulation of epithelial–mesenchymal transition (EMT) and ECM turnover [143]. In lung, liver, and intestinal fibrosis, MMP9 and –13 have been shown to contribute to ECM remodeling and can act as bidirectional regulators, at times accelerating fibrogenesis and in other settings facilitating ECM degradation and remodeling [143,144].

10. Vascular diseases

MMPs are key regulators of ECM remodeling in the heart and vasculature, and their dysregulation contributes directly to atherosclerosis, aneurysm formation, myocardial infarction (MI) remodeling, heart failure, and hypertension–mediated organ damage [5,124]. MMPs can be both pathogenic and adaptive or reparative. In cardiovascular tissues, MMPs are produced by endothelial cells, vascular smooth muscle cells (VSMCs), fibroblasts, cardiomyocytes, and infiltrating inflammatory cells [145,146]. Their expression is induced by hemodynamic stress, angiotensin II (Ang II), oxidative stress, and inflammatory cytokines, linking classic risk factors to structural remodeling [61,62,122,147,148]. MMPs participate in all stages of atherosclerosis, from lesion initiation to plaque rupture [149]. By degrading collagen and other ECM components in the fibrous cap, MMPs, over time, weaken plaques and increase the risk of acute coronary syndromes [150]. Elevated circulating MMP1, –3, and –12 correlate with atherosclerotic burden and future cardiovascular and cerebrovascular events in patients with carotid or coronary disease [149]. In plaques, MMP2, –9, and –12 derived from macrophages and VSMCs degrade cap collagen and elastin, promoting cap thinning, intraplaque hemorrhage, and rupture [145,146,150,151].

MMP-driven vascular wall degradation in abdominal aortic aneurysm (AAA) and arterial dissection represents another example of their detrimental functions [146]. High MMP activity erodes elastic lamellae, reduces VSMC content, and facilitates progressive aortic dilation and risk of rupture [152]. MMP2 and –9 activation is markedly increased in aneurysmal and dissected aortas and correlates with aneurysm growth and severity in both humans and animal models [146,152]. Pharmacologic MMP inhibition and blockade of upstream Ang II–TGF–β signaling reduce MMP–2/–9 activity, slow aneurysm development, and improve aortic structural integrity in experimental models [146,152]. However, it has only been demonstrated in murine models and not in humans [153,154].

Chronic hypertension activates MMPs, driving vascular hypertrophy, ECM rearrangement, and increased arterial stiffness, as well as left ventricular (LV) remodeling [153]. The resulting structural changes initially compensate for pressure overload but ultimately contribute to diastolic dysfunction, dilation, and heart failure [154]. In hypertension, altered MMP2 and –9 levels are associated with abnormal collagen turnover, reduced arterial compliance, and LV hypertrophy [152,154]. MMP2 is constitutively expressed, whereas inducible MMP9 reflects inflammatory cell infiltration [152,154].

After MI, coordinated MMP activation remodels the infarcted and remote myocardium. Excessive or prolonged activity favors adverse LV dilation and progression to heart failure [124,155]. MMPs degrade interstitial collagen and basement membranes, altering chamber geometry and wall thickness [124,155,156]. Interestingly, circulating MMPs and TIMPs are potentially emerging as biomarkers for cardiovascular risk stratification and remodeling assessment. Elevated plasma MMP2 and –9 associate with hypertension–mediated organ damage, heart failure, and risk of recurrent events; however, this has not been translated into clinical usage [124,155,156].

11. Central nervous system diseases

MMPs have crucial and context–dependent roles in central nervous system (CNS) diseases, where they regulate blood–brain barrier (BBB) integrity, neuroinflammation, demyelination, synaptic plasticity, and neurodegeneration [5,47]. Many CNS disorders share an MMP–linked triad of processes—BBB disruption, ECM remodeling, and inflammation—that can be either injurious or beneficial depending on timing, localization, and specific MMPs involved [5,47]. In an intact CNS, MMPs participate in physiological development, neurogenesis, and synaptic plasticity [157,158]. Controlled MMP activity shapes axonal pathfinding and dendritic remodeling and contributes to learning–related synaptic rearrangements [157,158]. MMP2 and –9 were shown to regulate neuronal circuit maturation and oligodendrocyte development during CNS development and adult neurogenesis [157,159].

Many CNS diseases feature increased MMP activity that disrupts BBB integrity, allowing peripheral immune cells, plasma proteins, and toxins to enter the brain and spinal cord [157,160]. BBB breakdown is a central pathogenic step in neuroinflammation, stroke, CNS infections, and CNS involvement of systemic cancers or leukemias [157,158,160]. In multiple sclerosis (MS) and related neuroinflammatory disorders, MMPs mediate leukocyte entry, demyelination, and vascular pathology, but also contribute to subsequent repair [47,160]. Excess proteolytic activity is detected in blood and cerebrospinal fluid (CSF) of patients with acute MS and in experimental autoimmune encephalomyelitis (EAE) models [47,161–165]. Overall, in CNS diseases, MMPs are derived from multiple cell types, including neurons, astrocytes, microglia, oligodendrocytes, endothelial cells, and infiltrating leukocytes [160,166]. Importantly, microglia are major producers of MMPs in neuroinflammatory diseases and gliomas, contributing to both BBB breakdown and local tissue remodeling. Typically, tight regulation by TIMPs is often lost in disease, leading to unchecked proteolysis, barrier failure, and bystander damage to myelin and neurons. Given the central role of MMPs, they remain putative therapeutic targets and candidate biomarkers in

CNS diseases, but their dual functions complicate direct inhibition strategies. Timing, specificity, and delivery across the BBB are critical considerations that will need to be incorporated in future studies.

12. Infectious diseases

During bacterial or viral infections, MMPs contribute to defending the host but also to tissue-damaging immunopathology that favors pathogen persistence and spread [5,47,167]. During infection, pattern-recognition receptors (PRRs) sensing pathogen-associated molecular patterns (PAMPs) typically rapidly upregulate MMP gene expression in leukocytes and stromal cells [167,168]. Proinflammatory cytokines such as TNFα, IL–1β, and granulocyte-macrophage colony-stimulating factor (GM–CSF) further amplify MMP production by monocytes and macrophages exposed to bacterial products like lipopolysaccharide (LPS) and heat shock proteins [167,168]. Controlled MMP activity supports effective immune responses to microbes through multiple mechanisms [167–169]. For example, some MMPs will cleave basement membranes to open paths for neutrophils and mononuclear cells to reach infected sites [167,168]. MMPs also cleave most, if not all, chemokines and cytokines to trigger a recruitment signal or an "eat-me" signal for phagocytes; in other words, they activate or inactivate chemokines and cytokines, thereby shaping gradients and intensity of inflammation [12,39,41–49,126,170]. In pulmonary *Cryptococcus* infection, broad MMP activity promotes inflammatory cell infiltration into lungs, enhancing antifungal immunity, illustrating a general role for MMPs in leukocyte trafficking in infection [168,169]. MMP12's C–terminal hemopexin domain has microbicidal activity, supporting pathogen killing by macrophages [171]. *Mycobacterium tuberculosis* and *M. bovis* induce robust MMP2 and -9 production from macrophages *in vitro* and are seen in infected tissues, which is regulated mainly via TNFα, IL–18, and IFNγ [172]. In human tuberculosis, MMP expression in inflammatory cells is associated with pulmonary tissue destruction and cavitation, which facilitates bacillary transmission while representing harmful immunopathology for the host [167,168]. In gram–negative bacterial infections and septic shock, multiple MMPs are upregulated and contribute to vascular permeability, hypotension, and organ dysfunction, yet their precise roles need to be further validated in humans [167,168]. After pathogen clearance, MMP-mediated ECM remodeling contributes to wound healing and restoration of tissue architecture, limiting chronic inflammation [167,168].

Viral infections also trigger pronounced and context-dependent changes in MMP expression and activity that shape inflammation, tissue damage, and infectivity of the virus [173–175]. For example, in mouse hepatitis virus infection of the CNS was shown to induce MMP3, –9, and –12, which facilitated leukocyte infiltration across the BBB [173]. These MMPs were also implicated in myelin protein degradation and generation of autoantigens, linking viral infection to demyelination and chronic neuroinflammatory disease; importantly, TIMP–1 induction partially counterbalances these effects [173]. During viral infections, IFNα is among the best characterized cytokines and has crucial roles in immunobiology, being involved in the pathogenesis of many autoimmune conditions as well as viral and bacterial infections, but the full complexity of its regulatory networks and signaling mechanisms is still not completely defined. In cells infected with coxsackievirus B3 (CVB3) or respiratory syncytial virus (RSV), MMP12 was shown to relocalize to the nucleus of infected cells, where it binds to the inhibitor kappa B alpha (IκBα) promoter shortly after viral entry and is required for the production and release of antiviral IFNα. In mice lacking Mmp12, infection with normally nonlethal doses of CVB3 or RSV failed to induce adequate IFNα secretion, leading to mortality rates exceeding 30 percent, underscoring the importance of this pathway for host survival. In addition to this intracellular role, MMP12 exerts extracellular functions during viral infection by participating in a negative feedback mechanism in which it selectively cleaves IFNα at the C–terminal region that engages the IFNα receptor (IFNαR2), thereby shutting down IFNα signaling and limiting systemic interferon–mediated toxicity. In *Mmp9*$^{-/-}$ mice, infection with CVB3 caused more extensive heart damage and larger infectious foci than in wild-types, whereas mice deficient in *Mmp8* did not differ from wild-type controls. The *Mmp9*$^{-/-}$ mice displayed a greater inflammatory cell infiltrate together with higher myocardial levels of IFNβ1 and IFNγ compared with wild-type counterparts [170]. Therefore, MMPs' regulation via precise cleavage of cytokines is critical for resolving some viral infections.

13. Techniques to study MMPs and identify their substrates

Many techniques have been developed to profile what substrates various MMPs can cleave and identified what site of cleavage at which this occurs [36,38,176]. Therefore, the terms degradomics [37] or

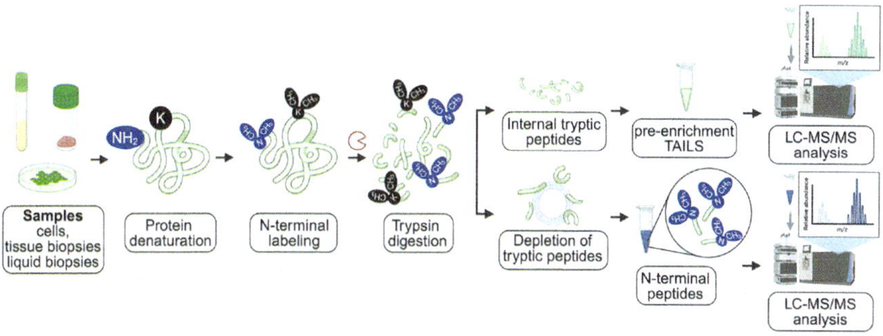

Fig. 3 Schematic representation of the N-terminomics/TAILS workflow. Cell culture, cells, tissue biopsies, or liquid biopsies can be analyzed. First, proteins from the samples are denatured, alkylated and the free cysteines are blocked. The N-termini are then blocked and labeled using reductive dimethylation. Trypsin is added to generated peptides. Samples are then split in two ways: 1) to analyze a pre-enrichment TAILS for peptide/protein abundance; and 2) to analyze the N-terminal peptides where a reactive polyaldehyde polymer is added for negative selection of tryptic peptides. Then, both samples are analyzed via liquid chromatography and tandem mass spectrometry (LC-MS/MS).

N-terminomics [36] were introduced to describe these techniques created to identified protease substrates. Several strategies now exist to isolate and analyze protein N-termini, typically relying on either positive or negative selection methods to achieve sufficient enrichment [Details covered in this review [36]]. One of the earliest positive enrichment methods, introduced in 2007 [177], used biotin–streptavidin chemistry. In this process, α-amines are first chemically modified through guanidination to block them from reacting further. Next, a reagent called sulfosuccinimidyl 2-(biotinamido)-ethyl-1,3-dithiopropionate (sulfo-NHS-SS-biotin) attaches a cleavable biotin label to the available amino groups. After digestion with trypsin, the newly generated N-terminal peptides are captured with streptavidin, and the linked peptides are then released by breaking the disulfide bonds. The added mass tag enables identification through liquid chromatography and tandem mass spectrometry (LC-MS/MS) analysis. This positive selection technique successfully enriched N-terminal peptides from a range of biological samples, including E. coli, yeast, mouse, human cell lysates, and even human serum. Since then, multiple N-terminomics approaches—using both enrichment strategies—have been developed, and several MMPs, but not all, have been profiled. TAILS or Terminal Amine Isotopic Labelling of Substrates has been used to profile multiple MMP substrates (Fig. 3). TAILS works via the study of peptide abundance and pinpoints specific cleavage events [38]. Traditional quantitative proteomics methods that rely on trypsin digestion face a limitation—trypsin cuts specifically after lysine (K) and arginine (R) residues. As a result, the resulting peptide mixture reflects this R/K specificity, making it difficult to differentiate between natural protease-generated fragments and those produced by trypsin itself during sample processing. TAILS overcame this challenge by allowing direct identification of cleavage substrates and their exact cut sites. Following the initial protease digestion, the key step involves chemically blocking and labelling all accessible primary amines (both α- and ε-amines) using isotopic reagents [38,176,178]. The labelled protein samples are then combined, precipitated, and digested with trypsin. Since the lysine ε-amines are already blocked, trypsin can only cleave after arginine residues, yielding peptides that show semi-ArgC specificity and thus help reveal newly formed N-termini. These samples can first be examined in a pre-enrichment or "shotgun" proteomics format via LC-MS/MS. Then, to enrich for peptides containing new N-termini, unlabeled internal peptides are selectively removed using a large aldehyde-reactive polymer in a negative selection step. Peptides that bind the TAILS polymer are trapped and separated by centrifugal filtration (typically using a 10 kDa cutoff filter). The peptides that pass through—enriched for new N-termini—are subsequently characterized by LC-MS/MS analysis. TAILS degradomics has shown that only about one quarter of validated MMP substrates are classical ECM proteins, whereas the majority are soluble or membrane proteins enriched for immune and inflammatory functions [5,37]. Non–ECM substrates include most CC chemokines, ELR+ CXC chemokines, complement components, coagulation factors, DAMPs, and cytokine-binding proteins, placing MMPs as higher-order regulators of leukocyte recruitment, barrier function, and resolution pathways. Key examples from chemokine biology illustrate this rheostat behavior: neutrophil MMP8 and MMP25 activate CXCL5 and CXCL8 and can relieve serpinA1–mediated inhibition of neutrophil elastase, creating a feed–forward loop for neutrophil recruitment [41,134]. In the context of acute inflammation, macrophage MMP12 subsequently inactivates all ELR+ CXC chemokines and converts multiple MCPs into broad CCR antagonists to terminate granulocyte and monocyte influx [43]. MMP2 and -9 also shape monocyte and lymphocyte trafficking by precisely truncating MCPs (CCL2/7/8/13), CCL15/23, CXCL12, and CX3CL1, toggling agonists to antagonists, shedding and inactivating fractalkine, and thereby coupling chemotactic gradients to tissue-specific resolution of inflammation and wound repair [39,55]. In the context of arthritis and peritonitis models,

MMP12 has been profiled using TAILS and hundreds of substrates linked to coagulation, complement activation, cytoskeleton, cell surface and many others were identified, demonstrating the broad range of substrates that some MMPs can cleave [126]. In tongue carcinoma, MMP8 substrates were profiled, and 36 new putative substrates were identified in this context [99].

14. Protease web and MMPs

Systematic *in vivo* studies and degradomics have reframed many MMPs from unidirectional tissue-destroying enzymes to nodes in an adaptive "protease web," where they activate other proteases, inactivate protease inhibitors, and thereby rewire entire cascades such as coagulation, complement, and kallikrein–kinin pathways [134,179]. For example, MMP2 cleavage of C1 inhibitor (serpin G1) in inflamed skin simultaneously enhances complement activation and bradykinin generation, increasing vascular permeability and exudation; genetic loss of MMP2 dampens exudation, lowers complement cleavage, and shifts the local inflammatory milieu [180].

Within this web, at least 10 of 24 murine MMPs display clear antitumorigenic or anti–inflammatory activities, making them bona fide drug "antitargets" whose inhibition can accelerate cancer, worsen autoimmunity, or impair host defence [84]. MMP2, -8, -9, -11, -12, -14, -19, -20, -25, and -28 all have validated beneficial roles in specific contexts (e.g., MMP8 restraining skin carcinogenesis and arthritis, MMP12 generating angiostatin and limiting lung metastasis, MMP2/9 controlling allergic lung inflammation and complement–driven exudation), such that broad inhibition risks blocking protective as well as pathogenic functions [84]. This spectrum of context–dependent "target" versus "antitarget" behavior is a major explanation for the failure of first–generation broad-spectrum hydroxamate inhibitors in oncology and has driven the field toward isoform–selective, exosite–directed, temporally constrained, or substrate-focused strategies. Therefore, MMP biology goes beyond a protease cleaving a single substrate and should be viewed as a protease web rippling effect that can impact substrates in a direct and indirect manners impact cascade of proteins and biological effects. Using proteomics and N-terminomics/degradomics technologies has revealed a much-extended role of MMP proteolysis in various disease processes. This makes the study of MMPs much more exciting than previously thought. There are still many MMPs that have never been profiled using these omics technologies, and still much work remains to understand tissue-specific MMP substrates and disease-specific MMP substrates.

15. Conclusion: MMPs as cell or immune rheostats

MMPs act as cell or immune rheostats by tuning the strength, duration, and quality of immune responses rather than simply turning inflammation "on or off." Through tightly regulated proteolysis of ECM and immune mediators, MMPs shape leukocyte development, migration, activation, and resolution of inflammation in health and disease. Interestingly, different MMPs perform vastly different functions, and there are still many understudied MMPs that need more in-depth characterization in diseases. The "neglected" MMPs include: MMP17, -19, -21, -23, -26, -27 and -28. Some MMPs have unique expression, such as MMP20, which is primarily expressed in developing tooth enamel by ameloblasts and odontoblasts, crucial for processing enamel matrix proteins to form hard, mineralized enamel [5,181]. MMP20 has also been identified in certain tumors but its function remains to be characterized [182].

The immune rheostat concept reflects how MMPs continuously adjust immune signaling thresholds in response to tissue context and inflammatory cues. Instead of behaving as purely destructive enzymes, many MMPs fine–tune cytokine gradients, cell–cell interactions, and ECM architecture, thereby calibrating effector responses and tissue repair [5,58,60,62,124]. MMP activity is controlled at multiple levels—transcription, zymogen activation, localization to specific microenvironments, inhibition by TIMPs, and post–translational modifications—allowing graded rather than binary control of immune outputs [10,85,101]. Because MMPs sit at key control points of immune regulation, their dysregulation contributes to diverse pathologies where the rheostat function is lost or mis-set. Viewing MMPs as cell or immune rheostats has important therapeutic implications, shifting emphasis from broad enzyme blockade to selective modulation of context-specific activities. Earlier generations of global MMP inhibitors caused off–target toxicities and impaired normal repair, reflecting disruption of both beneficial and harmful rheostat functions. This book is aimed at presenting details on the most promising approaches to inhibit MMPs, inhibitor development and how to test these inhibitors in preclinical models and clinical trials.

References

[1] J. Woessner, Catabolism of collagen and non-collagen protein in the rat uterus during post-partum involution, Biochem. J. 83 (2) (1962) 304–314, doi:10.1042/bj0830304.
[2] J. Gross, C.M. Lapiere, Collagenolytic activity in amphibian tissues: a tissue culture assay, Proc. Natl. Acad. Sci. 48 (6) (1962) 1014–1022, doi:10.1073/pnas.48.6.1014.

[3] Y. Nagai, C.M. Lapiere, J. Gross, Tadpole Collagenase. Preparation and Purification, Biochemistry 5 (10) (1966) 3123–3130, doi:10.1021/bi00874a007.
[4] I.E. Collier, S.M. Wilhelm, A.Z. Eisen, B.L. Marmer, G.A. Grant, J.L. Seltzer, et al., H-ras oncogene-transformed human bronchial epithelial cells (TBE-1) secrete a single metalloprotease capable of degrading basement membrane collagen, J. Biol. Chem. 263 (14) (1988) 6579–6587, doi:10.1016/S0021-9258(18)68680-6.
[5] L.G.N. de Almeida, H. Thode, Y. Eslambolchi, S. Chopra, D. Young, S. Gill, A.D. Laurent Devel, Matrix metalloproteinases: from molecular mechanisms to physiology, pathophysiology, and pharmacology, Pharmacol. Rev. 74 (3) (2022) 712–768, doi:10.1124/pharmrev.121.000349.
[6] Y. Okada, T. Morodomi, J.J. Enghild, K. Suzuki, A. Yasui, I. Nakanishi, et al., Matrix metalloproteinase 2 from human rheumatoid synovial fibroblasts. Purification and activation of the precursor and enzymic properties, Eur. J. Biochem. 194 (3) (1990) 721–730, doi:10.1111/j.1432-1033.1990.tb19462.x.
[7] W.A. Galloway, G. Murphy, J.D. Sandy, J. Gavrilovic, T.E. Cawston, J.J. Reynolds, Purification and characterization of a rabbit bone metalloproteinase that degrades proteoglycan and other connective-tissue components, Biochem. J. 209 (3) (1983) 741–752, doi:10.1042/bj2090741.
[8] J.R. Chin, G. Murphy, Z. Werb, Stromelysin, a connective tissue-degrading metalloendopeptidase secreted by stimulated rabbit synovial fibroblasts in parallel with collagenase. Biosynthesis, isolation, characterization, and substrates, J. Biol. Chem. 260 (22) (1985) 12367–12376, doi:10.1016/S0021-9258(17)39034-8.
[9] G. Murphy, M.I. Cockett, P.E. Stephens, B.J. Smith, A.J.P. Docherty, Stromelysin is an activator of procollagenase. A study with natural and recombinant enzymes, Biochem. J. 248 (1) (1987) 265–268, doi:10.1042/bj2480265.
[10] M.D. Sternlicht, Z. Werb, How matrix metalloproteinases regulate cell behavior, Annu. Rev. Cell Dev. Biol. 17 (1) (2001) 463–516, doi:10.1146/annurev.cellbio.17.1.463.
[11] N. Sela-Passwell, G. Rosenblum, T. Shoham, I. Sagi, Structural and functional bases for allosteric control of MMP activities: can it pave the path for selective inhibition? Biochim. Biophys. Acta. Mol. Cell Res. 1803 (1) (2010) 29–38, doi:10.1016/j.bbamcr.2009.04.010.
[12] A. Dufour, C.M. Overall, Subtracting matrix out of the equation: new key roles of matrix metalloproteinases in innate immunity and disease, in: Matrix Metalloproteinase Biology, John Wiley & Sons, Inc, Hoboken, NJ, USA, 2015, pp. 131–152, doi:10.1002/9781118772287.ch8.
[13] E. Morgunova, Structure of human pro-matrix metalloproteinase-2: activation mechanism revealed, Science 284 (5420) (1999) 1667–1670, doi:10.1126/science.284.5420.1667.
[14] S.H. Prior, Y.G. Fulcher, R.K. Koppisetti, A. Jurkevich, S.R. Van Doren, Charge-triggered membrane insertion of matrix metalloproteinase-7, supporter of innate immunity and tumors, Structure 23 (11) (2015) 2099–2110, doi:10.1016/j.str.2015.08.013.
[15] J. Li, P. Brick, M. O'Hare, T. Skarzynski, L. Lloyd, V. Curry, et al., Structure of full-length porcine synovial collagenase reveals a C-terminal domain containing a calcium-linked, four-bladed β-propeller, Structure 3 (6) (1995) 541–549, doi:10.1016/S0969-2126(01)00188-5.
[16] I. Bertini, V. Calderone, M. Fragai, R. Jaiswal, C. Luchinat, M. Melikian, et al., Evidence of reciprocal reorientation of the catalytic and hemopexin-like domains of full-length MMP-12, J. Am. Chem. Soc. 130 (22) (2008) 7011–7021, doi:10.1021/ja710491y.
[17] F. Mannello, V. Medda Nuclear localization of matrix metalloproteinases. Prog. Histochem. Cytochem.. 2012;47(1):27–58. doi:10.1016/j.proghi.2011.12.002

[18] H. Laronha, J. Caldeira Structure and function of human matrix metalloproteinases. Cells. 2020 ;9(5):1076. doi:10.3390/cells9051076
[19] E.A. Bauer, A.Z. Eisen, J.J. Jeffrey, Regulation of vertebrate collagenase activity in vivo and in vitro, J. Invest. Dermatol. 59 (1) (1972) 50–55, doi:10.1111/1523-1747.ep12625767.
[20] W.G. Stetler-Stevenson, P.D. Brown, M. Onisto, A.T. Levy, L.A. Liotta, Tissue inhibitor of metalloproteinases-2 (TIMP-2) mRNA expression in tumor cell lines and human tumor tissues, J. Biol. Chem. 265 (23) (1990) 13933–13938, doi:10.1016/S0021-9258(18)77438-3.
[21] S.S. Apte, M.G. Mattei, B.R. Olsen, Cloning of the cDNA encoding human tissue inhibitor of metalloproteinases-3 (TIMP-3) and mapping of the TIMP3 gene to chromosome 22, Genomics 19 (1) (1994) 86–90, doi:10.1006/geno.1994.1016.
[22] K.J. Leco, R. Khokha, N. Pavloff, S.P. Hawkes, D.R. Edwards, Tissue inhibitor of metalloproteinases-3 (TIMP-3) is an extracellular matrix-associated protein with a distinctive pattern of expression in mouse cells and tissues, J. Biol. Chem. 269 (12) (1994) 9352–9360, doi:10.1016/S0021-9258(17)37115-6.
[23] S.M. Silbiger, V.L. Jacobsen, R.L. Cupples, R.A. Koski, Cloning of cDNAs encoding human TIMP-3, a novel member of the tissue inhibitor of metalloproteinase family, Gene 141 (2) (1994) 293–297, doi:10.1016/0378-1119(94)90588-6.
[24] J.A. Uría, A.A. Ferrando, G. Velasco, J.M. Freije, C. López-Otín, Structure and expression in breast tumors of human TIMP-3, a new member of the metalloproteinase inhibitor family, Cancer Res. 54 (8) (1994) 2091–2094.
[25] C.G. Wilde, P.R. Hawkins, R.T. Coleman, W.B. Levine, A.M. Delegeane, P.M. Okamoto, et al., Cloning and characterization of human tissue inhibitor of metalloproteinases-3, DNA Cell Biol. 13 (7) (1994) 711–718, doi:10.1089/dna.1994.13.711.
[26] J. Greene, M. Wang, Y.E. Liu, L.A. Raymond, C. Rosen, Y.E. Shi, Molecular cloning and characterization of human tissue inhibitor of metalloproteinase 4, J. Biol. Chem. 271 (48) (1996), doi:10.1074/jbc.271.48.30375.
[27] R. Khokha, A. Murthy, A. Weiss, Metalloproteinases and their natural inhibitors in inflammation and immunity, Nat. Rev. Immunol. 13 (9) (2013) 649–665, doi:10.1038/nri3499.
[28] M.G. Masciantonio, C.K.S. Lee, V. Arpino, S. Mehta, S.E. Gill, The balance between metalloproteinases and TIMPs, Prog. Mol. Biol. Transl. Sci. 147 (2017) 101–131, doi:10.1016/bs.pmbts.2017.01.001.
[29] K. Brew, D. Dinakarpandian, H. Nagase, Tissue inhibitors of metalloproteinases: evolution, structure and function, Biochim. Biophy. Acta. Prot. Struct. Mol. Enzymol. 1477 (1–2) (2000) 267–283, doi:10.1016/S0167-4838(99)00279-4.
[30] K. Brew, H. Nagase, The tissue inhibitors of metalloproteinases (TIMPs): an ancient family with structural and functional diversity, Biochim. Biophys. Acta. Mol. Cell Res. 1803 (1) (2010) 55–71, doi:10.1016/j.bbamcr.2010.01.003.
[31] D.P. Jayawardena, N.P. Kulkarni, S.E. Gill, The role of tissue inhibitors of metalloproteinases in microvascular endothelial cell barrier dysfunction during sepsis, Metalloproteinases Med. 6 (2019) 1–12, doi:10.2147/MNM.S156245.
[32] H.W. Jackson, V. Defamie, P. Waterhouse, R. Khokha, TIMPs: versatile extracellular regulators in cancer, Nat. Rev. Cancer 17 (1) (2017) 38–53, doi:10.1038/nrc.2016.115.
[33] J. Oh, R. Takahashi, S. Kondo, A. Mizoguchi, E. Adachi, R.M. Sasahara, et al., The membrane-anchored MMP inhibitor RECK is a key regulator of extracellular matrix integrity and angiogenesis, Cell 107 (6) (2001) 789–800, doi:10.1016/S0092-8674(01)00597-9.

[34] R.P. Iyer, N.L. Patterson, G.B. Fields, M.L. Lindsey, The history of matrix metalloproteinases: milestones, myths, and misperceptions, Am. J. Physiol. Heart Circ. Physiol. 303 (8) (2012) H919–H930, doi:10.1152/ajpheart.00577.2012.
[35] A.R. Ziegler, N.E. Scott, L.E. Edgington-Mitchell, Advances in degradomics technologies to assess proteolytic cleavage events, Cell Chem. Biol. (2026), doi:10.1016/j.chembiol.2026.01.001.
[36] L. Wang, K. Main, H. Wang, O. Julien, A. Dufour, Biochemical tools for tracking proteolysis, J. Proteome Res. 20 (12) (2021) 5264–5279, doi:10.1021/acs.jproteome.1c00289.
[37] A. Dufour, Degradomics of matrix metalloproteinases in inflammatory diseases, Front. Biosci. 7 (1) (2015) 430, doi:10.2741/s430.
[38] O. Kleifeld, A. Doucet, U. auf dem Keller, A. Prudova, O. Schilling, R.K. Kainthan, et al., Isotopic labeling of terminal amines in complex samples identifies protein N-termini and protease cleavage products, Nat. Biotechnol. 28 (3) (2010) 281–288, doi:10.1038/nbt.1611.
[39] A.E. Starr, A. Dufour, J. Maier, C.M. Overall, Biochemical analysis of matrix metalloproteinase activation of chemokines CCL15 and CCL23 and increased glycosaminoglycan binding of CCL16, J. Biol. Chem. 287 (8) (2012) 5848–5860, doi:10.1074/jbc.M111.314609.
[40] T. Klein, U. Eckhard, A. Dufour, N. Solis, C.M. Overall, Proteolytic cleavage—mechanisms, function, and "Omic" approaches for a near-ubiquitous posttranslational modification, Chem. Rev. 118 (3) (2018) 1137–1168, doi:10.1021/acs.chemrev.7b00120.
[41] A.E. Starr, C.L. Bellac, A. Dufour, V. Goebeler, C.M. Overall, Biochemical characterization and N-terminomics analysis of leukolysin, the membrane-type 6 matrix metalloprotease (MMP25): chemokine and vimentin cleavages enhance cell migration and macrophage phagocytic activities, J. Biol. Chem. 287 (16) (2012) 13382–13395, doi:10.1074/jbc.M111.314179.
[42] J.H. Cox, A.E. Starr, R. Kappelhoff, R. Yan, C.R. Roberts, C.M. Overall, Matrix metalloproteinase 8 deficiency in mice exacerbates inflammatory arthritis through delayed neutrophil apoptosis and reduced caspase 11 expression, Arthritis Rheum. 62 (12) (2010) 3645–3655, doi:10.1002/art.27757.
[43] R.A. Dean, J.H. Cox, C.L. Bellac, A. Doucet, A.E. Starr, C.M. Overall, Macrophage-specific metalloelastase (MMP-12) truncates and inactivates ELR + CXC chemokines and generates CCL2, -7, -8, and -13 antagonists: potential role of the macrophage in terminating polymorphonuclear leukocyte influx, Blood 112 (8) (2008) 3455–3464, doi:10.1182/blood-2007-12-129080.
[44] G.A. McQuibban, G.S. Butler, J.H. Gong, L. Bendall, C. Power, I. Clark-Lewis, et al., Matrix metalloproteinase activity inactivates the CXC chemokine stromal cell-derived factor-1, J. Biol. Chem. 276 (47) (2001) 43503–43508, doi:10.1074/jbc.M107736200.
[45] D. Vergote, G.S. Butler, M. Ooms, J.H. Cox, C. Silva, M.D. Hollenberg, et al., Proteolytic processing of SDF-1 reveals a change in receptor specificity mediating HIV-associated neurodegeneration, Proc. Natl. Acad. Sci. 103 (50) (2006) 19182–19187, doi:10.1073/pnas.0604678103.
[46] A.M. Tester, J.H. Cox, A.R. Connor, A.E. Starr, R.A. Dean, X.S. Puente, et al., LPS responsiveness and neutrophil chemotaxis in vivo require PMN MMP-8 activity, PLoS One 2 (3) (2007) e312, doi:10.1371/journal.pone.0000312.
[47] S. Chopra, C.M. Overall, A. Dufour, Matrix metalloproteinases in the CNS: interferons get nervous, Cell. Mol. Life Sci. 76 (16) (2019) 3083–3095, doi:10.1007/s00018-019-03171-9.
[48] A. Dufour, C.L. Bellac, U. Eckhard, N. Solis, T. Klein, R. Kappelhoff, et al., C-terminal truncation of IFN-γ inhibits proinflammatory macrophage responses and

is deficient in autoimmune disease, Nat. Commun. 9 (1) (2018) 2416, doi:10.1038/s41467-018-04717-4.
[49] D.J. Marchant, C.L. Bellac, T.J. Moraes, S.J. Wadsworth, A. Dufour, G.S. Butler, et al., A new transcriptional role for matrix metalloproteinase-12 in antiviral immunity, Nat. Med. 20 (5) (2014) 493–502, doi:10.1038/nm.3508.
[50] G. Opdenakker, J. Van Damme, Cytokine-regulated proteases in autoimmune diseases, Immunol. Today 15 (3) (1994) 103–107, doi:10.1016/0167-5699(94)90151-1.
[51] P. Proost, S. Struyf, J. Van Damme, P. Fiten, E. Ugarte-Berzal, G. Opdenakker, Chemokine isoforms and processing in inflammation and immunity, J. Autoimmun. 85 (2017) 45–57, doi:10.1016/j.jaut.2017.06.009.
[52] K. Kessenbrock, V. Plaks, Z. Werb, Matrix metalloproteinases: regulators of the tumor microenvironment, Cell 141 (1) (2010) 52–67, doi:10.1016/j.cell.2010.03.015.
[53] B. Heissig, S. Rafii, H. Akiyama, Y. Ohki, Y. Sato, T. Rafael, et al., Low-dose irradiation promotes tissue revascularization through VEGF release from mast cells and MMP-9–mediated progenitor cell mobilization, J. Exp. Med. 202 (6) (2005) 739–750, doi:10.1084/jem.20050959.
[54] B. Heissig, K. Hattori, S. Dias, M. Friedrich, B. Ferris, N.R. Hackett, et al., Recruitment of stem and progenitor cells from the bone marrow niche requires MMP-9 mediated release of kit-ligand, Cell 109 (5) (2002) 625–637, doi:10.1016/S0092-8674(02)00754-7.
[55] D. Young, N. Das, A. Anowai, A. Dufour, Matrix metalloproteases as influencers of the cells' social media, Int. J. Mol. Sci. 20 (16) (2019) 3847, doi:10.3390/ijms20163847.
[56] V. Trivedi, A. Boire, B. Tchernychev, N.C. Kaneider, A.J. Leger, K. O'Callaghan, et al., Platelet matrix metalloprotease-1 mediates thrombogenesis by activating PAR1 at a cryptic ligand site, Cell 137 (2) (2009) 332–343, doi:10.1016/j.cell.2009.02.018.
[57] K.M. Austin, L. Covic, A. Kuliopulos Matrix metalloproteases and PAR1 activation. Blood. 2013;121(3):431–9. doi:10.1182/blood-2012-09-355958
[58] B. Cauwe, P.E. Van den Steen, G. Opdenakker, The biochemical, biological, and pathological kaleidoscope of cell surface substrates processed by matrix metalloproteinases, Crit. Rev. Biochem. Mol. Biol. 42 (3) (2007) 113–185, doi:10.1080/10409230701340019.
[59] A.R. Nelson, B. Fingleton, M.L. Rothenberg, L.M. Matrisian, Matrix metalloproteinases: biologic activity and clinical implications, J. Clin. Oncol. 18 (5) (2000) 1135, doi:10.1200/JCO.2000.18.5.1135.
[60] B. Cauwe, G. Opdenakker, Intracellular substrate cleavage: a novel dimension in the biochemistry, biology and pathology of matrix metalloproteinases, Crit. Rev. Biochem. Mol. Biol. 45 (5) (2010) 351–423, doi:10.3109/10409238.2010.501783.
[61] W. Wang, C.J. Schulze, W.L. Suarez-Pinzon, J.R.B. Dyck, G. Sawicki, R. Schulz, Intracellular action of matrix metalloproteinase-2 accounts for acute myocardial ischemia and reperfusion injury, Circulation 106 (12) (2002) 1543–1549, doi:10.1161/01.CIR.0000028818.33488.7B.
[62] W. Bassiouni, M.A.M. Ali, R. Schulz, Multifunctional intracellular matrix metalloproteinases: implications in disease, FEBS J. (febs) (2021) 15701, doi:10.1111/febs.15701.
[63] G.A. McQuibban, Inflammation dampened by gelatinase a cleavage of monocyte chemoattractant protein-3, Science 289 (5482) (2000) 1202–1206, doi:10.1126/science.289.5482.1202.
[64] P.E. Van den Steen, A. Wuyts, S.J. Husson, P. Proost, J. Van Damme, G. Opdenakker, Gelatinase B/MMP-9 and neutrophil collagenase/MMP-8 process the chemokines human GCP-2/CXCL6, ENA-78/CXCL5 and mouse GCP-2/LIX and modulate their physiological activities, Eur. J. Biochem. 270 (18) (2003) 3739–3749, doi:10.1046/j.1432-1033.2003.03760.x.

[65] M.A.M. Ali, A.K. Chow, A.D. Kandasamy, X. Fan, L.J. West, B.D. Crawford, et al., Mechanisms of cytosolic targeting of matrix metalloproteinase-2, J. Cell Physiol. 227 (10) (2012) 3397–3404, doi:10.1002/jcp.24040.
[66] D.H. Lovett, R. Mahimkar, R.L. Raffai, L. Cape, E. Maklashina, G. Cecchini, et al., A novel intracellular isoform of matrix metalloproteinase-2 induced by oxidative stress activates innate immunity, PLoS One 7 (4) (2012) e34177, doi:10.1371/journal.pone.0034177.
[67] D. Luo, B. Mari, I. Stoll, P. Anglard, Alternative splicing and promoter usage generates an intracellular stromelysin 3 isoform directly translated as an active matrix metalloproteinase, J. Biol. Chem. 277 (28) (2002) 25527–25536, doi:10.1074/jbc.M202494200.
[68] A. Szklarczyk, G. Oyler, R. McKay, C. Gerfen, K. Conant, Cleavage of neuronal synaptosomal-associated protein of 25 kDa by exogenous matrix metalloproteinase-7, J. Neurochem. 102 (4) (2007) 1256–1263, doi:10.1111/j.1471-4159.2007.04625.x.
[69] Z. Yang, D.K. Strickland, P. Bornstein, Extracellular matrix metalloproteinase 2 levels are regulated by the low density lipoprotein-related scavenger receptor and thrombospondin 2, J. Biol. Chem. 276 (11) (2001) 8403–8408, doi:10.1074/jbc.M008925200.
[70] L.J. Raggatt, S.C. Jefcoat, I. Choudhury, S. Williams, M. Tiku, N.C. Partridge Matrix metalloproteinase-13 influences ERK signalling in articular rabbit chondrocytes. Osteoarthr. Cartil. 2006;14(7):680–9. doi:10.1016/j.joca.2006.01.006
[71] P.E. Van den Steen, I. Van Aelst, V. Hvidberg, H. Piccard, P. Fiten, C. Jacobsen, et al., The hemopexin and O-glycosylated domains tune gelatinase B/MMP-9 bioavailability via inhibition and binding to cargo receptors, J. Biol. Chem. 281 (27) (2006) 18626–18637, doi:10.1074/jbc.M512308200.
[72] F. Tayyari, D. Marchant, T.J. Moraes, W. Duan, P. Mastrangelo, R.G. Hegele, Identification of nucleolin as a cellular receptor for human respiratory syncytial virus, Nat. Med. 17 (9) (2011) 1132–1135, doi:10.1038/nm.2444.
[73] T. Eguchi, S. Kubota, K. Kawata, Y. Mukudai, J. Uehara, T. Ohgawara, et al., Novel transcription factor-like function of human matrix metalloproteinase 3 regulating the CTGF/CCN2 gene, Mol. Cell Biol. 28 (7) (2008) 2391–2413, doi:10.1128/MCB.01288-07.
[74] K. Si-Tayeb, A. Monvoisin, C. Mazzocco, S. Lepreux, M. Decossas, G. Cubel, et al., Matrix metalloproteinase 3 is present in the cell nucleus and is involved in apoptosis, Am. J. Pathol. 169 (4) (2006) 1390–1401, doi:10.2353/ajpath.2006.060005.
[75] D.H. Choi, E.M. Kim, H.J. Son, T.H. Joh, Y.S. Kim, D. Kim, et al., A novel intracellular role of matrix metalloproteinase-3 during apoptosis of dopaminergic cells, J. Neurochem. 106 (1) (2008) 405–415, doi:10.1111/j.1471-4159.2008.05399.x.
[76] J.P. Miller, J. Holcomb, I. Al-Ramahi, M. de Haro, J. Gafni, N. Zhang, et al., Matrix metalloproteinases are modifiers of huntingtin proteolysis and toxicity in Huntington's disease, Neuron 67 (2) (2010) 199–212, doi:10.1016/j.neuron.2010.06.021.
[77] R. Shimizu-Hirota, W. Xiong, B.T. Baxter, S.L. Kunkel, I. Maillard, X.W. Chen, et al., MT1-MMP regulates the PI3Kδ·Mi-2/NuRD-dependent control of macrophage immune function, Genes Dev. 26 (4) (2012) 395–413, doi:10.1101/gad.178749.111.
[78] D.F. Quail, J.A. Joyce, Microenvironmental regulation of tumor progression and metastasis, Nat. Med. 19 (11) (2013) 1423–1437, doi:10.1038/nm.3394.
[79] I. Dagogo-Jack, A.T. Shaw, Tumour heterogeneity and resistance to cancer therapies, Nat. Rev. Clin. Oncol. 15 (2) (2018) 81–94, doi:10.1038/nrclinonc.2017.166.
[80] C.M. Overall, O. Kleifeld, Validating matrix metalloproteinases as drug targets and anti-targets for cancer therapy, Nat. Rev. Cancer 6 (3) (2006) 227–239.
[81] C.M. Overall, O. Kleifeld, Towards third generation matrix metalloproteinase inhibitors for cancer therapy, Br. J. Cancer 94 (7) (2006) 941–946.
[82] O.C. Olson, J.A. Joyce, Cysteine cathepsin proteases: regulators of cancer progression and therapeutic response, Nat. Rev. Cancer (2015), doi:10.1038/nrc4027.

[83] K.E. de Visser, J.A. Joyce, The evolving tumor microenvironment: from cancer initiation to metastatic outgrowth, Cancer Cell 41 (3) (2023) 374–403, doi:10.1016/j.ccell.2023.02.016.
[84] A. Dufour, C.M. Overall, Missing the target: matrix metalloproteinase antitargets in inflammation and cancer, Trends Pharmacol. Sci. 34 (4) (2013) 233–242, doi:10.1016/j.tips.2013.02.004.
[85] M. Egeblad, Z. Werb, New functions for the matrix metalloproteinases in cancer progression, Nat. Rev. Cancer 2 (3) (2002) 161–174, doi:10.1038/nrc745.
[86] Z. Piperigkou, K. Kyriakopoulou, C. Koutsakis, S. Mastronikolis, N.K. Karamanos Key matrix remodeling enzymes: functions and targeting in cancer. Cancers (Basel). 2021;13(6):1441. doi:10.3390/cancers13061441
[87] S. Zucker, J. Cao, W.T. Chen, Critical appraisal of the use of matrix metalloproteinase inhibitors in cancer treatment, Oncogene 19 (56) (2000) 6642–6650, doi:10.1038/sj.onc.1204097.
[88] I. Martínez-Reyes, N.S. Chandel, Cancer metabolism: looking forward, Nat. Rev. Cancer 21 (10) (2021) 669–680, doi:10.1038/s41568-021-00378-6.
[89] Z.L. Liu, H.H. Chen, L.L. Zheng, L.P. Sun, L. Shi, Angiogenic signaling pathways and anti-angiogenic therapy for cancer, Signal Transduct. Target. Ther. 8 (1) (2023) 198, doi:10.1038/s41392-023-01460-1.
[90] K. Kane, D. Edwards, J. Chen, The influence of endothelial metabolic reprogramming on the tumor microenvironment, Oncogene 44 (2) (2025) 51–63, doi:10.1038/s41388-024-03228-5.
[91] M. Potente, H. Gerhardt, P. Carmeliet, Basic and therapeutic aspects of angiogenesis, Cell 146 (6) (2011) 873–887, doi:10.1016/j.cell.2011.08.039.
[92] G. Bergers, Effects of angiogenesis inhibitors on multistage carcinogenesis in mice, Science 284 (5415) (1999) 808–812, doi:10.1126/science.284.5415.808.
[93] G. Bergers, R. Brekken, G. McMahon, T.H. Vu, T. Itoh, K. Tamaki, et al., Matrix metalloproteinase-9 triggers the angiogenic switch during carcinogenesis, Nat. Cell Biol. 2 (10) (2000) 737–744, doi:10.1038/35036374.
[94] M. De Palma, D. Biziato, T.V. Petrova, Microenvironmental regulation of tumour angiogenesis, Nat. Rev. Cancer 17 (8) (2017) 457–474, doi:10.1038/nrc.2017.51.
[95] R.V. Iozzo, J.J. Zoeller, A. Nyström, Basement membrane proteoglycans: modulators par excellence of cancer growth and angiogenesis, Mol. Cells 27 (5) (2009) 503–513, doi:10.1007/s10059-009-0069-0.
[96] J. Winkler, A. Abisoye-Ogunniyan, K.J. Metcalf, Z. Werb, Concepts of extracellular matrix remodelling in tumour progression and metastasis, Nat. Commun. 11 (1) (2020) 5120, doi:10.1038/s41467-020-18794-x.
[97] J.T. Korpi, V. Kervinen, H. Mäklin, A. Väänänen, M. Lahtinen, E. Läärä, et al., Collagenase-2 (matrix metalloproteinase-8) plays a protective role in tongue cancer, Br. J. Cancer 98 (4) (2008) 766–775, doi:10.1038/sj.bjc.6604239.
[98] B. Juurikka, N. Salo, Åström, The role of MMP8 in cancer: a systematic review, Int. J. Mol. Sci. 20 (18) (2019) 4506, doi:10.3390/ijms20184506.
[99] K. Juurikka, A. Dufour, K. Pehkonen, B. Mainoli, P. Campioni Rodrigues, N. Solis, et al. MMP8 increases tongue carcinoma cell–cell adhesion and diminishes migration via cleavage of anti-adhesive FXYD5. Oncogenesis. 2021;10(5):44. doi:10.1038/s41389-021-00334-x
[100] M. Balbín, A. Fueyo, A.M. Tester, A.M. Pendás, A.S. Pitiot, A. Astudillo, et al., Loss of collagenase-2 confers increased skin tumor susceptibility to male mice, Nat. Genet. 35 (3) (2003) 252–257, doi:10.1038/ng1249.
[101] J.D. Mott, Z. Werb, Regulation of matrix biology by matrix metalloproteinases, Curr. Opin. Cell Biol. 16 (5) (2004) 558–564, doi:10.1016/j.ceb.2004.07.010.

[102] C. Bonnans, J. Chou, Z. Werb, Remodelling the extracellular matrix in development and disease, Nat. Rev. Mol. Cell Biol. 15 (12) (2014) 786–801, doi:10.1038/nrm3904.
[103] R.G. Rowe, S.J. Weiss, Navigating ECM barriers at the invasive front: the cancer cell–stroma interface, Annu. Rev. Cell Dev. Biol. 25 (1) (2009) 567–595, doi:10.1146/annurev.cellbio.24.110707.175315.
[104] T.Y. Feinberg, R.G. Rowe, T.L. Saunders, S.J. Weiss, Functional roles of MMP14 and MMP15 in early postnatal mammary gland development, Development 143 (21) (2016), doi:10.1242/dev.136259.
[105] F. Sabeh, X.Y. Li, T.L. Saunders, R.G. Rowe, S.J. Weiss, Secreted versus membrane-anchored collagenases, J. Biol. Chem. 284 (34) (2009) 23001–23011, doi:10.1074/jbc.M109.002808.
[106] K. Holmbeck, P. Bianco, J. Caterina, S. Yamada, M. Kromer, S.A. Kuznetsov, et al., MT1-MMP-deficient mice develop dwarfism, osteopenia, arthritis, and connective tissue disease due to inadequate collagen turnover, Cell 99 (1) (1999) 81–92, doi:10.1016/S0092-8674(00)80064-1.
[107] K. Holmbeck, P. Bianco, S. Yamada, H. Birkedal-Hansen, MT1-MMP: a tethered collagenase, J. Cell Physiol. 200 (1) (2004) 11–19, doi:10.1002/jcp.20065.
[108] T.H. Chun, F. Sabeh, I. Ota, H. Murphy, K.T. McDonagh, K. Holmbeck, et al., MT1-MMP–dependent neovessel formation within the confines of the three-dimensional extracellular matrix, J. Cell Biol. 167 (4) (2004) 757–767, doi:10.1083/jcb.200405001.
[109] J. Shi, M.Y. Son, S. Yamada, L. Szabova, S. Kahan, K. Chrysovergis, et al., Membrane-type MMPs enable extracellular matrix permissiveness and mesenchymal cell proliferation during embryogenesis, Dev. Biol. 313 (1) (2008) 196–209, doi:10.1016/j.ydbio.2007.10.017.
[110] C.D. Paul, P. Mistriotis, K. Konstantopoulos, Cancer cell motility: lessons from migration in confined spaces, Nat. Rev. Cancer 17 (2) (2017) 131–140, doi:10.1038/nrc.2016.123.
[111] J. Redondo-Muñoz, E. Escobar-Diáz, R. Samaniego, M.J. Terol, J.A. Garciá-Marco, A. Garciá-Pardo, MMP-9 in B-cell chronic lymphocytic leukemia is up-regulated by $\alpha 4\beta 1$ integrin or CXCR4 engagement via distinct signaling pathways, localizes to podosomes, and is involved in cell invasion and migration, Blood 108 (9) (2006) 3143–3151, doi:10.1182/blood-2006-03-007294.
[112] J. Redondo-Muñoz, M. José Terol, J.A. García-Marco, A. García-Pardo, Matrix metalloproteinase-9 is up-regulated by CCL21/CCR7 interaction via extracellular signal-regulated kinase-1/2 signaling and is involved in CCL21-driven B-cell chronic lymphocytic leukemia cell invasion and migration, Blood 111 (1) (2008) 383–386, doi:10.1182/blood-2007-08-107300.
[113] J. Redondo-Muñoz, E. Ugarte-Berzal, J.A. García-Marco, M.H. del Cerro, P.E. Van den Steen, G. Opdenakker, et al., $\alpha 4\beta 1$ integrin and 190-kDa CD44v constitute a cell surface docking complex for gelatinase B/MMP-9 in chronic leukemic but not in normal B cells, Blood 112 (1) (2008) 169–178, doi:10.1182/blood-2007-08-109249.
[114] A. Dufour, N.S. Sampson, S. Zucker, J. Cao, Role of the hemopexin domain of matrix metalloproteinases in cell migration, J. Cell Physiol. 217 (3) (2008) 643–651, doi:10.1002/jcp.21535.
[115] A. Dufour, S. Zucker, N.S. Sampson, C. Kuscu, J. Cao, Role of matrix metalloproteinase-9 dimers in cell migration: design of inhibitory peptides, J. Biol. Chem. 285 (46) (2010) 35944–35956, doi:10.1074/jbc.M109.091769.
[116] K. Zarrabi, A. Dufour, J. Li, C. Kuscu, A. Pulkoski-Gross, J. Zhi, et al., Inhibition of matrix metalloproteinase 14 (MMP-14)-mediated cancer cell migration, J. Biol. Chem. 286 (38) (2011) 33167–33177.
[117] A. Dufour, N.S. Sampson, J. Li, C. Kuscu, R.C. Rizzo, J.L. DeLeon, et al., Small-molecule anticancer compounds selectively target the hemopexin domain of

matrix metalloproteinase-9, Cancer Res. 71 (14) (2011), doi:10.1158/0008-5472. CAN-10-4552.

[118] H.S. Lee, W.J. Kim, The role of matrix metalloproteinase in inflammation with a focus on infectious diseases, Int. J. Mol. Sci. 23 (18) (2022), doi:10.3390/ijms231810546.

[119] J.H.W. Distler, A. Jüngel, L.C. Huber, C.A. Seemayer, C.F. Reich, R.E. Gay, et al., The induction of matrix metalloproteinase and cytokine expression in synovial fibroblasts stimulated with immune cell microparticles, Proc. Natl. Acad. Sci. 102 (8) (2005) 2892–2897, doi:10.1073/pnas.0409781102.

[120] Y. Gong, E. Hart, A. Shchurin, J. Hoover-Plow, Inflammatory macrophage migration requires MMP-9 activation by plasminogen in mice, J. Clin. Invest. 118 (9) (2008) 3012–3024, doi:10.1172/JCI32750.

[121] B. Fingleton, Matrix metalloproteinases as regulators of inflammatory processes, Biochim. Biophys. Acta. Mol. Cell Res. 1864 (11) (2017) 2036–2042, doi:10.1016/j.bbamcr.2017.05.010.

[122] D.M. Tanase, E. Valasciuc, I.B. Anton, E.M. Gosav, N. Dima, A.I. Cucu, et al., Matrix metalloproteinases: pathophysiologic implications and potential therapeutic targets in cardiovascular disease, Biomolecules 15 (4) (2025), doi:10.3390/biom15040598.

[123] A.M. Manicone, J.K. McGuire, Matrix metalloproteinases as modulators of inflammation, Semin. Cell Dev. Biol. 19 (1) (2008) 34–41, doi:10.1016/j.semcdb.2007.07.003.

[124] J. Hu, P.E. Van den Steen, Q.X.A. Sang, G. Opdenakker, Matrix metalloproteinase inhibitors as therapy for inflammatory and vascular diseases, Nat. Rev. Drug Discov. 6 (6) (2007) 480–498, doi:10.1038/nrd2308.

[125] J. Vandooren, P.E. Van den Steen, G. Opdenakker, Biochemistry and molecular biology of gelatinase B or matrix metalloproteinase-9 (MMP-9): the next decade, Crit. Rev. Biochem. Mol. Biol. 48 (3) (2013) 222–272, doi:10.3109/10409238.2013.770819.

[126] C.L. Bellac, A. Dufour, M.J. Krisinger, A. Loonchanta, A.E. Starr, U. Auf dem Keller, et al., Macrophage matrix metalloproteinase-12 dampens inflammation and neutrophil influx in arthritis, Cell Rep. 9 (2) (2014) 618–632, doi:10.1016/j.celrep.2014.09.006.

[127] H. Haro, H.C. Crawford, B. Fingleton, K. Shinomiya, D.M. Spengler, L.M. Matrisian, Matrix metalloproteinase-7-dependent release of tumor necrosis factor-alpha in a model of herniated disc resorption, J. Clin. Invest. 105 (2) (2000) 143–150, doi:10.1172/JCI7091.

[128] A.J.H. Gearing, P. Beckett, M. Christodoulou, M. Churchill, J.M. Clements, M. Crimmin, et al., Matrix metalloproteinases and processing of pro-TNF-α, J. Leukoc. Biol. 57 (5) (1995) 774–777, doi:10.1002/jlb.57.5.774.

[129] B. Mallia-Milanes, A. Dufour, C. Philp, N. Solis, T. Klein, M. Fischer, et al., TAILS proteomics reveals dynamic changes in airway proteolysis controlling protease activity and innate immunity during COPD exacerbations, Am. J. Physiol. Lung Cell Mol. Physiol. 315 (6) (2018) L1003–L1014, doi:10.1152/ajplung.00175.2018.

[130] U. Chalise, M. Becirovic-Agic, S.R. Konfrst, J.R. Rodriguez-Paar, L.M. Cook, M.L. Lindsey, MMP-12 polarizes neutrophil signalome towards an apoptotic signature, J. Proteomics 264 (2022) 104636, doi:10.1016/j.jprot.2022.104636.

[131] S.E. Gill, S.T. Nadler, Q. Li, C.W. Frevert, P.W. Park, P. Chen, et al., Shedding of syndecan-1/CXCL1 complexes by matrix metalloproteinase 7 functions as an epithelial checkpoint of neutrophil activation, Am. J. Respir. Cell Mol. Biol. 55 (2) (2016) 243–251, doi:10.1165/rcmb.2015-0193OC.

[132] K.Y. Deleon-Pennell, R. Altara, A. Yabluchanskiy, A. Modesti, M.L. Lindsey, The circular relationship between matrix metalloproteinase-9 and inflammation following myocardial infarction, IUBMB Life 67 (8) (2015) 611–618, doi:10.1002/iub.1408.

[133] T. Kobayashi, H. Kim, X. Liu, H. Sugiura, T. Kohyama, Q. Fang, et al., Matrix metalloproteinase-9 activates TGF-β and stimulates fibroblast contraction of collagen

gels, Am. J. Physiol. Lung Cell Mol. Physiol. 306 (11) (2014) L1006–L1015, doi:10.1152/ajplung.00015.2014.
[134] N. Fortelny, J.H. Cox, R. Kappelhoff, A.E. Starr, P.F. Lange, P. Pavlidis, et al., Network analyses reveal pervasive functional regulation between proteases in the human protease web, PLoS Biol. 12 (5) (2014) e1001869, doi:10.1371/journal.pbio.1001869.
[135] S. O'Sullivan, J.F. Gilmer, C. Medina, Matrix metalloproteinases in inflammatory bowel disease: an update, Mediat. Inflamm. 2015 (2015) 964131, doi:10.1155/2015/964131.
[136] Y. Naito, T. Yoshikawa, Role of matrix metalloproteinases in inflammatory bowel disease, Mol. Aspects Med. 26 (4–5) (2005) 379–390, doi:10.1016/j.mam.2005.07.009.
[137] S. O'Sullivan, J.F. Gilmer, C. Medina, Matrix metalloproteinases in inflammatory bowel disease: an update, Mediat. Inflamm. 2015 (1) (2015), doi:10.1155/2015/964131.
[138] F.L. Koller, E.A. Dozier, K.T. Nam, M. Swee, T.P. Birkland, W.C. Parks, et al., Lack of MMP10 exacerbates experimental colitis and promotes development of inflammation-associated colonic dysplasia, Lab. Invest. 92 (12) (2012) 1749–1759, doi:10.1038/labinvest.2012.141.
[139] G. Murphy, H. Nagase, Progress in matrix metalloproteinase research, Mol. Aspects Med. 29 (5) (2008) 290–308, doi:10.1016/j.mam.2008.05.002.
[140] Y. Okada, H. Nagase, E.D. HArris, Matrix metalloproteinases 1, 2, and 3 from rheumatoid synovial cells are sufficient to destroy joints, J. Rheumatol. 14 (1987) 41–42.
[141] S.W. Manka, F. Carafoli, R. Visse, D. Bihan, N. Raynal, R.W. Farndale, et al., Structural insights into triple-helical collagen cleavage by matrix metalloproteinase 1, Proc. Natl. Acad. Sci. 109 (31) (2012) 12461–12466.
[142] N.H. Lim, E. Meinjohanns, G. Bou-Gharios, L.L. Gompels, E. Nuti, A. Rossello, et al., In vivo imaging of matrix metalloproteinase 12 and matrix metalloproteinase 13 activities in the mouse model of collagen-induced arthritis, Arthritis Rheumatol. 66 (3) (2014) 589–598, doi:10.1002/art.38295.
[143] G. Khalili-Tanha, E.S. Radisky, D.C. Radisky, A. Shoari, Matrix metalloproteinase-driven epithelial-mesenchymal transition: implications in health and disease, J. Transl. Med. 23 (1) (2025) 436, doi:10.1186/s12967-025-06447-w.
[144] Y. Wang, L. Jiao, C. Qiang, C. Chen, Z. Shen, F. Ding, et al., The role of matrix metalloproteinase 9 in fibrosis diseases and its molecular mechanisms, Biomed. Pharmacother. 171 (2024) 116116, doi:10.1016/j.biopha.2023.116116.
[145] A. Papazafiropoulou, N. Tentolouris, Matrix metalloproteinases and cardiovascular diseases, Hippokratia 13 (2) (2009) 76–82.
[146] G. Atkinson, R. Bianco, K. Di Gregoli, J.L. Johnson The contribution of matrix metalloproteinases and their inhibitors to the development, progression, and rupture of abdominal aortic aneurysms. Front. Cardiovasc. Med. 2023;10. doi:10.3389/fcvm.2023.1248561
[147] R. Schulz, Activation of MMP-2 as a key event in oxidative stress injury to the heart, Front. Biosci. 14 (2) (2009) 699–716, doi:10.2741/3274.
[148] M.A.M. Ali, X. Fan, R. Schulz, Cardiac sarcomeric proteins: novel intracellular targets of matrix metalloproteinase-2 in heart disease, Trends Cardiovasc. Med. 21 (4) (2011) 112–118, doi:10.1016/j.tcm.2012.03.008.
[149] H. Bräuninger, S. Krüger, L. Bacmeister, A. Nyström, K. Eyerich, D. Westermann, et al., Matrix metalloproteinases in coronary artery disease and myocardial infarction, Basic Res. Cardiol. 118 (1) (2023) 18, doi:10.1007/s00395-023-00987-2.
[150] G. Kremastiotis, I. Handa, C. Jackson, S. George, J. Johnson Disparate effects of MMP and TIMP modulation on coronary atherosclerosis and associated myocardial fibrosis. Sci. Rep. 2021;11(1):23081. doi:10.1038/s41598-021-02508-4
[151] J.L. Johnson, L. Devel, B. Czarny, S.J. George, C.L. Jackson, V. Rogakos, et al., A selective matrix metalloproteinase-12 inhibitor retards atherosclerotic plaque development in

apolipoprotein E-knockout mice, Arterioscler. Thromb. Vasc. Biol. 31 (3) (2011) 528–535, doi:10.1161/ATVBAHA.110.219147.
[152] M. Wang, S.H. Kim, R.E. Monticone, E.G. Lakatta, Matrix metalloproteinases promote arterial remodeling in aging, hypertension, and atherosclerosis, Hypertension 65 (4) (2015) 698–703, doi:10.1161/HYPERTENSIONAHA.114.03618.
[153] J. Parente, M. de Castro, Matrix metalloproteinase in the cardiovascular remodeling of hypertension: current insights and therapeutic potential, Metalloproteinases Med. 5 (2018) 1–11, doi:10.2147/MNM.S104793.
[154] X. Wang, E. Berry, S. Hernandez-Anzaldo, A. Takawale, Z. Kassiri, C. Fernandez-Patron, Matrix metalloproteinase-2 mediates a mechanism of metabolic cardioprotection consisting of negative regulation of the sterol regulatory element–binding protein-2/3-hydroxy-3-methylglutaryl-CoA reductase pathway in the heart, Hypertension 65 (4) (2015) 882–888, doi:10.1161/HYPERTENSIONAHA.114.04989.
[155] P. Liu, M. Sun, S. Sader, Matrix metalloproteinases in cardiovascular disease, Can. J. Cardiol. 22 (2006) 25B–30B, doi:10.1016/S0828-282X(06)70983-7.
[156] F.G. Spinale, Matrix Metalloproteinases: regulation and dysregulation in the failing heart, Circ. Res. 90 (5) (2002) 520–530, doi:10.1161/01.RES.0000013290.12884.A3.
[157] S. Agrawal, L. Lau, V. Yong, MMPs in the central nervous system: where the good guys go bad, Semin. Cell Dev. Biol. 19 (1) (2008) 42–51, doi:10.1016/j.semcdb.2007.06.003.
[158] H. Fujioka, Y. Dairyo, Y.K. Ichiro, K. Emoto, Neural functions of matrix metalloproteinases: plasticity, neurogenesis, and disease, Biochem. Res. Int. 2012 (2012) 1–8, doi:10.1155/2012/789083.
[159] M. Verslegers, K. Lemmens, I. Van Hove, L. Moons, Matrix metalloproteinase-2 and -9 as promising benefactors in development, plasticity and repair of the nervous system, Prog. Neurobiol. 105 (2013) 60–78, doi:10.1016/j.pneurobio.2013.03.004.
[160] R.G. Rempe, A.M.S. Hartz, B. Bauer, Matrix metalloproteinases in the brain and blood-brain barrier: versatile breakers and makers, J. Cereb. Blood Flow Metab. 36 (9) (2016) 1481–1507, doi:10.1177/0271678X16655551.
[161] M. Castellazzi, D. Ligi, E. Contaldi, D. Quartana, M. Fonderico, L. Borgatti, et al. Multiplex matrix metalloproteinases analysis in the cerebrospinal fluid reveals potential specific patterns in multiple sclerosis patients. Front. Neurol. 2018;9. doi:10.3389/fneur.2018.01080
[162] V.W. Yong, C. Power, P. Forsyth, D.R. Edwards, Metalloproteinases in biology and pathology of the nervous system, Nat. Rev. Neurosci. 2 (7) (2001) 502–511, doi:10.1038/35081571.
[163] A. Weaver, A.G. Da Silva, R.K. Nuttall, D.R. Edwards, S.D. Shapiro, S. Rivest, et al., An elevated matrix metalloproteinase (MMP) in an animal model of multiple sclerosis is protective by affecting Th1/Th2 polarization, FASEB J. 19 (12) (2005) 1668–1670, doi:10.1096/fj.04-2030fje.
[164] C. Camara-Lemarroy, L. Metz, J. Kuhle, D. Leppert, E. Willemse, D.K. Li, et al., Minocycline treatment in clinically isolated syndrome and serum NfL, GFAP, and metalloproteinase levels, Mult. Scler. 28 (13) (2022) 2081–2089, doi:10.1177/13524585221109761.
[165] L.M. Metz, D.K.B. Li, A.L. Traboulsee, P. Duquette, M. Eliasziw, G. Cerchiaro, et al., Trial of minocycline in a clinically isolated syndrome of multiple sclerosis, N. Engl. J. Med. 376 (22) (2017) 2122–2133, doi:10.1056/NEJMoa1608889.
[166] L.W. Lau, R. Cua, M.B. Keough, S. Haylock-Jacobs, V.W. Yong, Pathophysiology of the brain extracellular matrix: a new target for remyelination, Nat. Rev. Neurosci. 14 (10) (2013) 722–729, doi:10.1038/nrn3550.
[167] P.T.G. Elkington, C.M. O'Kane, J.S. Friedland, The paradox of matrix metalloproteinases in infectious disease, Clin. Exp. Immunol. 142 (1) (2005) 12–20, doi:10.1111/j.1365-2249.2005.02840.x.

[168] H.S. Lee, W.J. Kim, The role of matrix metalloproteinase in inflammation with a focus on infectious diseases, Int. J. Mol. Sci. 23 (18) (2022), doi:10.3390/ijms231810546.
[169] O. Supasorn, N. Sringkarin, P. Srimanote, P. Angkasekwinai, Matrix metalloproteinases contribute to the regulation of chemokine expression and pulmonary inflammation in *Cryptococcus* infection, Clin. Exp. Immunol. 183 (3) (2016) 431–440, doi:10.1111/cei.12725.
[170] C. Cheung, D. Marchant, E.K.Y. Walker, Z. Luo, J. Zhang, B. Yanagawa, et al., Ablation of matrix metalloproteinase-9 increases severity of viral myocarditis in mice, Circulation 117 (12) (2008) 1574–1582, doi:10.1161/CIRCULATIONAHA.107.733238.
[171] A.M. Houghton, W.O. Hartzell, C.S. Robbins, F.X. Gomis-Rüth, S.D. Shapiro, Macrophage elastase kills bacteria within murine macrophages, Nature 460 (7255) (2009) 637–641, doi:10.1038/nature08181.
[172] M. Quiding-Järbrink, D.A. Smith, G.J. Bancroft, Production of matrix metalloproteinases in response to mycobacterial infection, Infect. Immun. 69 (9) (2001) 5661–5670, doi:10.1128/IAI.69.9.5661-5670.2001.
[173] N.W. Marten, J. Zhou, The role of metalloproteinases in Corona Virus infection, in: Experimental Models of Multiple Sclerosis, Springer US, Boston, MA, 2005, pp. 839–848, doi:10.1007/0-387-25518-4_48.
[174] M. Wornle, M. Roeder, M. Sauter, A. Ribeiro, Role of matrix metalloproteinases in viral-associated glomerulonephritis, Nephrol. Dial. Transplant. 24 (4) (2008) 1113–1121, doi:10.1093/ndt/gfn627.
[175] S. Sinha, N.L. Rosin, R. Arora, E. Labit, A. Jaffer, L. Cao, et al., Dexamethasone modulates immature neutrophils and interferon programming in severe COVID-19, Nat. Med. 28 (1) (2022) 201–211, doi:10.1038/s41591-021-01576-3.
[176] O. Kleifeld, A. Doucet, A. Prudova, U. Auf Dem Keller, M. Gioia, J.N. Kizhakkedathu, et al., Identifying and quantifying proteolytic events and the natural N terminome by terminal amine isotopic labeling of substrates, Nat. Protoc. (2011), doi:10.1038/nprot.2011.382.
[177] J.C. Timmer, M. Enoksson, E. Wildfang, W. Zhu, Y. Igarashi, J.B. Denault, et al., Profiling constitutive proteolytic events in vivo, Biochem. J. 407 (1) (2007) 41–48, doi:10.1042/BJ20070775.
[178] A. Anowai, S. Chopra, B. Mainoli, D. Young, A. Dufour, N-Terminomics/TAILS of tissue and liquid biopsies, in: Methods Molecular Biology, 2022, pp. 85–94, doi:10.1007/978-1-0716-2124-0_7.
[179] W. Esser-Skala, N. Fortelny, The protease web, in: Proteolytic Signaling in Health and Disease, Elsevier, 2022, pp. 229–250, doi:10.1016/B978-0-323-85696-6.00001-9.
[180] U. Auf dem Keller, A. Prudova, U. Eckhard, B. Fingleton, C.M. Overall, Systems-level analysis of proteolytic events in increased vascular permeability and complement activation in skin inflammation, Sci. Signal. 6 (258) (2013) rs2, doi:10.1126/scisignal.2003512.
[181] J.P. Simmer, J.C.C. Hu, Expression, structure, and function of enamel proteinases, Connect Tissue Res. 43 (2–3) (2002) 441–449, doi:10.1080/03008200290001159.
[182] D. Kraus, J. Reckenbeil, S. Perner, J. Winter, R. Probstmeier, Expression pattern of matrix metalloproteinase 20 (MMP20) in human tumors, Anticancer Res. 36 (6) (2016) 2713–2718.

CHAPTER TWO

Activity-based probes to detect MMP activity

Konstantinos Patiniotis [a], Dimitris Georgiadis [a], Laurent Devel [b,*]
[a] Department of Chemistry, Laboratory of Organic Chemistry, National and Kapodistrian University of Athens, Panepistimiopolis, Zografou, Athens Greece
[b] CEA, INRAE, Médicaments et Technologies pour la Santé (MTS), SIMoS, Université Paris-Saclay, Gif-sur-Yvette, France
*Corresponding author. e-mail address: laurent.devel@cea.fr.

Contents

Abstract

Elucidating the precise biological functions of MMPs remains challenging because conventional analytical methods generally measure total protein abundance rather than enzymatic activity. Since only a fraction of the enzyme pool exists in an active form and proteolytic activity is the primary determinant of MMP function, approaches capable of selectively monitoring active MMPs provide a more accurate representation of their physiological and pathological roles. Recent advances in chemical biology have led to the development of molecular probes that directly interrogate MMP activity in complex biological environments. This chapter reviews two complementary classes of probes: substrate-based activatable probes and affinity-based probes. Activatable probes exploit MMP-mediated cleavage to generate detectable signals and enable sensitive monitoring of proteolytic activity in cells, tissues, biological fluids, and living organisms. In contrast, affinity-based probes selectively bind active MMPs and form covalent adducts that enable their direct visualization, enrichment, and unambiguous identification by proteomic methods. Particular emphasis is placed on probe design, substrate selectivity, analytical performance, and applications ranging from *in vivo* imaging to activity-based biomarker

Fungal and Protozoan Carbonic Anhydrases, Volume 60
ISSN 1874-6047
https://doi.org/10.1016/bs.enz.2026.06.003

discovery. Finally, current limitations and emerging opportunities are discussed, including machine learning-guided substrate design, synthetic biomarker technologies, and next-generation affinity-based profiling strategies, which collectively hold promise for advancing both fundamental MMP biology and translational diagnostic applications.

Keywords: Affinity-based probes, Diagnostic companion tools, Imaging, Substrate-based activatable probes, Synthetic biomarkers

1. Introduction

Understanding the precise roles of MMPs in physiological and pathological contexts remains challenging, largely because conventional analytical approaches frequently measure total protein abundance rather than enzymatic activity. Indeed, due to tight regulation at multiple levels, only a subset of the total MMP pool exists in an active form at any given time. Since the proteolytic function of MMPs is primarily responsible for their physiological and pathological effects, analytical strategies that specifically target the active fraction of these enzymes provide a more accurate representation of their biological role [1].

The presence of active MMP forms, combined with degradomics analyses, namely studies of MMPs substrates in complex biological matrices, provides critical insight into substrate repertoires and downstream signaling pathways [2–4]. These integrated approaches allow the identification not only of the MMPs responsible for specific proteolytic events but also of the functional consequences of these cleavages within a given pathological context. Such information is essential, particularly in diseases where multiple MMPs may exert opposing functions [1,5–7]. In addition, the activation patterns of MMPs are highly dynamic and can evolve over the course of disease progression. Temporal changes in MMP activation can therefore carry important biological and clinical information.

From a therapeutic perspective, both the ambivalent functions of MMPs and their dynamic activation patterns underscore the necessity of targeting the correct MMP at the appropriate stage of disease. In particular, precisely defining the therapeutic window within which it is necessary to selectively inhibit the MMPs that effectively support disease progression may enable more efficient and potentially safer interventions. This consideration is especially relevant given the historical challenges associated with MMP inhibitors used in cancer, which have produced undesirable side effects due to their lack of selectivity and their administration at inappropriate stages of disease [1,7,8]. In addition to therapeutic implications, activity-based detection of MMPs also offers promising opportunities for diagnostic applications. Because the pattern of MMP activation may reflect specific disease stages, monitoring these activation profiles throughout disease progression could contribute to the development of functional biomarkers directly associated with MMP activity, thereby improving disease stratification, prognosis, and treatment monitoring [9–12].

In this chapter, we focus on two major classes of chemical probes that enable the direct interrogation of MMPs activity in complex proteomic environments. The first class comprises activatable probes that incorporate peptide substrates selectively cleaved by MMPs. These probes are designed to remain silent until proteolytic processing occurs, at which point signal generation provides a direct and activity-dependent readout of MMP function in biological systems. The second class encompasses small-molecule affinity-based probes that are designed to selectively bind and react in proximity to the active site of MMPs. Upon interaction, these probes form covalent adducts with the target enzymes, enabling their visualization through gel-based methods or their enrichment and unambiguous identification using proteomic workflows.

Here, we examine key considerations underlying the design and application of these two probe classes, with particular emphasis on achieving selectivity in complex biological matrices and evaluating their potential as diagnostic companion tools.

2. Principle and design of MMP substrate-based activatable probes

Substrate-based activatable probes exploit one of the main advantages of enzymatic detection: catalytic turnover, whereby a single active MMP can cleave multiple probe molecules and thereby amplify the analytical signal. This is especially important for MMPs, which are often secreted or locally activated at low concentrations in complex extracellular environments. In their most common design, these probes contain a short MMP-cleavable peptide sequence flanked by reporter elements, typically a fluorophore–quencher pair or equivalent signal-generating module. Proteolysis separates these elements and converts a silent probe into a detectable signal (Fig. 1A). The peptide substrate is generally built around the scissile bond, with residues N-terminal to the cleavage site referred to as non-primed positions P_3–P_2–P_1, and residues C-terminal to it as primed positions P_1'–P_2'–P_3'; most

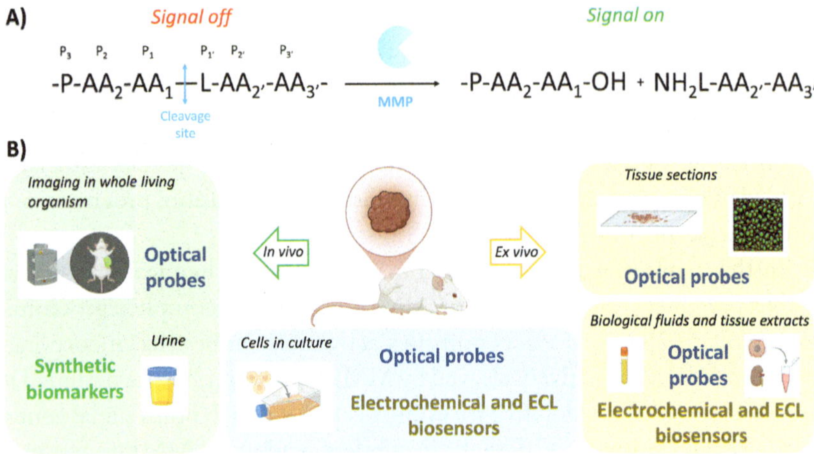

Fig. 1 Principle and applications of MMP substrate-based activatable probes.
(A) Representative peptide motifs commonly incorporated into activatable probes targeting MMPs, typically featuring a proline residue at the P_3 position and a leucine residue at the $P_{1'}$ position, which constitute important determinants for MMP substrate recognition and cleavage.
(B) MMP-activatable probes have been developed to monitor proteolytic activity across biological matrices of increasing complexity, including cultured cells (optical probes, electrochemical biosensors, and electrogenerated chemiluminescence (ECL) biosensors), tissue sections (*ex vivo* optical imaging and activatable zymography), biological fluids such as plasma or serum (optical, electrochemical, and ECL biosensors), and *in vivo* applications using optical imaging probes and synthetic biomarker approaches.

MMP probes therefore contain approximately three to five amino acids on each side of the cleavage site.

Early probe designs were largely inspired by known physiological substrates of MMPs, whereas later generations relied on systematic profiling approaches, including synthetic peptide libraries, mixture-based oriented peptide libraries [13], and substrate phage display [14–17]. More recent high-throughput substrate-profiling studies have further refined our understanding of MMP sequence preferences, revealing both strong redundancy and enzyme-specific signatures across the family [18,19]. Collectively, these studies indicate that MMP-mediated cleavage is largely driven by hydrophobic residues at the $P_{1'}$ position, such as leucine, isoleucine, or valine, with leucine emerging as a major specificity determinant. These residues are frequently found within consensus PXX#L/I/V-type motifs. In addition, a proline residue at the P_3 position is commonly favored, although alanine and other amino acids may be accommodated depending on the specific MMP. Residues at other positions can further modulate substrate recognition and contribute to partial selectivity within subsets of MMPs [18,19]. Nevertheless, achieving strict selectivity for a single MMP remains challenging because of the structural homology of MMP catalytic domains, overlapping substrate preferences, and the influence of the chemical groups flanking the peptide sequence. In this context, Table 1 summarizes representative peptide sequences commonly used for the development of MMP-activatable probes, together with the MMPs they target.

3. Activatable probes for *in vivo* and *ex vivo* multiscale profiling of protease activity

A broad variety of activatable probes has been developed for the profiling of MMP activity across biological matrices of increasing complexity, ranging from purified enzymes and cell supernatants to biological fluids, tissue sections, and whole living organisms (Fig. 1B). Owing to their high sensitivity and signal amplification properties, these probes are particularly well suited for monitoring the spatial and temporal regulation of proteolytic activity in complex biological environments. For *in vivo* applications, the main classes of activatable probes developed to date are imaging probes based on optical fluorescence [25,26,28,31,34,35,37,38,40,42–49,66,67,69,71,73–76,80,81,83–86,90–92,95–99,102,103,107–109], photoacoustic imaging [57,77,78], magnetic resonance imaging (MRI) [58,79], single-photon emission computed tomography (SPECT) [113]), or combinations of several imaging modalities within the same construct [29,30,32,33,36,39,41,72,106].

More recently, a new generation of activatable sensors, referred to as synthetic biomarkers, has emerged for the indirect monitoring of MMP proteolytic activity *in vivo* [9,20,22–24,60–65,89,112]. In these systems, MMP-cleavable peptide substrates are incorporated into nanoscale carriers designed to accumulate preferentially within pathological tissues. Proteolytic cleavage by active MMPs triggers the release of soluble reporter fragments that subsequently diffuse into the bloodstream and are ultimately recovered in urine for downstream analysis. These urinary reporters therefore provide a non-invasive and indirect readout of MMP activity occurring within diseased tissues.

Beyond *in vivo* imaging, fluorescent activatable probes have also found important applications in *ex vivo* tissue analysis, notably through the recent development of cleavage-based activatable zymography approaches

Table 1 Main substrate sequences used across activatable probe platforms, and their preferential protease targets.

Entry	Peptide sequence $P_3P_2P_1$# $P_{1'}P_{2'}P_{3'}$	Targeted MMPs and *other proteases*	Applications
a	**PLG#XRY,** with X = V, M or L and Y = G, Q, S, C, K or no aa	MMP-1/2/3/7/9/12/13/14/16 *Metalloprotease: ADAM-17* [20] *Serine protease: Trypsin* [21], *PRSS3/trypsin 3, KLK14 and FAP* [22–24]	Optical probes [25–49] Electrochemical sensors [50] Electrogenerated chemiluminescence biosensors [51–56] Photoacoustic probes [57] MRI probes [58] Mass-encoded [21,59] Synthetic biomarkers [20,22,60–65]
b	**PLG#XAY** With X = L, C(Me), M or I, and Y = G or R	MMP-2/9/13/14	Optical probes [37,66–76] Photoacoustic probes [77,78] MRI Probes [79]
c	**PLG#LEE**	MMP-12	Optical probes [80–82]
d	**PLG#LPG**	MMP-9	Optical probes [83,84]
e	**PLG#YLG**	MMP-2/9	Optical probes [85]
f	**PLG#LK(Dabcyl)A**	MMP-14	Optical probes [86]

Table 1 Main substrate sequences used across activatable probe platforms, and their preferential protease targets.—cont'd

Entry	Peptide sequence $P_3P_2P_1\# P_{1'}P_{2'}P_{3'}$	Targeted MMPs and *other proteases*	Applications
g	**PLG#MWS**	MMP-9	Electrochemical sensors [87] Electrogenerated chemiluminescence biosensors [88]
h	**PLG#Lp**-O-R	MMP-9	Synthetic biomarkers [89]
i	**PQG#IWG**	MMP-1/2/3/8/9/12/13/14 *Serine protease: FAP* [23,24]	Optical probes [90] Synthetic biomarkers [9,22–24,60]
j	**PKG#QTG**	MMP-13 *Serine protease: PRSS3* [22]	Synthetic biomarkers [22]
k	**PVG#LIG**	MMP-3/7/9/12/13 *Cysteine protease: Cathepsin K* [23]	Optical probes [91] Synthetic biomarkers [22–24,60]
l	**PFG#NLβA**	MMP-13	Optical probes [92]
m	**PWG#IWG**	MMP-2/7/9	Synthetic biomarkers [60]
n	**PMG#LLY**	MMP-7/12/13 *Serine protease: HGFAC, PRSS1, PRSS2* [24]	Synthetic biomarkers [24]

Continued

o	**PLS#LXM,** with X = T or V	MMP-1/2/3/7/9/10/11/12/13/14 *Metalloprotease: ADAM-17 and 28* [20] *Aspartic protease: Cathepsin D* [64,93], *E* [20], *and NAPSA/Napsin A* [22] *Cysteine protease: Cathepsin S/K/L* [20]	Optical probes [43,94–100] Electrogenerated chemiluminescence biosensors [101] Synthetic biomarkers [20,22,24,60,64]
b	**PLA#XWY** With X = L or Y, and Y = R or A	MMP-7/9	Optical probes [102–104] Mass-encoded probes [105]
k	**PLR#SWG**	MMP-14	Optical probes [106]
l	**PRS#LSG**	MMP-9	Optical probes [107]
m	**PRQ#ITA**	MMP-2/9	Optical probes [108,109]
n	**PMS#MRG**	MMP-1	Optical probes [110]
o	**PYA#YWM**	MMP-3	Mass-encoded probes [111]
p	**PIS#LIS**	MMP-1/3/12/13	Synthetic biomarkers [22]
q	**PFE#MSA**	MMP-3/10/12/13	Synthetic biomarkers [24]
r	**PAA#LRA**	MMP-1/2/3/7/8/9/10/11/12/13	Optical probes and synthetic biomarkers [112]
s	**IGF#LRT**	MMP-14	SPECT probes [113]

that enable direct visualization of localized protease activity in tissue sections with high spatial resolution [9,100,104]. In parallel, electrochemical [50,87,114,115] and electrogenerated chemiluminescence (ECL) [51–56,88,101] biosensors have emerged as attractive tools for *ex vivo* profiling of MMP activity in conditioned cell media and biological fluids, including pathological human serum samples, owing to their high analytical sensitivity, relatively simple instrumentation, and compatibility with quantitative diagnostic assays.

4. Optical activatable probes for *in vivo* MMP imaging

Most optical activatable probes developed for MMP *in vivo* imaging rely on fluorescence self-quenching or Förster resonance energy transfer (FRET) mechanisms. In these systems, the fluorophore signal is initially suppressed either through close proximity between identical fluorochromes (self-quenching probes) or through energy transfer between a donor fluorophore and an acceptor/quencher (FRET probes). Proteolytic cleavage of the MMP-sensitive peptide sequence induces the irreversible spatial separation of these groups, thereby restoring fluorescence emission and generating a measurable signal. When near-infrared (NIR) fluorophores are employed, such probes can be used for *in vivo* imaging owing to the improved tissue penetration and reduced autofluorescence associated with NIR wavelengths. The development of these probes nevertheless requires careful optimization of several parameters, including prolonged circulation time, efficient accumulation within pathological tissues, and minimization of signal attenuation caused by diffusion of the activated fluorescent fragments away from the proteolytic site.

A seminal contribution in this field was provided by the Weissleder group, which developed a polymer-based self-quenched fluorescent probe in which multiple fluorochromes were conjugated through MMP-cleavable peptide sequences to a stealth copolymer conferring prolonged blood circulation [25]. This probe architecture was subsequently validated in several mouse models of cancer [25,116], myocardial infarction [108], atherosclerosis [109], and cerebral ischemia [117]. Building on these pioneering studies, numerous nanoscale activatable probes with optimized pharmacokinetic properties and improved retention at pathological sites have been developed. These include probes based on dendrimers [71,102], ferritin cage-like nanostructures [28], metallic nanoparticles [30,32,37–40,46,72,84,104,118,119], organic nanoparticles [27,47,48,95–98,120], and nanocrystals [83,106]. Polyethylene glycol

(PEG)-based activatable probes have also been explored to tune circulation time and tissue distribution. Notably, a PEG12-containing construct was shown to provide a favorable balance between rapid pharmacokinetics and efficient targeting of MMP activity, enabling detectable *in vivo* fluorescence within 30 min after administration [31]. The performance of these PEGylated probes has since been further optimized in several pathological models [34,35,121].

To further limit diffusion of the fluorescent products away from the site of proteolysis, several elegant retention strategies have been developed. Tsien and co-workers designed activatable FRET probes in which an MMP-cleavable sequence was inserted between complementary polycationic and polyanionic domains [66]. Upon MMP-mediated cleavage, a polyarginine-based cell-penetrating peptide (CPP) bearing an NIR fluorophore is released and rapidly internalized into nearby cells, thereby selectively concentrating the fluorescent signal at MMP-active sites. This strategy has been successfully applied in several mouse cancer models, notably for fluorescence-guided surgery [66,69,122]. Using a comparable concept, probes incorporating highly cell-permeable boron dipyrromethene (BODIPY) fluorophores were shown to undergo rapid intracellular accumulation after extracellular activation by MMPs, thereby improving local signal retention [68]. Schultz and colleagues also reported an MMP-12-selective FRET probe containing an N-terminal palmitic acid moiety that enabled anchoring of the cleaved fluorescent fragment onto cell membranes, thus limiting signal diffusion [80], although this approach has not yet been validated *in vivo*. More recently, supramolecular aggregation-induced emission (AIE) strategies have emerged to enhance probe retention and fluorescence amplification. In these systems, the cleaved probe fragments self-assemble into fluorescent nanostructures such as nanofibers [85] or nanodots [74].

Beyond qualitative imaging, several activatable optical probes have also been engineered for ratiometric quantification of MMP activity in pathological tissues [41,46,48,69,84,119,122–124]. In this context, a recent study introduced a ratiometric afterglow luminescence (AGL)-based probe incorporating a self-calibration function for quantitative *in vivo* imaging of MMP activity [48]. Many of these optical probes have additionally been integrated into multimodal imaging platforms to improve spatial resolution and anatomical localization, notably through combinations with MRI [30,70,71,106,125], photoacoustic imaging [32,39,41,72], PET [33], or SPECT imaging [44]. Finally, recent years have witnessed the emergence of dual-responsive MMP-activatable probes capable of integrating proteolytic activity with secondary pathological triggers such as local pH variations, thereby improving specificity and contextual imaging of diseased tissues [37,84].

5. Synthetic biomarkers for the indirect interrogation of MMP activity *in vivo*

The seminal work of Bhatia and co-workers introduced a new paradigm in which MMP activity occurring within pathological tissues is converted into the appearance of synthetic biomarkers in urine [60]. In this approach, activatable probes are composed of mass-encoded reporter peptides conjugated to nanostructures through protease-cleavable peptide linkers. Following systemic administration and accumulation at diseased sites, proteolytic cleavage by active MMPs releases soluble reporters that subsequently diffuse into the circulation and are ultimately excreted in urine, where they can be detected by mass spectrometry. This strategy enables the multiplexed monitoring of protease activities, including those of several MMPs, through minimally invasive sampling. Since these pioneering studies, the concept has been successfully extended to multiple animal models [61–63], with sustained efforts directed toward improving its translational potential. In particular, simplified readout platforms such as lateral-flow assays [24,61] and colorimetric detection systems [63] have been developed to facilitate biomarker detection directly from urine samples.

Several complementary strategies have also been explored to improve the specificity and sensitivity of these synthetic biomarker systems. To enhance tissue specificity and reduce background activation, local delivery approaches involving pulmonary instillation [22,23] or nebulization [24] have been implemented for lung-associated diseases. In parallel, signal sensitivity has been markedly increased through the incorporation of DNA-barcoded urinary reporters that can be amplified and detected using CRISPR-Cas-based assays directly in unprocessed urine samples [20,24]. More recently, the concept of synthetic biomarkers has been further refined through the development of AND-gated protease-responsive nano sensors [112]. In this system, urinary reporter release is conditioned by the simultaneous presence of two independent proteolytic activities within the pathological tissue microenvironment, namely granzyme B secreted by activated CD8+ T cells and MMP activity associated with tumor cells, thereby substantially improving disease specificity.

Beyond urinary biomarkers, MMP activity has very recently been exploited to generate exogenous volatile reporters that function as synthetic breath biomarkers [89]. In this study, a self-immolative linker selectively cleaved by MMPs was developed, resulting in the release of alcohol-derived volatile compounds that could subsequently be detected in exhaled breath by mass spectrometry. This innovative approach opens new perspectives for the non-invasive monitoring of proteolytic activity through breath analysis.

6. *Ex vivo* profiling of MMP activity

Beyond *in vivo* imaging applications, activatable probes have also been extensively developed for the *ex vivo* profiling of MMP activity in living cells, tissue sections, and biological fluids. In living cells, optical activatable probes have notably been designed for the simultaneous and multiplexed detection of MMPs together with other proteolytic activities, including thrombin [45,126], urokinase-type plasminogen activator (uPA) [43,118], cathepsin B [43], ADAM proteases [107], and additional serine proteases [45]. More sophisticated systems have also been engineered to integrate secondary stimuli such as pH variations [127] or hydrogen sulfide (H_2S) concentrations [49], thereby increasing the contextual specificity of probe activation. In parallel, fluorescence detection has recently been combined with CRISPR-based signal amplification strategies to further improve analytical sensitivity [103].

Optical activatable probes have similarly been employed for the profiling of MMP activity directly on tissue sections [9,92,100,104,126,128]. Among the most significant recent advances in this field are the cleavage-based *in situ* zymography approaches [9,100,104]. Inspired by the original activatable cell-penetrating peptide strategy [66], these probes incorporate an MMP-cleavable peptide sequence flanked by complementary polyglutamate and polyarginine domains. Upon proteolytic cleavage, the NIR-labelled polyarginine fragment is released and strongly associates with surrounding cellular membranes, thereby minimizing diffusion within the tissue section and substantially improving the signal-to-noise ratio. These approaches now provide powerful tools for the spatially resolved visualization of proteolytic activity directly within pathological tissue samples.

In biological fluids such as plasma and serum, several activatable sensing strategies have also been reported, including optical probes [45,73,98,107], electrochemical biosensors [50,87,114,115], and electrogenerated chemiluminescence (ECL)-based platforms [51–56,88,101]. These systems are generally intended to selectively detect the circulating active fraction of MMPs with very high analytical sensitivity, thereby potentially complementing conventional immunoassays such as ELISA, which do not discriminate between active, latent, or inhibited enzyme forms. However, the actual clinical relevance of detecting active circulating MMPs remains open to discussion, since a substantial fraction of MMPs present in blood likely corresponds to enzymes already trapped and inactivated by α2-macroglobulin as a downstream footprint of proteolytic activation occurring within tissues. In this context, activatable probes may only provide a significant diagnostic advantage if MMP–α2-macroglobulin complexes partially escape detection by antibody-based assays or by highly sensitive affinity-based technologies such as proximity extension assays (PEA) coupled to next-generation sequencing [129]. The relatively simple instrumentation required for the implementation of electrochemical and electrogenerated chemiluminescence biosensors may also constitute an important advantage for their future development as routine diagnostic devices.

7. Selective targeting of MMPs in complex biological matrices

The selectivity of activatable probes targeting MMPs can be considered at several distinct levels, including selective accumulation within pathological tissues, selective cleavage by MMPs relative to other protease families, and, ultimately, selective detection of a single MMP within the highly homologous MMP family.

At the tissue level, several *in vivo* co-targeting strategies have been developed to simultaneously direct probes toward pathological cells while reporting local MMP activity, thereby improving both tissue specificity and signal-to-background ratio. In this context, activatable optical probes have notably been engineered to combine an MMP-cleavable peptide sequence with ligands recognizing cell-surface biomarkers associated with MMP-rich pathological environments, including integrin $\alpha v\beta 3$ [40,44,48,85,123,130], gastric cancer-associated antigens [83], glucose transporters [37], folate receptors [84], EGFR [46], and CD13 [119].

Selectivity at the proteolytic level is most commonly addressed through the design of peptide substrates incorporating sequence motifs preferentially recognized by MMPs, notably through the introduction of a proline residue at the P_3 position (Table 1). The specificity of such probes' activation is generally evaluated using scrambled peptide controls [25,41,66,95,96,106], D-amino acid-containing peptides [100], replacement of the scissile bond by non-cleavable Gly/Gly motifs [74], or inhibition assays. Regarding this last point, many inhibition studies rely on broad-spectrum metalloprotease inhibitors that also target ADAMs and related metalloproteases, thereby complicating interpretation of the observed signal suppression. Only a limited number of studies have employed selective MMP inhibitors to validate probe specificity [81,123]. Importantly, recent studies have demonstrated that the presence of a P_3 proline does not guarantee exclusive MMP-dependent cleavage. For example, the widely used PLG#XRY motif (Table 1, entry a) can also be processed by ADAM-17 [20] and several serine proteases [21–24]. In particular, the presence of an arginine residue near the cleavage site favors hydrolysis by trypsin-like proteases, although this off-target cleavage can be partially reduced by inversion of the arginine stereochemistry [35]. Similarly, the broadly used PLS#LXM motif (Table 1, entry o) can be cleaved not only by MMPs but also by ADAM-17 and ADAM-28 [20], aspartyl proteases [20,22,64,93], and cysteine proteases [20]). More recently, the simplified PAA#LRA sequence (Table 1, entry r) was incorporated into activatable optical probes and showed improved selectivity toward several MMPs while sparing metalloprotease (ADAMTS-1) and many non-protease families, including serine, threonine, and aspartyl proteases [112]. Nevertheless, cleavage by trypsin-like enzymes was not specifically evaluated in this study.

Achieving selectivity for a single MMP isoform remains considerably more challenging because of the strong structural homology and overlapping substrate preferences shared across the MMP family. Only a limited number of peptide substrates currently display partial selectivity for individual MMPs (Table 1). For example, the PLG#LEE sequence (Table 1, entry c) appears relatively selective for MMP-12 [80,81,110], although it has not been systematically evaluated against a broad panel of proteases. Likewise, the PFG#NLβA sequence (Table 1, entry l) has shown promising selectivity toward MMP-13 [92]. Beyond peptide substrate optimization alone, improved selectivity may also be achieved by combining an MMP-cleavable sequence (Table 1, entry f) with an allosteric recognition element directed toward the targeted enzyme. Such a strategy has notably been validated for the selective targeting of MMP-14 *in vivo* through the incorporation of an MMP-14 allosteric binder within the probe architecture [86].

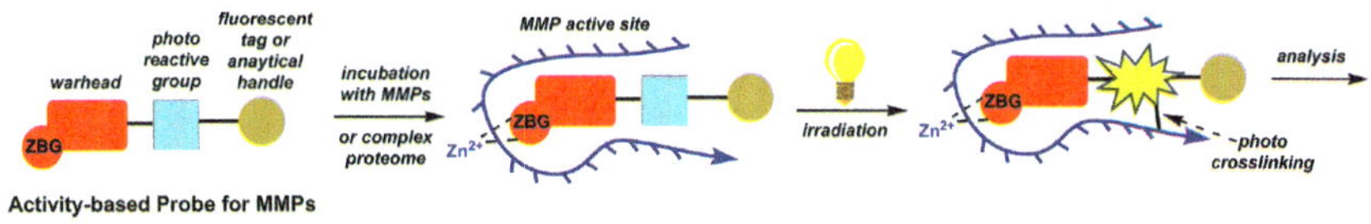

Fig. 2 Generic structure of a photoactivatable AfBP for MMPs and mechanism of action.

8. The evolution of activity-based probes (ABPs) for MMPs: light-triggered crosslinking of affinity-based probes (A*f*BPs) to MMPs

The covalent binding of molecular probes to target proteins forms the foundation of a groundbreaking technology introduced in the late 1990s, which unlocked new possibilities for profiling the activity of diverse enzymatic classes within complex proteomes [131,132]. Unlike the substrate-based probes described above, the recognition warheads of ABPs are derived from enzymatic inhibitors. In this context, mechanism-based inhibitors bearing reactive units capable of targeting a conserved nucleophile of the active site enabled covalent crosslinking with cysteine and serine proteases, however this approach was unsuitable for zinc and aspartyl proteases. To address this limitation, seminal studies by the research groups of Cravatt [133] and Yao [134] demonstrated that labelling of MPs is feasible even with non-covalent inhibitors, provided that a post-binding photo-crosslinking event would mediate the covalent binding of the probe on the protein scaffold (Fig. 2).

Specifically, Yao and co-workers developed affinity-based probes (A*f*BPs) comprising a hydroxamate zinc-binding group (ZBG) appended to a peptidyl segment, a diazirine photoactivatable group targeting the non-primed subsites of MPs' active site, and a Cy3 fluorescent reporter tag (Fig. 3, probe **1**). Although inherently unselective due to the non-optimized peptidyl sequence, probe **1** was nonetheless capable of efficiently labelling MMP-9 with a detection limit comparable to that of a more MMP-selective probe (Fig. 3, probe **2**) developed concurrently by Cravatt and co-workers [133]. The design of probes of type **2** was based on the structure of ilomastat, a broad-spectrum inhibitor of MMPs which occupies the primed subsites of the active site. This approach was intended to enhance the specificity of the resulting probes for MMPs which tend to form a more extensive network of interactions in their primed region. By replacing the hydrophobic indole ring of ilomastat with a photoactivatable benzophenone and appending

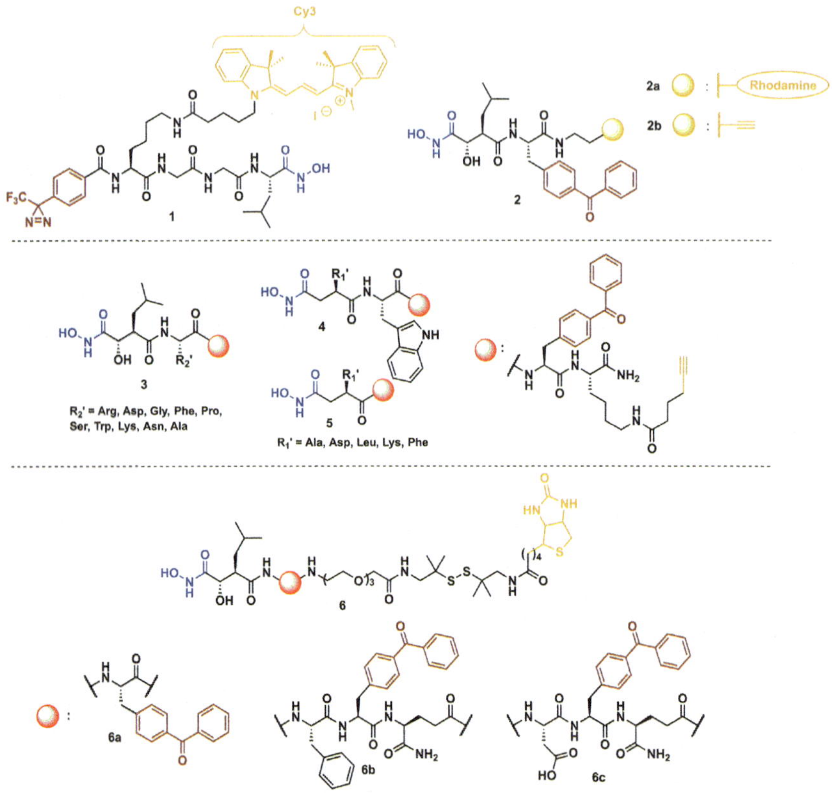

Fig. 3 Chemical structures of probes developed by the research groups of Yao (probe 1, [134]), Cravatt (probes 2–5, [133,136]) and Tannenbaum/Griffith (probes 6a-c, [137,138]. The ZBG, photoreactive group and fluorescent/analytical tag are colored in blue, brown and orange, respectively.

a rhodamine tag at the C-terminus (probe **2a**), the researchers achieved detection limits as low as 3 nM for MMP-2, with no observed labelling of pro-MMP-2. The utility of this probe in *in vivo* labelling was demonstrated through the spatial detection of gelatinolytic activity in zebrafish embryos and developing tadpoles [135]. The same group extended their approach to a broader range of MMPs and other MPs by using cocktails of photoactivatable affinity probes with varying specificity, incorporating a hydroxamate as a zinc-binding group and an alkyne handle to facilitate post-irradiation fluorescent labelling or avidin/biotin enrichment via click chemistry and subsequent LC-MS analysis (Fig. 3, probes **3**–**5**) [136]. By applying this cocktail of probes to complex proteomes, the authors managed to label and detect more than 20 metalloproteases, including 6 MMPs.

More recently, Ravidra and co-workers extended the results of Cravatt on the benzophenone probe series of type **2** with the aim of improving sensitivity and enabling the detection of endogenous MMPs. This challenge is hampered by the very low concentrations at which active MMPs are present in biological samples, which lie below the detection limits of existing A*f*BP-based analytical methods [137]. To this respect, they developed probe **6a**, which retains the same warhead and photolabile group as the established A*f*BP **2**, alongside probes **6b** and **6c** that could potentially target a larger variety of MMPs and ADAMTSs. All type **6** probes incorporate a biotin tag to promote enrichment through avidin bead capture, and a cleavable disulfide linker protected by vicinal dimethyl groups. These dimethyl groups are able to stabilize the disulfide linker against premature reduction by free cysteine thiols or even DTT, while still permitting reduction-mediated release from magnetic beads. This design substantially reduces background signal arising from non-specific absorption of proteins, a major source of noise in bead-based capture workflows, thus increasing the overall sensitivity. With this platform in hand, the authors successfully applied this strategy (ABPP-MudPIT, Activity-Based Protein Profiling-Multidimensional Identification Technology) to the identification of seven endogenous MMPs in an *in vitro* model of early-stage cartilage degradation [137].

Despite the success of the aforementioned technique in enabling efficient activity-based profiling of MPs in complex proteomes, a more refined, MMP-dedicated approach was proposed a year later by Dive and co-workers [139], based on the use of optimized MMP inhibitors able to potently and selectively target this class of enzymes (Fig. 4, probe **7**). The authors recognized that lowering the detection limit was essential for accurately recording the functional state of MMPs in complex proteomes, where MMPs are predominantly present either as zymogens or in TIMP-bound forms. To this end, the improvement of cross-linking yields was identified as a key priority, prompting the redesign of photoactivatable A*f*BPs. More specifically, guided by advances in inhibitor development and relevant crystallographic data, the photoreactive group was strategically positioned in the P_1' side-chain of a tight-binding phosphinic pseudopeptide inhibitor. This placement would minimize solvent-induced quenching of reactive nitrene intermediates generated upon irradiation of the parent azide photoreactive group, leading to increased probe/protein proximity and crosslinking yields. Indeed, by utilizing probe **7** and radioactive labelling, 8 different MMPs

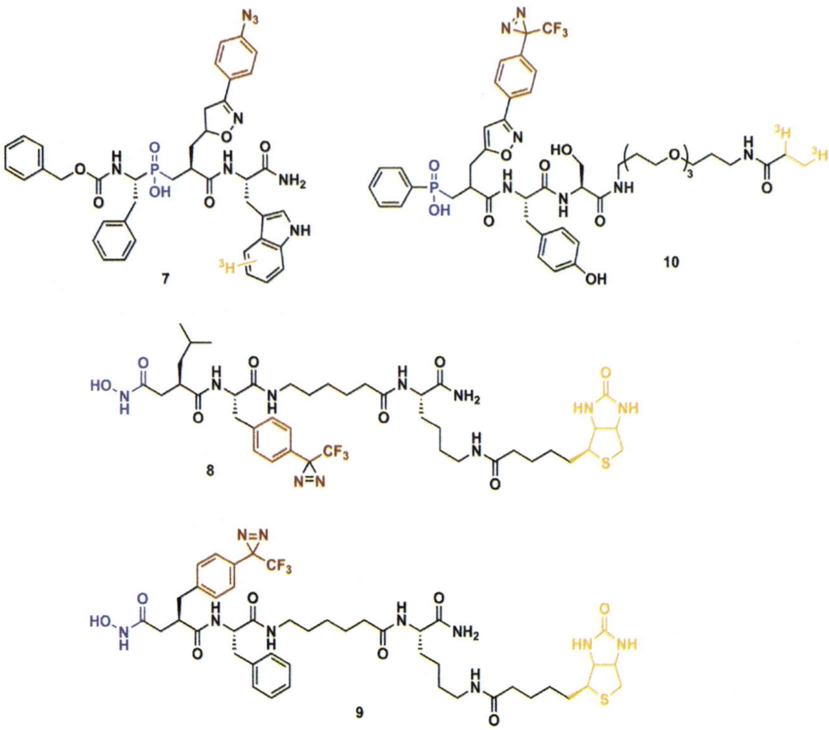

Fig. 4 Chemical structures of probes developed by the research groups of Dive (probes 7 and 10, [139,142]) and Overkleeft/Bischoff (probes 8 and 9, [143,144]). The ZBG, photoreactive group and radioactive/analytical tag are colored in blue, brown and orange, respectively.

were detected with varying sensitivity, reaching a detection limit of 2.5 fmol for MMP-12. Notably, crosslinking efficacy ranged from 1 percent for MMP-8 to 42 percent for MMP-12, indicating that probe sensitivity is strongly influenced by the microenvironment of the photoreactive group within the S_1' cavity. Subsequent studies attributed this preference for MMP-12 to covalent modification of the ε-NH_2 of Lys241 residue located in its S_1' loop, a residue not conserved across MMPs, explaining the marked differences in crosslinking yields between probe **7** and other family members [140]. At higher pH, labelling of MMP-12 as well as other MMPs bearing nucleophilic residues at an equivalent position (e.g., His in MMP-3), was furthered enhanced underscoring the importance of an appropriately positioned nucleophile in trapping nitrene intermediates generated upon irradiation [141].

The dense network of P_1'/S_1' interactions that has been well-documented in crystallographic studies [145], likely creates the conditions for proximity-driven, faster crosslinking compared to other primed subsites of MMPs where less defined, opportunistic interactions may allow diffusion effects to lead to quenching of reactive intermediates by the solvent. To this respect, Overkleeft, Bischoff and co-workers compared two hydroxamate peptidyl probes (Fig. 4, probes **8** and **9**) bearing a diazirine photoreactive group either at P_2' (**8**) or at P_1' (**9**) and observed that A*f*BP **8** was markedly more efficient across a panel of nine MMP family members [143,144]. Beyond positional effects, the broader efficiency of diazirine-based probes can be partially attributed to the chemical versatility of the carbene intermediates generated upon nitrogen expulsion from trifluoromethyl diazirines, which can insert into a wide range of X-H bonds (ranging from O–H bonds to typically inert C–H bonds) in contrast to electrophilic nitrene intermediates, which react exclusively with nucleophilic residues. This was further demonstrated in the case of phosphinic probe **10** (Fig. 4) bearing a diazirine photoreactive group at the terminus of a long, hydrophobic P_1' side chain [142]. Using a broad-spectrum phosphinic MMP inhibitor as a structural template, probe **10** proved to be a pan-MMP activity probe exhibiting high labelling efficacy across 7 MMPs tested with crosslinking yields ranging from 11 to 58 percent, which are substantially improved compared to probe **7**.

Interestingly, labelling preferences between the azide probe **7** and the diazirine probe **10** (Fig. 4) were not equivalent. For instance, probe **7** yields negligible labelling of MMP-3 (~1 percent) and strong labelling of MMP-12 (42 percent), whereas the opposite pattern is observed with diazirine probe **10** (42 percent for MMP-3 and 11 percent for MMP-12). These discrepancies emphasize that crosslinking yield depends not only on the mechanism of crosslinking reaction itself, but also on subtle spatial parameters that govern the proximity and orientation of the reactive group relative to the active site. Moreover, estimations on crosslinking yields cannot be derived from comparison of K_is as these two parameters must not be directly correlated. Geurink et al. observed that although probe **8** is 25-fold more potent inhibitor for ADAM-17 than probe **9**, both are equally efficient in labelling ADAM-17. Similarly, comparison of K_is of probe **7** with crosslinking yield denotes that these two properties must not be directly correlated. For instance, probe **10** labels MMP-3 and MMP-12 with similar efficiency although probe **10** is a far weaker inhibitor of MMP-3 (K_i = 527 nM) than of MMP-12 (K_i = 1.0 nM). This dissociation between inhibitory

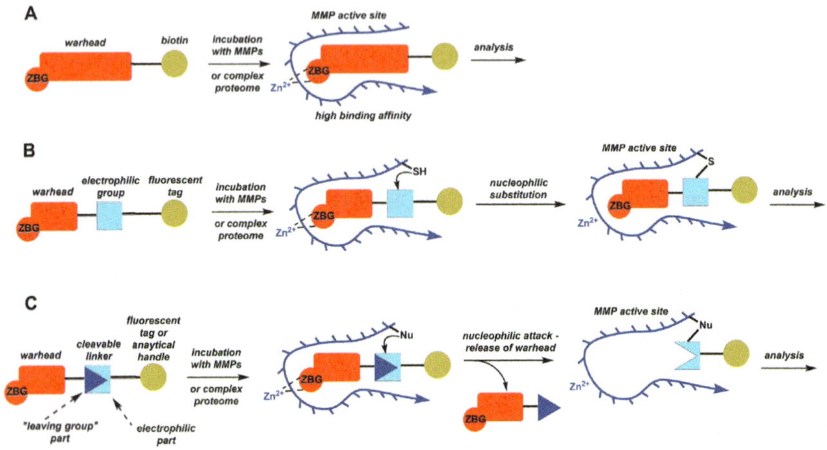

Fig. 5 Mechanism of action of alternative probes of MMPs that do not rely on photo-crosslinking: A) affinity probe, developed by Bregant et al. [146], B) proximity-induced labelling by electrophilic probes, developed by Bogyo and co-workers [147], and C) proximity-induced labelling by electrophilic probes bearing cleavable linkers, developed by Devel and co-workers [148–150].

potency and labelling efficiency reflects the influence of residence time on crosslinking yield which is better associated to K_{off} and not K_{i}. Factors affecting K_{off} will modulate the lifetime of enzyme-probe complex and therefore the probability of productive photo-crosslinking, independently of overall binding affinity.

9. Next generation of affinity-based probes for MMPs: Novel technologies beyond photoactivation

The first period of development of A*f*BPs for MMPs was defined by the need of including an additional irradiation step to mediate covalent binding between a photoreactive probe and the target metalloprotease. However, limitations associated with the large variability in crosslinking yields across MMP family members, the incompatibility of irradiation with use in living organisms and the technical demands of the photoactivation step itself, motivated research efforts towards alternative, next-generation probes that would not depend on irradiation. From a mechanistic perspective, these approaches are summarized in Fig. 5.

The first approach (Fig. 5A) relies exclusively on binding affinity, rather than covalent crosslinking, with the prerequisite that the probe achieves

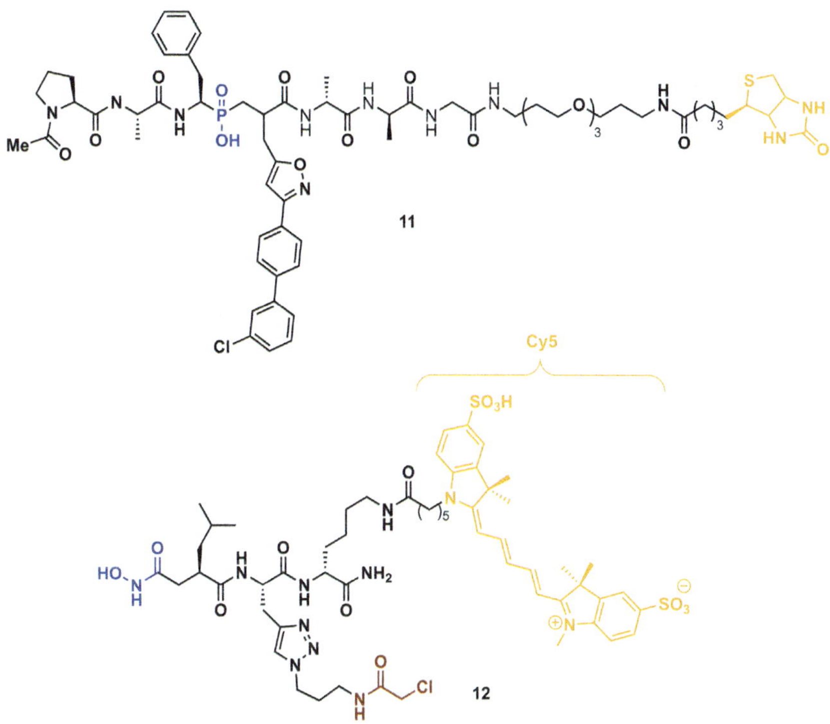

Fig. 6 Chemical structures of probes developed by the research groups of Dive (probes 11, [146]) and Bogyo (probe 12,[147]). The ZBG, photoreactive group and fluorescent/analytical tag are colored in blue, brown and orange, respectively.

subnanomolar binding affinity for MMPs. Therefore, probe **11** (Fig. 6) was developed that was found superior compared to a structurally related photoreactive probe in capturing hMMP-12 and hMMP-8 in tumor extracts using streptavidin magnetic beads [146]. Although the absence of covalent bond formation might raise concerns regarding capture efficiency, the exceptional potency of probe **11** (K_i values of 23 and 11 pM toward hMMP-8 and hMMP-12, respectively), confers remarkable stability to the MMP/probe **11** complex, corresponding to a dissociation half-life of about 2.7 h. For hMMP-12, quantitative capture and detection limits as low as 25 femtomoles were reported.

The second approach, depicted in Fig. 5B, was developed by the Bogyo group and relies on proximity-induced labelling of engineered proteases with electrophilic probes bearing an α-chloroacetamide group at their P_2' position [147]. Specifically, the researchers introduced a Cys residue on MMP-12 and MMP-14 generating mutants with properties identical to those of wild-type enzymes both *in vitro* and *in cellulo*. These engineered MMPs could be efficiently labelled by probe **12** (Fig. 6) through a mechanism involving nucleophilic attack of the engineered Cys thiol to the α-chloroacetamide thus forming a stable covalent complex. While this approach is not applicable to native enzymes in standard protein profiling settings, its extension to *in vivo* functional studies on specific MMPs would require gene-editing techniques to generate stable cell lines that express the engineered variant. Indeed, the same group successfully applied this strategy to profile activity of MMP-14 in several cell types as well as in breast cancer tumors [151].

Recent advances in irradiation-free technologies for the chemical labelling of native proteins pioneered by Hamachi's group [152,153] inspired the development of next-generation A*f*BPs for MMPs. In particular, our group recently demonstrated that ligand-directed chemistry based on acylimidazole and *N*-acyl-*N*-alkyl sulphonamide (NASA) cleavable linkers can be successfully applied to the covalent modification of the active site of MMPs without any external trigger [148]. These probes were built on the structural scaffold of a potent and selective inhibitor of MMP-12 (RXP470.1, [154]) with the cleavable linker introduced at the P_3' position (Fig. 7, probes **13a** and **13b**), targeting a Lys residue within the S_3' region. Although maximum labelling efficiency, comparable to that of photoactivatable probes, was achieved with probes **13** for MMP-12, weaker labelling was also observed for MMP-3, MMP-9 and MMP-13, which was attributed to the presence of reactive Lys residues with flexible side chains in their S_3' region. As illustrated in Fig. 5C, the warhead is released upon labelling, enabling the modified protein to restore its proteolytic activity, as experimentally demonstrated in the case of MMP-12 labelled with probe **13a**. This unique feature unlocks possibilities for investigating the localization of labelled proteins while preserving their functionality. However, probe selectivity as well as undesired off-target labelling are strongly governed by the reactivity of the cleavable linker, rendering careful tuning of this property essential for overall probe optimization.

Building on this framework, our group developed a second generation of probes incorporating a modified NASA cleavable linker with moderate electrophilicity, aiming to improve selectivity and reduce non-specific reactivity in complex biological matrices [150]. In these probes (Fig. 7, probes **14** and **15**), the cleavable linker was repositioned to S_3 subsite, which according to structural and sequence alignment data harbors several lysine residues

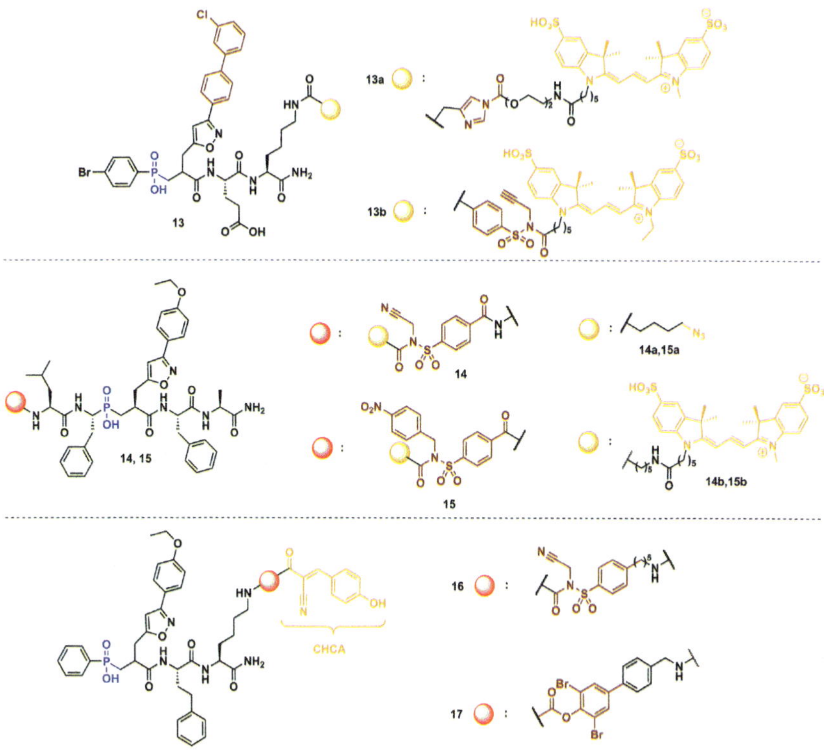

Fig. 7 Chemical structures of MMP probes 13–17 bearing cleavable linkers for proximity-induced labelling, developed by Devel and co-workers [148,150,155]. The ZBG, photoreactive group and fluorescent/analytical tag are colored in blue, brown and orange, respectively.

across a broader range of MMPs, thereby increasing crosslinking probability and enabling broader-spectrum MMP profiling. Comparative analysis of probe **14** and the less reactive probe **15** revealed that reduced electrophilicity lowers labelling yield but also substantially suppresses non-specific off-target labelling of abundant proteins in cell lysates, such as serum albumin.

In the same context, the most recent contribution from our group provides a practical framework for enrichment-free MALDI-detection of MMPs, achieved by transferring an α-cyano-4-hydroxycinammic acid (CHCA) mass tag onto MMPs using A*f*BPs **16** and **17** (Fig. 7) [155]. These A*f*BPs incorporated either NASA (probe **16**) or dibromophenyl benzoate (BB, probe **17**) as the cleavable linker. Probe **16** was able to label a broader range of MMPs whereas probe **17** containing the less reactive BB linker was highly selective for MMP-12. The substantial improvement in signal-to-noise ratio facilitated by the CHCA mass tag enables detection of multiple targets in complex proteomes at nanomolar concentration, without the need for enrichment and with selectivity tunable through the electrophilicity of the cleavable linker. Future directions include the development of CHCA-derived AfBPs targeting other Lys-rich regions, such as S_3 region to broaden MMP coverage, as well as the identification of alternative nucleophilic residues that could be potentially targeted within the active site of MMPs. Undoubtedly, the development of optimized probes in the future will require a thorough understanding of the key parameters governing protein-probe interactions, rigorous assessment of current knowledge on factors affecting labelling yield and selectivity and precise modulation of chemical reactivity to suppress non-specific labelling and extend probe applicability to cellular and *in vivo* settings [156].

10. Perspectives for activity-based probes targeting MMPs

One of the major future directions for activatable MMP probes will be the development of peptide substrates displaying improved selectivity toward MMPs while minimizing off-target cleavage by other protease families. In this context, recent advances in computational prediction tools and machine learning approaches are expected to considerably accelerate substrate optimization. Deep learning-based algorithms have notably emerged as powerful tools to predict protease cleavage sites and anticipate potential off-target proteolytic events [157,158]. Beyond cleavage prediction alone, very recent studies have further demonstrated that artificial intelligence-guided approaches can be used to generate novel peptide substrates with enhanced selectivity not only for MMPs as a family, but even for individual MMP isoforms [159]. Importantly, substrate selectivity does not solely depend on the peptide sequence itself, but is also strongly influenced by the chemical environment flanking both sides of the MMP-cleavable motif, including the nature of fluorophores, quenchers, polymers, nanoparticles, or anchoring groups incorporated within the probe architecture.

Once experimentally validated, highly selective peptide substrates, whether targeting a subset of MMPs or a single MMP isoform, could be incorporated into multiple classes of activatable constructs adapted to distinct biomedical applications. In particular, MMP-selective fluorescent probes could become highly valuable tools for *in situ* zymography on pathological tissue sections obtained from patients, with the aim of monitoring changes in spatial MMP activation patterns during disease progression or according to pathological grade. In this setting, achieving high selectivity toward individual MMPs may prove critical to properly interpret the biological significance of the detected signals. Such approaches could additionally be strengthened through parallel validation using selective MMP inhibitors.

Another highly promising application concerns fluorescence-guided surgery and intraoperative imaging. In this regard, AVB-620, a ratiometric protease-activatable fluorescent peptide probe primarily targeting MMP-2 and MMP-9, was evaluated in a first-in-human Phase I clinical study performed in patients undergoing breast cancer surgery (NCT02391194), where it demonstrated acceptable safety together with detectable tumor-versus-normal tissue contrast. A Phase II clinical study in women undergoing surgery for primary nonrecurrent nonmetastatic breast cancer has also recently been completed (NCT03113825). Similarly, Pegulicianine (Lumicell, LUM015, LumiSystem), a clinically advanced protease-activatable fluorescent imaging agent responsive to both MMPs and cathepsins depending on tissue context, has been evaluated for intraoperative detection of residual tumor tissue during breast cancer surgery, notably in the INSITE trial (NCT03686215). While AVB-620 currently represents one of the most direct examples of an MMP substrate-activatable optical probe translated into clinical imaging studies, Pegulicianine illustrates the broader clinical potential of protease-responsive imaging systems integrating multiple proteolytic inputs.

Among the different classes of activatable probes, synthetic biomarkers currently appear as one of the most clinically disruptive evolutions of protease-responsive systems. In contrast to imaging probes, these approaches do not require sophisticated imaging infrastructure and instead convert local protease activity into easily detectable urinary or breath reporters. Importantly, they also enable the simultaneous monitoring of multiple proteases, an aspect that is particularly relevant in biomarker research, where complex pathological states are generally better characterized through multidimensional biomarker signatures rather than isolated individual markers. In this perspective, multiplexed protease profiling may provide a more accurate representation of disease state, progression, or therapeutic response. Nevertheless, further efforts will be required to simplify the downstream detection methods to facilitate routine clinical implementation. In this regard, the development of lateral-flow assays, colorimetric tests, CRISPR-amplified readouts, and more recently synthetic breath biomarkers represents an important step toward simple, rapid, and potentially point-of-care-compatible diagnostic platforms.

Although affinity-based probes (AfBPs) enable the strict identification of active MMPs within complex proteomes, their application has so far remained restricted to a limited number of biological matrices, with only one study reporting their use in clinically relevant biological fluids [138]. This limitation mainly arises from the complexity of the analytical workflow, which generally involves a photoaffinity labelling step requiring dedicated photo-irradiation equipment, followed by enrichment and proteomic identification by mass spectrometry. Recently, the development of a new generation of AfBPs capable of forming covalent MMP–probe adducts without photoactivation has considerably simplified the labelling process [148–150]. Moreover, the extension of this photoactivation-free strategy into mass-encoded activity-based protein profiling workflows has further streamlined the analytical pipeline [155]. In particular, the direct identification and quantification of active MMPs through MALDI-based activity fingerprints provides a scalable and multiplexable platform for investigating MMP proteolytic activity across diverse native biological environments, including human biological fluids containing active MMP species or MMP–α2-macroglobulin complexes.

Beyond liquid biopsies, the intrinsic compatibility of MALDI with mass spectrometry imaging should also enable spatially resolved mapping of MMP activity directly within tissue sections. Furthermore, combining activatable probes with mass-encoded activity-based protein profiling could potentially leverage the complementary strengths of both approaches, namely the high sensitivity and signal amplification associated with substrate-based activatable probes together with the strict molecular identification enabled by AfBPs. Such hybrid strategies could ultimately facilitate the discovery of functional MMP activity biomarkers and accelerate the translational development of activity-based probes targeting MMPs.

References

[1] L.G.N. de Almeida, H. Thode, Y. Eslambolchi, S. Chopra, D. Young, S. Gill, L. Devel, A. Dufour, Matrix metalloproteinases: from molecular mechanisms to physiology, pathophysiology, and pharmacology, Pharmacol. Rev. 74 (3) (2022) 712–768, doi:10.1124/pharmrev.121.000349.

[2] G.S Butler, C.M. Overall, Updated biological roles for matrix metalloproteinases and new « intracellular » substrates revealed by degradomics, Biochemistry 48 (46) (2009) 10830–10845, doi:10.1021/bi901656f.

[3] P. Schlage, U. auf dem Keller, Proteomic approaches to uncover MMP function, Matrix Biol. 44–46 (2015) 232–238, doi:10.1016/j.matbio.2015.01.003.

[4] O. Kollet, A. Das, N. Karamanos, U. Auf dem Keller, I. Sagi, Redefining metalloproteases specificity through network proteolysis, Trends Mol. Med. 30 (2) (2024) 147–163, doi:10.1016/j.molmed.2023.11.001.

[5] A. Dufour, C.M. Overall, Missing the target: matrix metalloproteinase antitargets in inflammation and cancer, Trends Pharmacol. Sci. 34 (4) (2013) 233–242, doi:10.1016/j.tips.2013.02.004.

[6] R.E Vandenbroucke, C. Libert, Is there new hope for therapeutic matrix metalloproteinase inhibition? Nat. Rev. Drug Discov. 13 (12) (2014) 904–927, doi:10.1038/nrd4390.

[7] C.M Overall, O. Kleifeld, Tumour microenvironment—opinion: validating matrix metalloproteinases as drug targets and anti-targets for cancer therapy, Nat. Rev. Cancer 6 (3) (2006) 227–239, doi:10.1038/nrc1821.

[8] C.M Overall, O. Kleifeld, Towards third generation matrix metalloproteinase inhibitors for cancer therapy, Br. J. Cancer 94 (7) (2006) 941–946, doi:10.1038/sj.bjc.6603043.

[9] A.P Amini, J.D Kirkpatrick, C.S Wang, A.M Jaeger, S. Su, S. Naranjo, Q. Zhong, C.M Cabana, T. Jacks, S.N. Bhatia, Multiscale profiling of protease activity in cancer, Nat. Commun. 13 (1) (2022) 5745, doi:10.1038/s41467-022-32988-5.

[10] E.S. Radisky, Extracellular proteolysis in cancer: proteases, substrates, and mechanisms in tumor progression and metastasis, J. Biol. Chem. 300 (6) (2024) 107347, doi:10.1016/j.jbc.2024.107347.

[11] J.L Montoya Mira, A. Quentel, R.K Patel, D. Keith, M. Sousa, J. Minnier, B.R Kingston, L. David, S.C Esener, R.C Sears, C.D Lopez, B.C Sheppard, U. Demirci, M.H Wong, J.M. Fischer, Early detection of pancreatic cancer by a high-throughput protease-activated nanosensor assay, Sci. Transl. Med. 17 (785) (2025) eadq3110, doi:10.1126/scitranslmed.adq3110.

[12] B. Grillet, R.VS. Pereira, J. Van Damme, A. Abu El-Asrar, P. Proost, G. Opdenakker, Matrix metalloproteinases in arthritis: towards precision medicine, Nat. Rev. Rheumatol. 19 (6) (2023) 363–377, doi:10.1038/s41584-023-00966-w.

[13] B.E Turk, L.L Huang, E.T Piro, L.C. Cantley, Determination of protease cleavage site motifs using mixture-based oriented peptide libraries, Nat. Biotechnol. 19 (7) (2001) 661–667, doi:10.1038/90273.

[14] S.J Kridel, E. Chen, L.P Kotra, E.W Howard, S. Mobashery, J.W. Smith, Substrate hydrolysis by matrix metalloproteinase-9, J. Biol. Chem. 276 (23) (2001) 20572–20578, doi:10.1074/jbc.M100900200.

[15] M.M Smith, L. Shi, M. Navre, Rapid identification of highly active and selective substrates for stromelysin and matrilysin using bacteriophage peptide display libraries, J. Biol. Chem. 270 (12) (1995) 6440–6449, doi:10.1074/jbc.270.12.6440.

[16] S. Ohkubo, K. Miyadera, Y. Sugimoto, K. Matsuo, K. Wierzba, Y. Yamada, Identification of substrate sequences for membrane type-1 matrix metalloproteinase using bacteriophage peptide display library, Biochem. Biophys. Res. Commun. 266 (2) (1999) 308–313, doi:10.1006/bbrc.1999.1816.

[17] S.J Deng, D.M Bickett, J.L Mitchell, M.H Lambert, R.K Blackburn, H.L Carter, J. Neugebauer, G. Pahel, M.P Weiner, M.L. Moss, Substrate specificity of human collagenase 3 assessed using a phage-displayed peptide library, J. Biol. Chem. 275 (40) (2000) 31422–31427, doi:10.1074/jbc.M004538200.

[18] U. Eckhard, P.F Huesgen, O. Schilling, C.L Bellac, G.S Butler, J.H Cox, A. Dufour, V. Goebeler, R. Kappelhoff, K.UA. dem, T. Klein, P.F Lange, G. Marino, C.J Morrison, A. Prudova, D. Rodriguez, A.E Starr, Y. Wang, C.M Overall, Active site specificity profiling of the matrix metalloproteinase family: proteomic identification of 4300 cleavage sites by nine MMPs explored with structural and synthetic peptide cleavage analyses, Matrix Biol. 49 (2016) 37–60, doi:10.1016/j.matbio.2015.09.003.

[19] M. Kukreja, S.A Shiryaev, P. Cieplak, N. Muranaka, D.A Routenberg, A.V Chernov, S. Kumar, A.G Remacle, J.W Smith, I.A Kozlov, A.Y. Strongin, High-throughput multiplexed peptide-centric profiling illustrates both substrate cleavage redundancy and specificity in the MMP family, Chem. Biol. 22 (8) (2015) 1122–1133, doi:10.1016/j.chembiol.2015.07.008.
[20] L. Hao, R.T Zhao, N.L Welch, E.KW. Tan, Q. Zhong, N.S Harzallah, C. Ngambenjawong, H. Ko, H.E Fleming, P.C Sabeti, S.N. Bhatia, CRISPR-Cas-amplified urinary biomarkers for multiplexed and portable cancer diagnostics, Nat. Nanotechnol. 18 (7) (2023) 798–807, doi:10.1038/s41565-023-01372-9.
[21] J. Hu, F. Liu, Y. Chen, J. Fu, G. Shangguan, H. Ju, Mass-encoded suspension array for multiplex detection of matrix metalloproteinase activities, Anal. Chem. 94 (16) (2022) 6380–6386, doi:10.1021/acs.analchem.2c00854.
[22] J.D Kirkpatrick, A.D Warren, A.P Soleimany, P.MK. Westcott, J.C Voog, C. Martin-Alonso, H.E Fleming, T. Tammela, T. Jacks, S.N. Bhatia, Urinary detection of lung cancer in mice via noninvasive pulmonary protease profiling, Sci. Transl. Med. 12 (537) (2020) eaaw0262, doi:10.1126/scitranslmed.aaw0262.
[23] J.D Kirkpatrick, A.P Soleimany, J.S Dudani, H.J Liu, H.C Lam, C. Priolo, E.P Henske, S.N. Bhatia, Protease activity sensors enable real-time treatment response monitoring in lymphangioleiomyomatosis, Eur. Respir. J. 59 (4) (2022) 2100664, doi:10.1183/13993003.00664-2021.
[24] Q. Zhong, E.KW. Tan, C. Martin-Alonso, T. Parisi, L. Hao, J.D Kirkpatrick, T. Fadel, H.E Fleming, T. Jacks, S.N. Bhatia, Inhalable point-of-care urinary diagnostic platform, Sci. Adv. 10 (1) (2024) eadj9591, doi:10.1126/sciadv.adj9591.
[25] C. Bremer, C.H Tung, R. Weissleder, In vivo molecular target assessment of matrix metalloproteinase inhibition, Nat. Med. 7 (6) (2001) 743–748, doi:10.1038/89126.
[26] S. Lee, K. Park, S.Y Lee, J.H Ryu, J.W Park, H.J Ahn, I.C Kwon, I.C Youn, K. Kim, K. Choi, Dark quenched matrix metalloproteinase fluorogenic probe for imaging osteoarthritis development in vivo, Bioconjug. Chem. 19 (9) (2008) 1743–1747, doi:10.1021/bc800264z.
[27] S. Lee, J.H Ryu, K. Park, A. Lee, S.Y Lee, I.C Youn, C.H Ahn, S.M Yoon, S.J Myung, D.H Moon, X. Chen, K. Choi, I.C Kwon, K. Kim, Polymeric nanoparticle-based activatable near-infrared nanosensor for protease determination in vivo, Nano Lett. 9 (12) (2009) 4412–4416, doi:10.1021/nl902709m.
[28] X. Lin, J. Xie, L. Zhu, S. Lee, G. Niu, Y. Ma, K. Kim, X. Chen, Hybrid ferritin nanoparticles as activatable probes for tumor imaging, Angew. Chem. Int. Ed Engl. 50 (7) (2011) 1569–1572, doi:10.1002/anie.201006757.
[29] I.C Sun, D.K Eun, H. Koo, C.Y Ko, H.S Kim, D.K Yi, K. Choi, I.C Kwon, K. Kim, C.H Ahn, Tumor-targeting gold particles for dual computed tomography/optical cancer imaging, Angew. Chem. Int. Ed Engl. 50 (40) (2011) 9348–9351, doi:10.1002/anie.201102892.
[30] E.J Cha, E.S Jang, I.C Sun, I.J Lee, J.H Ko, Y.I Kim, I.C Kwon, K. Kim, C.H. Ahn, Development of MRI/NIRF « activatable » multimodal imaging probe based on iron oxide nanoparticles, J. Control Release 155 (2) (2011) 152–158, doi:10.1016/j.jconrel.2011.07.019.
[31] L. Zhu, J. Xie, M. Swierczewska, F. Zhang, Q. Quan, Y. Ma, X. Fang, K. Kim, S. Lee, X. Chen, Real-time video imaging of protease expression in vivo, Theranostics 1 (2011) 18–27, doi:10.7150/thno/v01p0018.
[32] X. Xia, M. Yang, L.K Oetjen, Y. Zhang, Q. Li, J. Chen, Y. Xia, An enzyme-sensitive probe for photoacoustic imaging and fluorescence detection of protease activity, Nanoscale 3 (3) (2011) 950–953, doi:10.1039/c0nr00874e.

[33] C.W Huang, Z. Li, P.S. Conti, Radioactive smart probe for potential corrected matrix metalloproteinase imaging, Bioconjug. Chem. 23 (11) (2012) 2159–2167, doi:10.1021/bc3001968.
[34] Y. Cai, L. Zhu, F. Zhang, G. Niu, S. Lee, S. Kimura, X. Chen, Noninvasive monitoring of pulmonary fibrosis by targeting matrix metalloproteinases (MMPs), Mol. Pharm. 10 (6) (2013) 2237–2247, doi:10.1021/mp300613x.
[35] L. Zhu, Y. Ma, D.O Kiesewetter, Y. Wang, L. Lang, S. Lee, G. Niu, X. Chen, Rational design of matrix metalloproteinase-13 activatable probes for enhanced specificity, ACS Chem. Biol. 9 (2) (2014) 510–516, doi:10.1021/cb400698s.
[36] S. Lee, S.W Kang, J.H Ryu, J.H Na, D.E Lee, S.J Han, C.M Kang, Y.S Choe, K.C Lee, J.F Leary, K. Choi, K.H Lee, K. Kim, Tumor-homing glycol chitosan-based optical/PET dual imaging nanoprobe for cancer diagnosis, Bioconjug. Chem. 25 (3) (2014) 601–610, doi:10.1021/bc500020g.
[37] X. Zhao, C.X Yang, L.G Chen, X.P. Yan, Dual-stimuli responsive and reversibly activatable theranostic nanoprobe for precision tumor-targeting and fluorescence-guided photothermal therapy, Nat. Commun. 8 (2017) 14998, doi:10.1038/ncomms14998.
[38] L. Sun, S. Xie, J. Qi, E. Liu, D. Liu, Q. Liu, S. Chen, H. He, V.C. Yang, Cell-permeable, MMP-2 activatable, nickel ferrite and his-tagged fusion protein self-assembled fluorescent nanoprobe for tumor magnetic-targeting and imaging, ACS Appl. Mater. Interfaces 9 (45) (2017) 39209–39222, doi:10.1021/acsami.7b12918.
[39] C. Liu, S. Li, Y. Gu, H. Xiong, W.T Wong, L. Sun, Multispectral photoacoustic imaging of tumor protease activity with a gold nanocage-based activatable probe, Mol. Imaging Biol. 20 (6) (2018) 919–929, doi:10.1007/s11307-018-1203-1.
[40] H. Shi, Y. Sun, R. Yan, S. Liu, L. Zhu, S. Liu, Y. Feng, P. Wang, J. He, Z. Zhou, D. Ye, Magnetic semiconductor Gd-doping CuS nanoparticles as activatable nanoprobes for bimodal imaging and targeted photothermal therapy of gastric tumors, Nano Lett. 19 (2) (2019) 937–947, doi:10.1021/acs.nanolett.8b04179.
[41] L. Yin, H. Sun, H. Zhang, L. He, L. Qiu, J. Lin, H. Xia, Y. Zhang, S. Ji, H. Shi, M. Gao, Quantitatively visualizing tumor-related protease activity in vivo using a ratiometric photoacoustic probe, J. Am. Chem. Soc. 141 (7) (2019) 3265–3273, doi:10.1021/jacs.8b13628.
[42] N. Hananya, O. Press, A. Das, A. Scomparin, R. Satchi-Fainaro, I. Sagi, D. Shabat, Persistent chemiluminescent glow of phenoxy-dioxetane luminophore enables unique CRET-based detection of proteases, Chemistry 25 (64) (2019) 14679–14687, doi:10.1002/chem.201903489.
[43] J. Xu, L. Fang, M. Shi, Y. Huang, L. Yao, S. Zhao, L. Zhang, H. Liang, A peptide-based four-color fluorescent polydopamine nanoprobe for multiplexed sensing and imaging of proteases in living cells, Chem. Commun. (Camb) 55 (11) (2019) 1651–1654, doi:10.1039/c8cc09359h.
[44] L. Yin, H. Sun, M. Zhao, A. Wang, S. Qiu, Y. Gao, J. Ding, S.J Ji, H. Shi, M. Gao, Rational design and synthesis of a metalloproteinase-activatable probe for dual-modality imaging of metastatic lymph nodes in vivo, J. Org. Chem. 84 (10) (2019) 6126–6133, doi:10.1021/acs.joc.9b00331.
[45] F. Liu, R. Chen, W. Song, L. Li, C. Lei, Z. Nie, Modular combination of proteolysis-responsive transcription and spherical nucleic acids for smartphone-based colorimetric detection of protease biomarkers, Anal. Chem. 93 (7) (2021) 3517–3525, doi:10.1021/acs.analchem.0c04894.
[46] Y. Fang, Y. Li, Y. Li, R. He, Y. Zhang, X. Zhang, Y. Liu, H. Ju, In situ protease secretion visualization and metastatic lymph nodes imaging via a cell membrane-anchored upconversion nanoprobe, Anal. Chem. 93 (19) (2021) 7258–7265, doi:10.1021/acs.analchem.1c00469.

[47] W. Zeng, L. Wu, Y. Sun, Y. Wang, J. Wang, D. Ye, Ratiometric imaging of MMP-2 activity facilitates tumor detection using activatable near-infrared fluorescent semiconducting polymer nanoparticles, Small 17 (36) (2021) e2101924, doi:10.1002/smll.202101924.
[48] W. Huang, W. Zeng, Z. Huang, D. Fang, H. Liu, M. Feng, L. Mao, D. Ye, Ratiometric afterglow luminescent imaging of matrix metalloproteinase-2 activity via an energy diversion process, Angew. Chem. Int. Ed Engl. 63 (26) (2024) e202404244, doi:10.1002/anie.202404244.
[49] L. Wu, Q. Tong, X. Cao, D. Zhang, F. Yang, H. Lin, Q. Fan, Dual-locked near-infrared fluorescent probe for real-time imaging of hydrogen sulfide/matrix metallopeptidase-2 in vivo, ACS Nano 19 (7) (2025) 7294–7305, doi:10.1021/acsnano.4c17799.
[50] L. Wang, H. Xie, X. Zhou, Y. Lin, Y. Qin, J. Yang, J. Zhao, G. Li, An electrochemical biosensor to identify the phenotype of aggressive breast cancer cells, Chem. Commun. (Camb) 59 (26) (2023) 3890–3893, doi:10.1039/d3cc00611e.
[51] Q. Dang, H. Gao, Z. Li, H. Qi, Q. Gao, C. Zhang, Simple and sensitive electrogenerated chemiluminescence peptide-based biosensor for detection of matrix metalloproteinase 2 released from living cells, Anal. Bioanal. Chem. 408 (25) (2016) 7067–7075, doi:10.1007/s00216-016-9360-z.
[52] X. Fan, S. Wang, H. Liu, Z. Li, Q. Sun, Y. Wang, X. Fan, A sensitive electrochemiluminescence biosensor for assay of cancer biomarker (MMP-2) based on NGQDs-Ru@SiO2 luminophore, Talanta 236 (2022) 122830, doi:10.1016/j.talanta.2021.122830.
[53] M. Qian, H. Huang, D. Zhang, C. Zhang, H. Qi, Coordination-based site-specific labeling strategy for electrogenerated chemiluminescence biosensing of matrix metalloproteinase 2, Anal. Chem. 96 (49) (2024) 19504–19510, doi:10.1021/acs.analchem.4c04078.
[54] C. Wang, M. Qian, X. Liu, X. Yang, C. Zhang, M. Li, H. Qi, Peptide sequence modulating the analytical performance of electrogenerated chemiluminescence peptide-based biosensors for matrix metalloproteinase 2, Anal. Chem. 97 (37) (2025) 20013–20019, doi:10.1021/acs.analchem.5c04487.
[55] Q. Zhu, X. Sun, Y. Du, X. Zhang, X. Ren, H. Ma, D. Wu, H. Ju, Q. Wei, Detection of matrix metalloproteinase 2 by a polypeptide cleavable split-type sensor based on atomically precise bimetallic nanocluster Au3Ag5(MSA)3, Anal. Chem. 97 (27) (2025) 14574–14583, doi:10.1021/acs.analchem.5c02007.
[56] Q. Luo, C. Zheng, Y. Huang, L. Liang, Y. Gong, J. Zhang, Q. Tang, K. Zhang, X. Liao, Polydopamine-stabilized CsPbBr3 enables toxicity-reduced ECL detection of MMP-2 via CRISPR/Cas12a trans-cleavage, Biosens. Bioelectron. 293 (2026) 118193, doi:10.1016/j.bios.2025.118193.
[57] K. Yang, L. Zhu, L. Nie, X. Sun, L. Cheng, C. Wu, G. Niu, X. Chen, Z. Liu, Visualization of protease activity in vivo using an activatable photo-acoustic imaging probe based on CuS nanoparticles, Theranostics 4 (2) (2014) 134–141, doi:10.7150/thno.7217.
[58] X. Yue, Z. Wang, L. Zhu, Y. Wang, C. Qian, Y. Ma, D.O Kiesewetter, G. Niu, X. Chen, Novel 19F activatable probe for the detection of matrix metalloprotease-2 activity by MRI/MRS, Mol. Pharm. 11 (11) (2014) 4208–4217, doi:10.1021/mp500443x.
[59] J. Hu, F. Liu, Y. Chen, J. Fu, H. Ju, Signal-on mass spectrometric biosensing of multiplex matrix metalloproteinases with a phospholipid-structured mass-encoded microplate, Anal. Chem. 95 (23) (2023) 8974–8981, doi:10.1021/acs.analchem.3c01039.
[60] G.A Kwong, G. von Maltzahn, G. Murugappan, O. Abudayyeh, S. Mo, I.A Papayannopoulos, D.Y Sverdlov, S.B Liu, A.D Warren, Y. Popov, D. Schuppan, S.N. Bhatia, Mass-encoded synthetic biomarkers for multiplexed urinary monitoring of disease, Nat. Biotechnol. 31 (1) (2013) 63–70, doi:10.1038/nbt.2464.

[61] A.D Warren, G.A Kwong, D.K Wood, K.Y Lin, S.N. Bhatia, Point-of-care diagnostics for noncommunicable diseases using synthetic urinary biomarkers and paper microfluidics, Proc. Natl. Acad. Sci. U.S.A. 111 (10) (2014) 3671–3676, doi:10.1073/pnas.1314651111.
[62] J.S Dudani, M. Ibrahim, J. Kirkpatrick, A.D Warren, S.N. Bhatia, Classification of prostate cancer using a protease activity nanosensor library, Proc. Natl. Acad. Sci. U.S.A. 115 (36) (2018) 8954–8959, doi:10.1073/pnas.1805337115.
[63] C.N Loynachan, A.P Soleimany, J.S Dudani, Y. Lin, A. Najer, A. Bekdemir, Q. Chen, S.N Bhatia, M.M. Stevens, Renal clearable catalytic gold nanoclusters for in vivo disease monitoring, Nat. Nanotechnol. 14 (9) (2019) 883–890, doi:10.1038/s41565-019-0527-6.
[64] L. Hao, N. Rohani, R.T Zhao, E.M Pulver, H. Mak, O.J Kelada, H. Ko, H.E Fleming, F.B Gertler, S.N. Bhatia, Microenvironment-triggered multimodal precision diagnostics, Nat. Mater. 20 (10) (2021) 1440–1448, doi:10.1038/s41563-021-01042-y.
[65] Z. Wu, R. Liu, J. Chen, X. Cai, J. Yi, J. Wang, D. Wang, M. Hu, An enzymatic cleavage-triggered minimally invasive nanosensor for urine-based detection of early atherosclerosis, Sci. Adv. 11 (11) (2025) eadu7614, doi:10.1126/sciadv.adu7614.
[66] T. Jiang, E.S Olson, Q.T Nguyen, M. Roy, P.A Jennings, R.Y. Tsien, Tumor imaging by means of proteolytic activation of cell-penetrating peptides, Proc. Natl. Acad. Sci. U.S.A. 101 (51) (2004) 17867–17872, doi:10.1073/pnas.0408191101.
[67] Q.T Nguyen, E.S Olson, T.A Aguilera, T. Jiang, M. Scadeng, L.G Ellies, R.Y. Tsien, Surgery with molecular fluorescence imaging using activatable cell-penetrating peptides decreases residual cancer and improves survival, Proc. Natl. Acad. Sci. U.S.A. 107 (9) (2010) 4317–4322, doi:10.1073/pnas.0910261107.
[68] T. Myochin, K. Hanaoka, T. Komatsu, T. Terai, T. Nagano, Design strategy for a near-infrared fluorescence probe for matrix metalloproteinase utilizing highly cell permeable boron dipyrromethene, J. Am. Chem. Soc. 134 (33) (2012) 13730–13737, doi:10.1021/ja303931b.
[69] E.N Savariar, C.N Felsen, N. Nashi, T. Jiang, L.G Ellies, P. Steinbach, R.Y Tsien, Q.T. Nguyen, Real-time in vivo molecular detection of primary tumors and metastases with ratiometric activatable cell-penetrating peptides, Cancer Res. 73 (2) (2013) 855–864, doi:10.1158/0008-5472.CAN-12-2969.
[70] N. Hua, F. Baik, T. Pham, A. Phinikaridou, N. Giordano, B. Friedman, M. Whitney, Q.T Nguyen, R.Y Tsien, J.A. Hamilton, Identification of high-risk plaques by MRI and fluorescence imaging in a rabbit model of atherothrombosis, PLoS One 10 (10) (2015) e0139833, doi:10.1371/journal.pone.0139833.
[71] S. Chen, J. Cui, T. Jiang, E.S Olson, Q.Y Cai, M. Yang, W. Wu, J.M Guthrie, J.D Robertson, S.A Lipton, L. Ma, R.Y Tsien, Z. Gu, Gelatinase activity imaged by activatable cell-penetrating peptides in cell-based and in vivo models of stroke, J. Cereb. Blood Flow Metab. 37 (1) (2017) 188–200, doi:10.1177/0271678X15621573.
[72] F. Xia, J. Niu, Y. Hong, C. Li, W. Cao, L. Wang, W. Hou, Y. Liu, D. Cui, Matrix metallopeptidase 2 targeted delivery of gold nanostars decorated with IR-780 iodide for dual-modal imaging and enhanced photothermal/photodynamic therapy, Acta Biomater. 89 (2019) 289–299, doi:10.1016/j.actbio.2019.03.008.
[73] Q. Li, Y. Wang, G. Yu, Y. Liu, K. Tang, C. Ding, H. Chen, S. Yu, Fluorescent polymer dots and graphene oxide based nanocomplexes for « off-on » detection of metalloproteinase-9, Nanoscale 11 (43) (2019) 20903–20909, doi:10.1039/c9nr06557a.
[74] X. Chen, H. Gao, Y. Deng, Q. Jin, J. Ji, D. Ding, Supramolecular aggregation-induced emission nanodots with programmed tumor microenvironment responsiveness for image-guided orthotopic pancreatic cancer therapy, ACS Nano 14 (4) (2020) 5121–5134, doi:10.1021/acsnano.0c02197.

[75] F. Wu, Y. Huang, X. Yang, J.J Hu, X. Lou, F. Xia, Y. Song, L. Jiang, Tunning intermolecular interaction of peptide-conjugated AIEgen in nano-confined space for quantitative detection of tumor marker secreted from cells, Anal. Chem. 93 (48) (2021) 16257–16263, doi:10.1021/acs.analchem.1c04422.
[76] K. Nagai, T. Sato, C. Kojima, Design of a dendrimer with a matrix metalloproteinase-responsive fluorescence probe and a tumor-homing peptide for metastatic tumor cell imaging in the lymph node, Bioorg. Med. Chem. Lett. 33 (2021) 127726, doi:10.1016/j.bmcl.2020.127726.
[77] J. Levi, S.R Kothapalli, T.J Ma, K. Hartman, B.T Khuri-Yakub, S.S. Gambhir, Design, synthesis, and imaging of an activatable photoacoustic probe, J. Am. Chem. Soc. 132 (32) (2010) 11264–11269, doi:10.1021/ja104000a.
[78] K. Miki, N. Imaizumi, K. Nogita, M. Oe, H. Mu, W. Huo, H. Harada, K. Ohe, MMP-2-activatable photoacoustic tumor imaging probes based on Al- and Si-naphthalocyanines, Bioconjug. Chem. 32 (8) (2021) 1773–1781, doi:10.1021/acs.bioconjchem.1c00266.
[79] X. Zhu, H. Lin, L. Wang, X. Tang, L. Ma, Z. Chen, J. Gao, Activatable T1 relaxivity recovery nanoconjugates for kinetic and sensitive analysis of matrix metalloprotease 2, ACS Appl. Mater. Interfaces 9 (26) (2017) 21688–21696, doi:10.1021/acsami.7b05389.
[80] A. Cobos-Correa, J.B Trojanek, S. Diemer, M.A Mall, C. Schultz, Membrane-bound FRET probe visualizes MMP12 activity in pulmonary inflammation, Nat. Chem. Biol. 5 (9) (2009) 628–630, doi:10.1038/nchembio.196.
[81] N.H Lim, E. Meinjohanns, G. Bou-Gharios, L.L Gompels, E. Nuti, A. Rossello, L. Devel, V. Dive, M. Meldal, H. Nagase, In vivo imaging of matrix metalloproteinase 12 and matrix metalloproteinase 13 activities in the mouse model of collagen-induced arthritis, Arthritis Rheumatol. 66 (3) (2014) 589–598, doi:10.1002/art.38295.
[82] Z. Chang, K. Wang, W. Hao, M. Ouyang, Y. Tang, X. Gao, B. Tang, Simultaneous detection of MMP-1 and MMP-12 using Au-Se nanoprobe for advancing NSCLC diagnosis, Anal. Chem. 97 (8) (2025) 4452–4460, doi:10.1021/acs.analchem.4c05846.
[83] Y. Hou, J. Zhou, Z. Gao, X. Sun, C. Liu, D. Shangguan, W. Yang, M. Gao, Protease-activated ratiometric fluorescent probe for pH mapping of malignant tumors, ACS Nano 9 (3) (2015) 3199–3205, doi:10.1021/acsnano.5b00276.
[84] T. Ma, Y. Hou, J. Zeng, C. Liu, P. Zhang, L. Jing, D. Shangguan, M. Gao, Dual-ratiometric target-triggered fluorescent probe for simultaneous quantitative visualization of tumor microenvironment protease activity and pH in vivo, J. Am. Chem. Soc. 140 (1) (2018) 211–218, doi:10.1021/jacs.7b08900.
[85] H.W An, D. Hou, R. Zheng, M.D Wang, X.Z Zeng, W.Y Xiao, T.D Yan, J.Q Wang, C.H Zhao, L.M Cheng, J.M Zhang, L. Wang, Z.Q Wang, H. Wang, W. Xu, A near-infrared peptide probe with tumor-specific excretion-retarded effect for image-guided surgery of renal cell carcinoma, ACS Nano 14 (1) (2020) 927–936, doi:10.1021/acsnano.9b08209.
[86] X. Ji, S. Xie, Y. Jiao, X. Zhang, D. Sun, V.C Yang, M. Wang, H. He, L. Sun, MT1-MMP activatable fluorogenic probes with enhanced specificity via high-affinity peptide conjugation for tumor imaging, Biomater. Sci. 8 (8) (2020) 2308–2317, doi:10.1039/c9bm02007a.
[87] D.S Shin, Y. Liu, Y. Gao, T. Kwa, Z. Matharu, A. Revzin, Micropatterned surfaces functionalized with electroactive peptides for detecting protease release from cells, Anal. Chem. 85 (1) (2013) 220–227, doi:10.1021/ac302547p.
[88] Y. Du, H. Geng, J. Chen, R. Feng, T. Wu, X. Zhang, Q. Wei, Supramolecular organic framework-enabled aggregation-enhanced electrochemiluminescence for ultrasensitive detection of MMP 9, Anal. Chem. 97 (38) (2025) 21060–21070, doi:10.1021/acs.analchem.5c04283.

[89] S.T Wang, M. Anahtar, D.M Kim, T.S Samad, C.M Zhang, S. Patel, H. Ko, C. Ngambenjawong, C.S Wang, J.D Kirkpatrick, V. Kumar, H.E Fleming, S.N. Bhatia, Engineering multiplexed synthetic breath biomarkers as diagnostic probes, bioRxiv (2025), doi:10.1101/2024.12.30.630769.
[90] Z. Yan, T. Kavanagh, S. Harrabi R da, L. ST, C. Tang, R. Beavil, M.M Müller, A. Beavil, S. Ameer-Beg, R.MP. da Silva, E. Gentleman, FRET sensor-modified synthetic hydrogels for real-time monitoring of cell-derived matrix metalloproteinase activity using fluorescence lifetime imaging, Adv. Funct. Mater. 34 (21) (2024) adfm.202309711, doi:10.1002/adfm.202309711.
[91] Z. Cheng, Y. Jin, J. Li, G. Shi, L. Yu, B. Shao, J. Tian, Y. Du, Z. Yuan, Fibronectin-targeting and metalloproteinase-activatable smart imaging probe for fluorescence imaging and image-guided surgery of breast cancer, J. Nanobiotechnol. 21 (1) (2023) 112, doi:10.1186/s12951-023-01868-5.
[92] B. Mills, D. Norberg, K. Dhaliwal, A.R Akram, M. Bradley, A. Megia-Fernandez, A matrix metalloproteinase activation probe for painting human tumours, Chem. Commun. (Camb) 56 (69) (2020) 9962–9965, doi:10.1039/d0cc03886e.
[93] L.W Chan, M.N Anahtar, T.H Ong, K.E Hern, R.R Kunz, S.N. Bhatia, Engineering synthetic breath biomarkers for respiratory disease, Nat. Nanotechnol. 15 (9) (2020) 792–800, doi:10.1038/s41565-020-0723-4.
[94] W. Pham, Y. Choi, R. Weissleder, C.H. Tung, Developing a peptide-based near-infrared molecular probe for protease sensing, Bioconjug. Chem. 15 (6) (2004) 1403–1407, doi:10.1021/bc049924s.
[95] S.M Yoon, S.J Myung, I.W Kim, E.J Do, B.D Ye, J.H Ryu, K. Park, K. Kim, I.C Kwon, M.J Kim, D.H Moon, D.H Yang, K.J Kim, J.S Byeon, S.K Yang, J.H. Kim, Application of near-infrared fluorescence imaging using a polymeric nanoparticle-based probe for the diagnosis and therapeutic monitoring of colon cancer, Dig. Dis. Sci. 56 (10) (2011) 3005–3013, doi:10.1007/s10620-011-1685-z.
[96] J.H Ryu, A. Lee, J.U Chu, H. Koo, C.Y Ko, H.S Kim, S.Y Yoon, B.S Kim, K. Choi, I.C Kwon, K. Kim, I. Youn, Early diagnosis of arthritis in mice with collagen-induced arthritis, using a fluorogenic matrix metalloproteinase 3-specific polymeric probe, Arthritis Rheum. 63 (12) (2011) 3824–3832, doi:10.1002/art.30628.
[97] A. Lee, K. Park, S.J Choi, D.H Seo, K. Kim, H.S Kim, K. Choi, I.C Kwon, S.Y Yoon, I. Youn, Prediction of antiarthritic drug efficacies by monitoring active matrix metalloproteinase-3 (MMP-3) levels in collagen-induced arthritic mice using the MMP-3 probe, Mol. Pharm. 11 (5) (2014) 1450–1458, doi:10.1021/mp400622q.
[98] R. Lee, S.J Choi, K.C Moon, J.W Park, K. Kim, S.Y Yoon, I. Youn, Fluorogenic probe for detecting active matrix metalloproteinase-3 (MMP-3) in plasma and peripheral blood neutrophils to indicate the severity of rheumatoid arthritis, ACS Biomater. Sci. Eng. 5 (6) (2019) 3039–3048, doi:10.1021/acsbiomaterials.9b00084.
[99] Y. Li, W. Liu, Q. Xu, J. Hu, C.Y. Zhang, Construction of a sensitive protease sensor with DNA-peptide conjugates for single-molecule detection of multiple matrix metalloproteinases, Biosens. Bioelectron. 169 (2020) 112647, doi:10.1016/j.bios.2020.112647.
[100] A.P Soleimany, J.D Kirkpatrick, S. Su, J.S Dudani, Q. Zhong, A. Bekdemir, S.N. Bhatia, Activatable zymography probes enable in situ localization of protease dysregulation in cancer, Cancer Res. 81 (1) (2021) 213–224, doi:10.1158/0008-5472.CAN-20-2410.
[101] M. Qian, Y. Zeng, M. Li, Q. Gao, C. Zhang, H. Qi, Electrogenerated chemiluminescence biosensor for quantization of matrix metalloproteinase-3 in serum via target-induced cleavage of oligopeptide, Biosensors (Basel) 14 (4) (2024) 181, doi:10.3390/bios14040181.
[102] R.L Scherer, M.N VanSaun, J.O McIntyre, L.M. Matrisian, Optical imaging of matrix metalloproteinase-7 activity in vivo using a proteolytic nanobeacon, Mol. Imaging 7 (3) (2008) 118–131.

[103] D.M Snider, M.L Coffin, B.J Armijo, R. Khetan, M.W Duchow, A. Capasso, D. Samanta, Conformationally locked peptide-DNA nanostructures for CRISPR-amplified activity-based sensing, Angew. Chem. Int. Ed Engl. 64 (25) (2025) e202500649, doi:10.1002/anie.202500649.

[104] S. Kaushal, H.TN. Nguyen, M. Fisher, H.C Kuo, Z.D Schultz, Y. Ren, Matrix metalloproteinase-9 (MMP-9) activatable gold nanoparticles for in situ zymography and diagnostics of neurofibromatosis type 2 (NF2) tumors, ACS Appl. Nano Mater. 9 (5) (2026) 2245–2257, doi:10.1021/acsanm.5c04657.

[105] J. Hu, F. Liu, Y. Chen, J. Fu, H. Ju, Signal-on mass spectrometric biosensing of multiplex matrix metalloproteinases with a phospholipid-structured mass-encoded microplate, Anal. Chem. 95 (23) (2023) 8974–8981, doi:10.1021/acs.analchem.3c01039.

[106] J. Park, J. Yang, E.K Lim, E. Kim, J. Choi, J.K Ryu, N.H Kim, J.S Suh, J.I Yook, Y.M Huh, S. Haam, Anchored proteinase-targetable optomagnetic nanoprobes for molecular imaging of invasive cancer cells, Angew. Chem. Int. Ed Engl. 51 (4) (2012) 945–948, doi:10.1002/anie.201106758.

[107] Y. Zhang, X. Chen, S. Yuan, L. Wang, X. Guan, Joint entropy-assisted graphene oxide-based multiplexing biosensing platform for simultaneous detection of multiple proteases, Anal. Chem. 92 (22) (2020) 15042–15049, doi:10.1021/acs.analchem.0c03007.

[108] J. Chen, C.H Tung, J.R Allport, S. Chen, R. Weissleder, P.L. Huang, Near-infrared fluorescent imaging of matrix metalloproteinase activity after myocardial infarction, Circulation 111 (14) (2005) 1800–1805, doi:10.1161/01.CIR.0000160936.91849.9F.

[109] J.o Deguchi, M. Aikawa, C.H Tung, E. Aikawa, D.E Kim, V. Ntziachristos, R. Weissleder, P. Libby, Inflammation in atherosclerosis: visualizing matrix metalloproteinase action in macrophages in vivo, Circulation 114 (1) (2006) 55–62, doi:10.1161/CIRCULATIONAHA.106.619056.

[110] Z. Chang, K. Wang, W. Hao, M. Ouyang, Y. Tang, X. Gao, B. Tang, Simultaneous detection of MMP-1 and MMP-12 using au-se nanoprobe for advancing NSCLC diagnosis, Anal. Chem. 97 (8) (2025) 4452–4460, doi:10.1021/acs.analchem.4c05846.

[111] J. Hu, F. Liu, Y. Chen, J. Fu, H. Ju, Signal-on mass spectrometric biosensing of multiplex matrix metalloproteinases with a phospholipid-structured mass-encoded microplate, Anal. Chem. 95 (23) (2023) 8974–8981, doi:10.1021/acs.analchem.3c01039.

[112] A. Sivakumar, H. Phuengkham, H. Rajesh, Q.D Mac, L.C Rogers, A.D Silva Trenkle, S.S Bawage, R. Hincapie, Z. Li, S. Vainikos, I. Lee, M. Xue, P. Qiu, M.G Finn, G.A Kwong, AND-gated protease-activated nanosensors for programmable detection of anti-tumour immunity, Nat. Nanotechnol. 20 (3) (2025) 441–450, doi:10.1038/s41565-024-01834-8.

[113] G.A Watkins, E.F Jones, M. Scott Shell, H.F VanBrocklin, M.H Pan, S.M Hanrahan, J.J Feng, J. He, N.E Sounni, K.A Dill, C.H Contag, L.M Coussens, B.L Franc, Development of an optimized activatable MMP-14 targeted SPECT imaging probe, Bioorg. Med. Chem. 17 (2) (2009) 653–659, doi:10.1016/j.bmc.2008.11.078.

[114] H. Wang, Z. Ma, H. Han, A novel impedance enhancer for amperometric biosensor based ultrasensitive detection of matrix metalloproteinase-2, Bioelectrochemistry 130 (2019) 107324, doi:10.1016/j.bioelechem.2019.06.009.

[115] L. Wang, H. Li, L. Shi, L. Li, F. Jia, T. Gao, G. Li, In situ peptide self-assembly on ionic nanochannel for dynamic monitoring of MMPs in extracellular matrix, Biosens. Bioelectron. 195 (2022) 113671, doi:10.1016/j.bios.2021.113671.

[116] C. Bremer, S. Bredow, U. Mahmood, R. Weissleder, C.H. Tung, Optical imaging of matrix metalloproteinase-2 activity in tumors: feasibility study in a mouse model, Radiology 221 (2) (2001) 523–529, doi:10.1148/radiol.2212010368.

[117] J. Klohs, N. Baeva, J. Steinbrink, R. Bourayou, C. Boettcher, G. Royl, D. Megow, U. Dirnagl, J. Priller, A. Wunder, In vivo near-infrared fluorescence imaging of matrix metalloproteinase activity after cerebral ischemia, J. Cereb. Blood Flow Metab. 29 (7) (2009) 1284–1292, doi:10.1038/jcbfm.2009.51.

[118] R. Zhan, X. Li, L. Zang, K. Xu, An Au-Se nanoprobe for the evaluation of the invasive potential of breast cancer cells via imaging the sequential activation of uPA and MMP-2, Analyst 145 (3) (2020) 1008–1013, doi:10.1039/c9an01830a.
[119] H. Chen, H. Zhang, Z. Wang, A ratiometric fluorescent probe based on peptide modified MnFe2O4 nanoparticles for matrix metalloproteinase-7 activity detection in vitro and in vivo, Analyst 147 (8) (2022) 1581–1588, doi:10.1039/d2an00212d.
[120] A. Lee, S.J Choi, K. Park, J.W Park, K. Kim, K. Choi, S.Y Yoon, I. Youn, Detection of active matrix metalloproteinase-3 in serum and fibroblast-like synoviocytes of collagen-induced arthritis mice, Bioconjug. Chem. 24 (6) (2013) 1068–1074, doi:10.1021/bc4001273.
[121] Y. Cai, S. Leng, Y. Ma, T. Xu, D. Chang, S. Ju, Dynamic change of MMP-9 in diabetic stroke visualized by optical imaging and treated with CD28 superagonist, Biomater. Sci. 9 (7) (2021) 2562–2570, doi:10.1039/d0bm02014a.
[122] M. Miampamba, J. Liu, A. Harootunian, A.J Gale, S. Baird, S.L Chen, Q.T Nguyen, R.Y Tsien, J.E. González, Sensitive in vivo visualization of breast cancer using ratiometric protease-activatable fluorescent imaging agent, AVB-620, Theranostics 7 (13) (2017) 3369–3386, doi:10.7150/thno.20678.
[123] W. Zeng, L. Wu, Y. Sun, Y. Wang, J. Wang, D. Ye, Ratiometric imaging of MMP-2 activity facilitates tumor detection using activatable near-infrared fluorescent semiconducting polymer nanoparticles, Small 17 (36) (2021) e2101924, doi:10.1002/smll.202101924.
[124] X. Li, B. Chen, M. He, B. Hu, A dual-functional magnetic microsphere for ICP-MS quantification and fluorescence imaging of matrix metalloproteinase 2 in cell secretion, Anal. Chim. Acta 1161 (2021) 338479, doi:10.1016/j.aca.2021.338479.
[125] E. Schellenberger, F. Rudloff, C. Warmuth, M. Taupitz, B. Hamm, J. Schnorr, Protease-specific nanosensors for magnetic resonance imaging, Bioconjug. Chem. 19 (12) (2008) 2440–2445, doi:10.1021/bc800330k.
[126] A. Megia-Fernandez, B. Mills, C. Michels, S.V Chankeshwara, N. Krstajić, C. Haslett, K. Dhaliwal, M. Bradley, Bimodal fluorogenic sensing of matrix proteolytic signatures in lung cancer, Org. Biomol. Chem. 16 (43) (2018) 8056–8063, doi:10.1039/c8ob01790e.
[127] S. Rainu, S. Parameswaran, S. Krishnakumar, N. Singh, Dual-sensitive fluorescent nanoprobes for detection of matrix metalloproteinases and low pH in a 3D tumor microenvironment, J. Mater. Chem. B 10 (28) (2022) 5388–5401, doi:10.1039/d2tb00519k.
[128] B.M Wallis de Vries, J.L Hillebrands, G.M van Dam, R.A Tio, J.S de Jong, R. Slart, C.J Zeebregts, Images in cardiovascular medicine. Multispectral near-infrared fluorescence molecular imaging of matrix metalloproteinases in a human carotid plaque using a matrix-degrading metalloproteinase-sensitive activatable fluorescent probe, Circulation 119 (20) (2009) e534–e536, doi:10.1161/CIRCULATIONAHA.108.821389.
[129] L. Wik, N. Nordberg, J. Broberg, J. Björkesten, E. Assarsson, S. Henriksson, I. Grundberg, E. Pettersson, C. Westerberg, E. Liljeroth, A. Falck, M. Lundberg, Proximity extension assay in combination with next-generation sequencing for high-throughput proteome-wide analysis, Mol. Cell. Proteomics 20 (2021) 100168, doi:10.1016/j.mcpro.2021.100168.
[130] J.L Crisp, E.N Savariar, H.L Glasgow, L.G Ellies, M.A Whitney, R.Y. Tsien, Dual targeting of integrin $\alpha v\beta 3$ and matrix metalloproteinase-2 for optical imaging of tumors and chemotherapeutic delivery, Mol. Cancer Ther. 13 (6) (2014) 1514–1525, doi:10.1158/1535-7163.MCT-13-1067.
[131] M. Fonović, M. Bogyo, Activity based probes for proteases: applications to biomarker discovery, molecular imaging and drug screening, Curr. Pharm. Des. 13 (3) (2007) 253–261, doi:10.2174/138161207779313623.

[132] L.E Edgington, M. Verdoes, M. Bogyo, Functional imaging of proteases: recent advances in the design and application of substrate-based and activity-based probes, Curr. Opin. Chem. Biol. 15 (6) (2011) 798–805, doi:10.1016/j.cbpa.2011.10.012.
[133] A. Saghatelian, N. Jessani, A. Joseph, M. Humphrey, B.F. Cravatt, Activity-based probes for the proteomic profiling of metalloproteases, Proc. Natl. Acad. Sci. U.S.A. 101 (27) (2004) 10000–10005, doi:10.1073/pnas.0402784101.
[134] E.WS. Chan, S. Chattopadhaya, R.C Panicker, X. Huang, S.Q. Yao, Developing photoactive affinity probes for proteomic profiling: hydroxamate-based probes for metalloproteases, J. Am. Chem. Soc. 126 (44) (2004) 14435–14446, doi:10.1021/ja047044i.
[135] J.Y Keow, E.D Pond, J.S Cisar, B.F Cravatt, B.D. Crawford, Activity-based labeling of matrix metalloproteinases in living vertebrate embryos, PLoS One 7 (8) (2012) e43434, doi:10.1371/journal.pone.0043434.
[136] S.A Sieber, S. Niessen, H.S Hoover, B.F. Cravatt, Proteomic profiling of metalloprotease activities with cocktails of active-site probes, Nat. Chem. Biol. 2 (5) (2006) 274–281, doi:10.1038/nchembio781.
[137] K.C Ravindra, C.C Ahrens, Y. Wang, J.Y Ramseier, J.S Wishnok, L.G Griffith, A.J Grodzinsky, S.R. Tannenbaum, Chemoproteomics of matrix metalloproteases in a model of cartilage degeneration suggests functional biomarkers associated with posttraumatic osteoarthritis, J. Biol. Chem. 293 (29) (2018) 11459–11469, doi:10.1074/jbc.M117.818542.
[138] C.C Ahrens, E.L Chiswick, K.C Ravindra, M.A Miller, J.Y Ramseier, K.B Isaacson, D.A Lauffenburger, L.G. Griffith, Development and application of the metalloprotease activity multiplexed bead-based immunoassay (MAMBI), Biochemistry 58 (38) (2019) 3938–3942, doi:10.1021/acs.biochem.9b00584.
[139] A. David, D. Steer, S. Bregant, L. Devel, A. Makaritis, F. Beau, A. Yiotakis, V. Dive, Cross-linking yield variation of a potent matrix metalloproteinase photoaffinity probe and consequences for functional proteomics, Angew. Chem. Int. Ed Engl. 46 (18) (2007) 3275–3277, doi:10.1002/anie.200604408.
[140] A.S Dabert-Gay, B. Czarny, L. Devel, F. Beau, E. Lajeunesse, S. Bregant, R. Thai, A. Yiotakis, V. Dive, Molecular determinants of matrix metalloproteinase-12 covalent modification by a photoaffinity probe: insights into activity-based probe development and conformational variability of matrix metalloproteinases, J. Biol. Chem. 283 (45) (2008) 31058–31067, doi:10.1074/jbc.M805795200.
[141] A.S Dabert-Gay, B. Czarny, E. Lajeunesse, R. Thai, H. Nagase, V. Dive, Covalent modification of matrix metalloproteinases by a photoaffinity probe: influence of nucleophilicity and flexibility of the residue in position 241, Bioconjug. Chem. 20 (2) (2009) 367–375, doi:10.1021/bc800478b.
[142] C. Nury, B. Czarny, E. Cassar-Lajeunesse, D. Georgiadis, S. Bregant, V. Dive, A pan photoaffinity probe for detecting active forms of matrix metalloproteinases, ChemBioChem 14 (1) (2013) 107–114, doi:10.1002/cbic.201200583.
[143] M.A Leeuwenburgh, P.P Geurink, T. Klein, H.F Kauffman, G.A van der Marel, R. Bischoff, H.S. Overkleeft, Solid-phase synthesis of succinylhydroxamate peptides: functionalized matrix metalloproteinase inhibitors, Org. Lett. 8 (8) (2006) 1705–1708, doi:10.1021/ol060409e.
[144] Design of peptide hydroxamate-based photoreactive activity-based probes of zinc-dependent metalloproteases—Geurink Eur. J. Organ. Chem., 2010 https://chemistry-europe.onlinelibrary.wiley.com/doi/full/10.1002/ejoc.200901385.
[145] L. Devel, B. Czarny, F. Beau, D. Georgiadis, E. Stura, V. Dive, Third generation of matrix metalloprotease inhibitors: gain in selectivity by targeting the depth of the S1' cavity, Biochimie 92 (11) (2010) 1501–1508, doi:10.1016/j.biochi.2010.07.017.

[146] S. Bregant, C. Huillet, L. Devel, A.S Dabert-Gay, F. Beau, R. Thai, B. Czarny, A. Yiotakis, V. Dive, Detection of matrix metalloproteinase active forms in complex proteomes: evaluation of affinity versus photoaffinity capture, J. Proteome Res. 8 (5) (2009) 2484–2494, doi:10.1021/pr801069c.

[147] M. Morell, T. Nguyen Duc, A.L Willis, S. Syed, J. Lee, E. Deu, Y. Deng, J. Xiao, B.E Turk, J.R Jessen, S.J Weiss, M. Bogyo, Coupling protein engineering with probe design to inhibit and image matrix metalloproteinases with controlled specificity, J. Am. Chem. Soc. 135 (24) (2013) 9139–9148, doi:10.1021/ja403523p.

[148] M. Kaminska, P. Bruyat, C. Malgorn, M. Doladilhe, E. Cassar-Lajeunesse, C. Fruchart Gaillard, M. De Souza, F. Beau, R. Thai, I. Correia, A. Galat, D. Georgiadis, O. Lequin, V. Dive, S. Bregant, L. Devel, Ligand-directed modification of active matrix metalloproteases: activity-based probes with no Photolabile Group, Angew. Chem. Int. Ed Engl. 60 (33) (2021) 18272–18279, doi:10.1002/anie.202106117.

[149] C. Malgorn, F. Becher, P. Bruyat, C. Fruchart-Gaillard, F. Beau, S. Bregant, L. Devel, A new affinity-based probe to profile MMP active forms, Methods Mol. Biol. 2747 (2024) 29–39, doi:10.1007/978-1-0716-3589-6_3.

[150] L. Devel, C. Malgorn, R.W Tohon, M. Launay, K. Patiniotis, M. Sejalon-Cipolla, F. Beau, R. Thai, P. Bruyat, A. Bonino, S. Bregant, G. Subra, S. Cantel, D. Georgiadis, Covalent labeling of matrix metalloproteases with affinity-based probes containing tuned reactive N-Acyl-N-Alkyl sulfonamide cleavable linkers, ChemBioChem 25 (23) (2024) e202400441, doi:10.1002/cbic.202400441.

[151] N. Amara, M. Tholen, M. Bogyo, Chemical tools for selective activity profiling of endogenously expressed MMP-14 in multicellular models, ACS Chem. Biol. 13 (9) (2018) 2645–2654, doi:10.1021/acschembio.8b00562.

[152] T. Tamura, M. Kawano, I. Hamachi, Targeted covalent modification strategies for drugging the undruggable targets, Chem. Rev. 125 (2) (2025) 1191–1253, doi:10.1021/acs.chemrev.4c00745.

[153] T. Tamura, I. Hamachi, Chemistry for covalent modification of endogenous/native proteins: from test tubes to complex biological systems, J. Am. Chem. Soc. 141 (7) (2019) 2782–2799, doi:10.1021/jacs.8b11747.

[154] L. Devel, V. Rogakos, A. David, A. Makaritis, F. Beau, P. Cuniasse, A. Yiotakis, V. Dive, Development of selective inhibitors and substrate of matrix metalloproteinase-12, J. Biol. Chem. 281 (16) (2006) 11152–11160, doi:10.1074/jbc.M600222200.

[155] L. Berthy, U. Pasco, M. Sejalon-Cipolla, P. Bruyat, M. Kaminska, C. Malgorn, F. Beau, M. Osolin, M. Novak, B. Breznik, D. Georgiadis, I. Correia, E. Groult, O. Lequin, S. Bregant, Y.M Pers, D. Noel, R. Thai, G. Subra, S. Cantel, L. Devel, Proximity-induced transfer of a mass tag enables direct profiling of active matrix metalloproteases, Angew. Chem. Int. Ed Engl. (2026) e23132, doi:10.1002/anie.202523132.

[156] M. Launay, L. Berthy, S.L Rizo, M.A Balachouti, D. Georgiadis, L. Devel, Guidelines for the design of custom affinity-based probes for metalloproteases, Methods Mol. Biol. 3007 (2026) 183–194, doi:10.1007/978-1-0716-5092-9_11.

[157] F. Li, C. Wang, X. Guo, T. Akutsu, G.I Webb, L.JM. Coin, L. Kurgan, J. Song, Prosperous-Plus: a one-stop and comprehensive platform for accurate protease-specific substrate cleavage prediction and machine-learning model construction, Brief Bioinform. 24 (6) (2023) bbad372, doi:10.1093/bib/bbad372.

[158] F. Li, J. Chen, A. Leier, T. Marquez-Lago, Q. Liu, Y. Wang, J. Revote, A.I Smith, T. Akutsu, G.I Webb, L. Kurgan, J. Song, DeepCleave: a deep learning predictor for caspase and matrix metalloprotease substrates and cleavage sites, Bioinformatics 36 (4) (2020) 1057–1065, doi:10.1093/bioinformatics/btz721.

[159] C. Martin-Alonso, S. Alamdari, T.S Samad, K.K Yang, S.N Bhatia, A.P. Amini, Deep learning guided design of protease substrates, Nat. Commun. 17 (1) (2026) 54, doi:10.1038/s41467-025-67226-1.

CHAPTER THREE

Classical MMP inhibitors

Andrea Trabocchi*
Department of Chemistry "Ugo Schiff", University of Florence, Sesto Fiorentino, Florence, Italy
*Corresponding author. e-mail address: andrea.trabocchi@unifi.it.

Contents

Abstract

Considering the pivotal roles of matrix metalloproteinases (MMPs) in tumor progression, MMP inhibitors (MMPIs) have emerged as promising therapeutic agents, resulting in the development of several generations of inhibitors and candidate cancer drugs. First-generation inhibitors were conceived as peptide-based inhibitors and hydroxamate-based peptidomimetics such as batimastat and marimastat, demonstrating at preclinical level the strong anticancer activity by targeting the catalytic Zn^{2+} ion of MMPs. Nevertheless, their clinical success was limited by poor bioavailability and selectivity, resulting to severe side-effects. Second-generation MMPIs consisted of alternative zinc-binding groups, such as thiol, carboxylic acid, pyrimidine, and hydroxypyrone moieties compounds, resulting in failure in clinical trials of improved candidate drugs such as prinomastat and tanomastat due to low oral bioavailability and safety concerns. A major issue has been the non-specific inhibition across multiple MMP family members due to high conservation of the MMP catalytic site. Herein is summarized the development of MMP inhibitors and are described those compounds that reached clinical trials. It is expected that advances in structural biology highlighting the structural differences among MMPs may facilitate the development of more selective inhibitors with improved efficacy and reduced toxicity and enable new MMPI as candidate cancer drugs.

Keywords: Cancer, Drug discovery, Enzyme inhibition, Hydroxamic acid, Metalloproteases, Peptides, Peptidomimetics

Fungal and Protozoan Carbonic Anhydrases, Volume 60
ISSN 1874-6047
https://doi.org/10.1016/bs.enz.2026.07.003

1. Introduction

Matrix metalloproteases are a family of zinc-dependent endopeptidases normally produced in most cells of the body that play an important role in the turnover and remodeling of the extracellular matrix (ECM) in both physiological and pathological processes [1–3]. They exhibit specific proteolytic activity against ECM and various non-ECM proteins, such as collagen, elastin, gelatin, protoglycans, and matrix glycoproteins [4]. Also, taking advantage of their action, tumor cells are able to release several pro-angiogenic factors like VEGF and TGF-beta. The remodeling of the ECM is a regulated process in normal physiological events such as wound healing and bone remodeling. This is particularly effective at the molecular level through the fine and balanced tuning of the expression and activation of MMPs and their endogenous inhibitors, such as TIMPs (tissue inhibitors of metalloproteases). Nevertheless, in cancer and other pathological conditions such regulated process is altered, particularly resulting in an increased MMP activity to degrade the ECM, which promotes tumor invasion, intravasation, extravasation, and metastatic spread. These proteolytic enzymes are produced at the invasive front of tumor cells, where tumor-derived cytokines stimulate protease production by stromal cells. In addition, MMPs facilitate tumor growth through the release of matrix-bound growth factors and contribute to angiogenesis by enabling new blood vessel formation [5]. There is clinical evidence documenting the increased expression of MMPs in numerous solid tumors, such as lung, pancreatic, breast, ovarian, and colorectal cancer [6]. Also, as the role of MMPs is key in other numerous cellular processes, such as apoptosis, migration, cell proliferation, differentiation, and their dysfunction is relevant to other pathologies beyond cancer in the context of cardiovascular, neurological and autoimmune diseases.

2. Structure and classification of MMPs

Twenty-six sub-families of metalloproteases have been identified, 23 of which are present in humans [7–9]. These have been classified into six categories based on their function in the body, their location, and their secondary structure [10,11]:

- Collagenases (MMP1, MMP8, MMP13)
- Gelatinases (MMP2, MMP9)
- Stromelysins (MMP3, MMP10, MMP11, MMP12)
- Membrane-type (MMP14, MMP17, MMP24, MMP25)

- Matrilysins (MMP7, MMP26)
- Other

All metalloproteases have well-defined common structural domains, namely:

- a *N*-terminal signal peptide sequence that directs the enzyme to the endoplasmic reticulum;
- a catalytic domain of approximately 170 amino acids with a Zn^{2+} binding site chelated by three histidine residues in a conserved HExxHxxGxxH peptide sequence;
- a *C*-terminal hemopexin-like domain of approximately 200 amino acids, with a helical structure, which mediates protein-protein interactions and contributes to enzyme activation and substrate recognition. This region is connected to the catalytic domain through a 75-amino acid linker, which does not have a well-defined structure.

MMPs are synthesized as inactive pro-enzymes, also called zymogens, thanks to the presence of a cysteine residue in the catalytic pro-domain that coordinates the zinc ion that serves to inhibit its catalytic activity. The *N*-terminal pro-domain is then removed during transport to the endoplasmic reticulum, thus activating the enzyme and performing a cellular signaling function. Loss of the pro-domain removes the coordination of the cysteine thiol group, thus allowing water molecules, that are responsible for proteolytic activity, to enter the active site. The catalytic mechanism of MMPs involves the cleavage of an amino acid sequence through the nucleophilic attack by a water molecule at neutral pH, that is activated as a nucleophile by the interaction with a glutamic acid and the zinc ion of the MMP (Fig. 1). The peptide substrate to be cleaved enters the enzyme's active site, forming a Michaelis complex thanks to interactions with the S1 and S1'subsites and the bulge edge sequence [12,13]. Subsequently, the zinc ion coordinates the catalytic water molecule in a pentameric geometry with the three histidine residues and the oxygen atom of the amide carbonyl group.

The catalytic site consists of a Zn^{2+} ion characterized by left- and right-handed subsites, also termed "unprimed" and "primed" pockets, such as S3, S2, and S1 and S1', S2', and S3', respectively. The most important pocket among the MMPs is the S1' subsite, which varies in shape, amino acid composition and depth, whereas the shallower S2' and S3' pockets are more solvent exposed. Accordingly, MMPs are generally classified based on the depth of the S1' pocket into three groups, namely shallow, intermediate, and deep pocket MMPs [14]. Shallow S1' pockets are found in MMP-1 and −7,

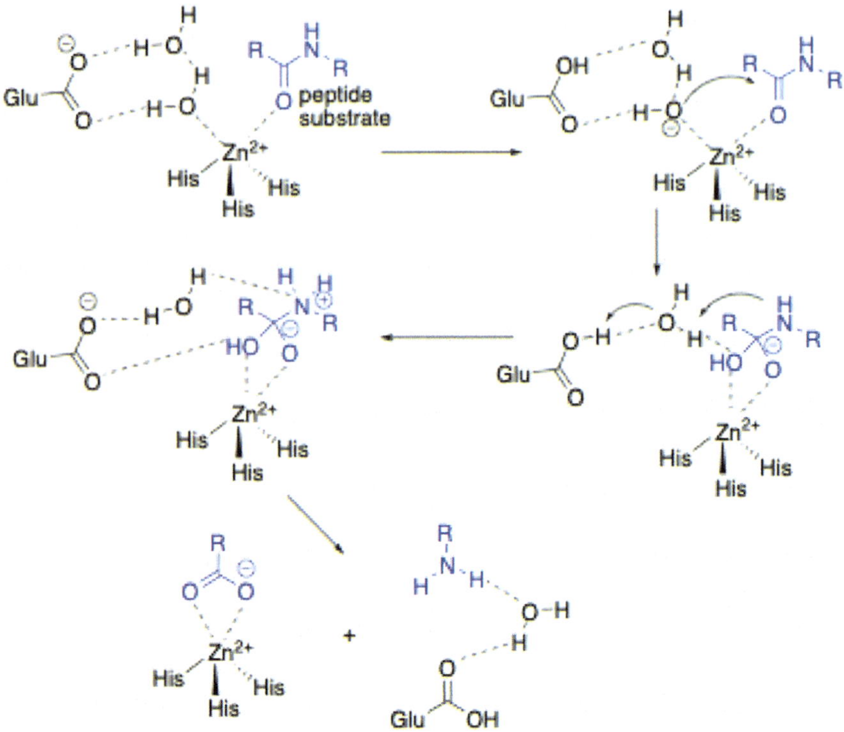

Fig. 1 Proposed reaction mechanism of MMP with the participation of two water molecules.

intermediate S1' subsites are found in MMP-2, −8, and −9, and deep pockets are found in MMP-3, −11, −12, −13, and −14. This is the catalytic pocket that has been studied and exploited most to develop selective inhibitors. In fact, it is known that the active site is conserved among MMPs except for the S1' pocket region, which varies slightly among MMPs in sequence, shape, and depth [15]. The shape of the S1' pocket is characterized by a tunnel formed by different amino acids and delimited by the W-loop. An additional feature of the S1' pocket is the diversity of certain residues in the pocket, such as the basic Arg218 in MMP-1 that is replaced with non-polar leucine in MMP-2, −3, −12, and −14 and with the aromatic residue tyrosine in MMP-7.

3. MMP inhibition

Matrix metalloproteinases have attracted considerable attention as therapeutic targets since their identification as regulators of the ECM remodeling and tumor progression [16]. Specifically, the capability of these zinc-dependent endopeptidases to degrade the structural proteins of the ECM, and their involvement in angiogenesis, tumor invasion, and metastasis, stimulated the development of MMP inhibitors as a promising strategy for cancer therapy Accordingly, several generations of inhibitors have been developed over the years, almost all designed to chelate the zinc ion present in the enzyme's active site, significantly reducing the activity of the proteolytic enzymes through competitive inhibition [17]. This approach was particularly studied between the 1990s and early 2000s, as a strategy to interfere with multiple stages of tumor progression rather than targeting the proliferation of tumor cells, thus resulting in the development of more than fifty MMP inhibitors that were advanced to preclinical evaluation and clinical trials. Nevertheless, despite encouraging results in experimental models, none of these compounds demonstrated sufficient efficacy and safety to obtain final approval for the pharmacological treatment of cancer. Such failure highlighted the complexity of selectively targeting this enzyme family [18].

The initial strategy for MMP inhibition within first-generation MMPI was largely based on the knowledge of the highly conserved catalytic activity characteristic of all MMPs [19], characterized by a Zn^{2+} ion coordinated by three histidine residues activating a water molecule to perform a nucleophilic attack on the specific amide bond of the substrate adjacent to the zinc ion [20]. As the catalytic zinc ion is essential for the enzymatic activity, early inhibitor design focused on developing molecules capable of directly coordinating the metal ion to prevent catalysis [21]. This approach was achieved through the generation of inhibitors possessing a zinc-binding group (ZBG), which became the distinguishing structural feature of the first generation of MMP inhibitors. Besides replacing the catalytic water molecule, the ZBG was particularly effective as a chelating moiety capable to stabilizing the inhibitor within the active site and correctly orienting the other portions of the molecule toward the enzyme pockets [22–24]. Over the years, numerous zinc-chelating groups were investigated, including hydroxamates, carboxylates, thiols, phosphonates, aminocarboxylates, hydrazides, *N*-hydroxyureas, and hydroxypyrones (Fig. 2) [25].

Nevertheless, hydroxamic acid rapidly emerged as the preferred zinc-binding motif because of its straightforward synthesis and the high affinity for Zn^{2+} ions through its chelating activity, resulting to excellent inhibitory potency against a broad range of MMPs [26–28]. Specifically, hydroxamic acids proved to establish highly stable complexes with the zinc ion through

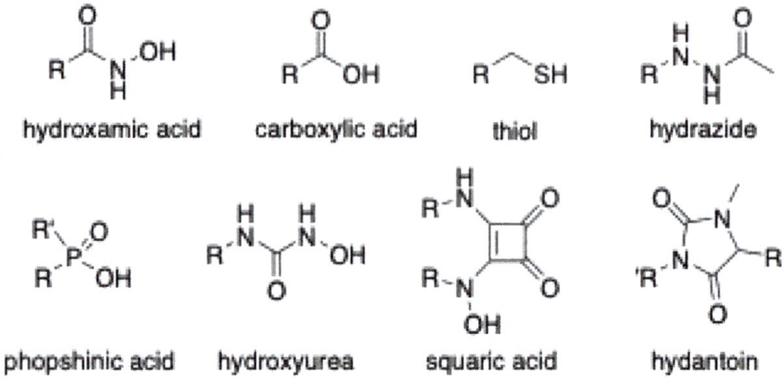

Fig. 2 Chemical structures of representative zinc-binding groups (ZBGs).

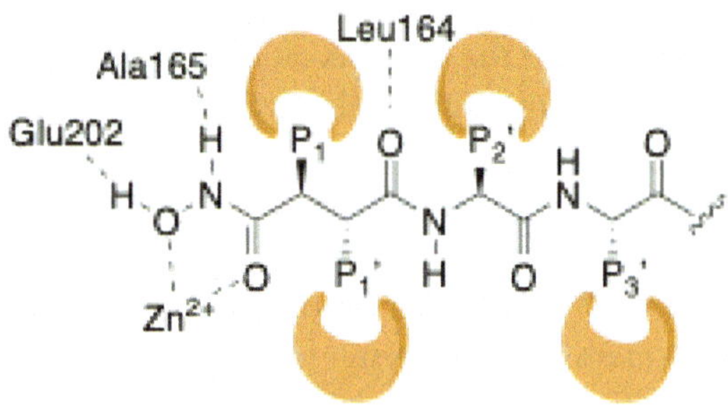

Fig. 3 Canonical binding mode of hydroxamic acid-based MMPI.

the simultaneous coordination of the carbonyl and hydroxyl oxygen atoms, thus acting as monoanionic bidentate ligands (Fig. 3) [29]. Also, their strong interaction with the catalytic metal ion is stabilized by hydrogen-bonding interactions with highly conserved amino acid residues within the active site. Structure-activity relationship (SAR) studies demonstrated that the replacement of the hydroxamic acid moiety with less efficient chelators, such as the corresponding carboxylic acid, resulted in a dramatic reduction of the inhibitory potency of about two to three orders of magnitude.

Earliest MMP inhibitors were designed by mimicking the peptide substrates naturally recognized by MMPs, thus displaying a peptide character. Specifically, as collagen constitutes one of the principal physiological substrates of several MMPs, substrate-derived peptide analogues were initially developed to reproduce the amino acid sequence flanking the scissile bond while incorporating a hydroxamic acid group to coordinate the catalytic zinc ion. Although these peptide-based inhibitors proved to successfully bind to

MMPs, taking advantage of the reproduction of substrate recognition, they displayed several pharmacological disadvantages, including high conformational flexibility, rapid metabolic degradation, poor membrane permeability, and limited oral bioavailability. Such unfavorable features significantly restricted their therapeutic potential despite their ability to inhibit the MMP activity *in vitro* [22]. To overcome these limitations, medicinal chemistry efforts were conceived to improve the structure of MMPI through the development of peptidomimetics, in which structural modifications preserved the key interactions with the catalytic site while improving the metabolic stability and pharmacokinetic properties through local modification of the peptide core. The incorporation of hydroxamic acid into peptidomimetic scaffolds generated the first clinically relevant MMP inhibitors, characterized by high potency but limited selectivity. Because the catalytic domains of MMPs exhibit a remarkable degree of structural conservation, particularly around the zinc-binding region, these compounds generally inhibited multiple MMP subtypes simultaneously. While broad-spectrum inhibition was initially considered advantageous during drug discovery of MMPI as cancer drugs, it later became evident that such lack of selectivity represented one of the major obstacles to the achievement of approved drug through successful clinical evaluation.

Among the first developed hydroxamate-based peptidomimetics, batimastat (BB-94) [30] and marimastat (BB-2516) [31] represented the typical broad-spectrum MMPI. Batimastat exhibited potent inhibition of numerous MMP subtypes at nanomolar concentration and demonstrated remarkable antitumor activity in preclinical models, displaying the suppression of key tumor phenotypes such as tumor invasion, angiogenesis, and metastatic dissemination. Nevertheless, poor aqueous solubility and oral bioavailability impaired the possibility of achieving an oral drug, thus requiring intraperitoneal administration and severely limiting its clinical applicability and widespread therapeutic use. This was particularly disappointing despite early clinical studies confirmed its biological activity. Marimastat (BB-2516) was successively developed as an orally bioavailable analogue of batimastat, maintaining the hydroxamate zinc-binding group and the broad-spectrum inhibitory profile. This MMPI displayed improved pharmacokinetic properties and was advanced into phase II and phase III clinical trials for the treatment of several cancers, such as pancreatic, gastric, ovarian, breast, and non-small-cell lung cancer. Nevertheless, all these trials failed to meet the desired endpoints for survival or disease progression. Also, musculoskeletal syndrome was found as a frequent and severe side-effect, that forced to reconsider dosage and ultimately contributed to the discontinuation of its clinical development. At the molecular level, the limited performance of these hydroxamate-based MMPI in terms of side effects was due to the strong zinc-chelating ability not being restricted to MMPs but also to several other zinc-dependent metalloproteins, such as ADAM (A Disintegrin and Metalloproteinase) and ADAMTS (A Disintegrin and Metalloproteinase with Thrombospondin Motifs) enzyme classes. Concerning poor bioavailability, this was due to pharmacokinetic issues including solubility and metabolic instability, the latter due to the conversion of hydroxamic acid to the corresponding less active carboxylic acid derivative through enzymatic cleavage, which reduced the effective drug concentrations *in vivo* [32].

Concerning the selectivity among MMP subtypes, advances in structural biology provided important insights to the molecular basis of MMP inhibition. Specifically, crystallographic studies of MMPs, such as MMP-1 and MMP-3, together with extensive structure-activity relationship analyses, revealed the organization of the catalytic cleft into multiple substrate-recognition subsites positioned on either side of the catalytic zinc ion. Among these binding pockets, the S1′ subsite emerged as the key element of diversity across different MMP isoforms, displaying considerable variability in both amino acid composition and cavity depth, thus was exploited to improve inhibitor selectivity in subsequent generations of MMPI. As discussed, MMPs are broadly classified according to the morphology of S1' subsite into shallow-pocket enzymes, including MMP-1 and MMP-7; intermediate-pocket enzymes, such as MMP-2, MMP-8, and MMP-9; and deep-pocket enzymes, including MMP-3, MMP-11, MMP-12, MMP-13, and membrane-type MMPs. These differences provided the first convincing approach to develop selective inhibitors capable of discriminating between individual MMP family members.

Also, advances in structural biology and X-ray crystallographic studies enabled the development of other classes of MMPI not necessarily derived from the structural features of substrates to achieve competitive inhibitors, but facilitating the development of smaller and more rigid compounds with improved pharmacological properties and enhanced selectivity profiles.

Concerning the improvements connected to the inhibitory activity associated to the use of hydroxamic acid moiety and to reduce the related limitations, alternative zinc-binding groups were identified, as previously illustrated (Fig. 2). It was ascertained that the strong metal-chelating properties of hydroxamic acid, though a key driver for its high potency, also restricted the conformational flexibility within the active site, thus contributing to broad inhibition of several Zn-metalloproteins. As a consequence, it was postulated that weaker zinc-chelating functionalities could preserve sufficient affinity for the catalytic zinc ion and enable the establishment of additional interactions with less conserved substrate-binding pockets to elicit inhibitor selectivity. Among alternative zinc-binding groups [33], thiol-containing inhibitors represented one of the earliest classes, as sulfhydryl groups proved to maintain the ability to coordinate the catalytic metal ion and to display lower affinity than hydroxamates. Similarly, carboxylate- and phosphinate-based inhibitors were developed to exploit weaker metal coordination and to enable the establishment of specific interactions within the catalytic cleft. Pyrimidine-derived compounds, hydroxypyrones, aminocarboxylates, hydrazides, *N*-hydroxyureas, and squaric acid analogues were subsequently introduced as additional alternatives to hydroxamic acid. *N*-hydroxyureas attracted particular attention because of their structural similarity to hydroxamic acids and the favorable oral bioavailability previously demonstrated by *N*-hydroxyurea-derived lipoxygenase inhibitors. Nevertheless, direct comparisons with hydroxamate analogues revealed lower inhibitory potency against MMPs. Similarly, hydrazide-based zinc-binding groups, despite being isosteric with hydroxamic acids and capable of displaying bidentate metal coordination, generally exhibited substantially reduced activity as compared to well-established hydroxamate inhibitors. Finally, squaric acid derivatives were also investigated as hydroxamate bioisosteres capable of forming six-membered chelate rings with the catalytic zinc ion instead of the five-membered complexes characteristic of hydroxamic acids [34]. Although these compounds generally exhibited reduced inhibitory potency as compared with hydroxamate derivatives, their weaker zinc-binding properties often translated into lower off-target inhibition and improved pharmacokinetic behavior. Such structural improvements of MMPI at the molecular level enabled the transition from the first generation of broad-spectrum hydroxamate-based inhibitors toward a second generation of MMP inhibitors, ultimately improving potency, selectivity, pharmacokinetic properties and safety.

The availability of crystallographic data encouraged the transition from peptidomimetic to non-peptidomimetic inhibitors, that provided greater conformational rigidity, improved metabolic stability and enhanced oral bioavailability. Among representative compounds developed using this strategy, prinomastat (AG3340) became one of the most extensively investigated second-generation inhibitors [35–37]. Nevertheless, phase III clinical studies in patients with advanced non-small-cell lung cancer and other solid tumors were disappointing. Similar results were obtained for numerous other inhibitors developed during the same period, that were particularly disappointing concerning the therapeutic treatment of patients with metastatic or late-stage disease. This outcome suggested that the knowledge regarding MMP biology was incomplete, and that early therapeutic strategies had been based on the erroneous hypothesis that the inhibition of extracellular matrix degradation would directly translate into the suppression of tumor invasion and metastasis. Subsequent experimental evidence demonstrated that MMPs participate in a broad range of physiological and pathological processes, confirming the evidence of these enzymes as challenging targets. For example, as well as displaying ECM-degrading potential, MMPs regulate the bioavailability of cytokines, chemokines, growth factors, cell-surface receptors, and adhesion molecules, thus displaying a direct effect to inflammation, angiogenesis, immune cell recruitment, wound healing, and tissue remodeling. Such evidence clarified that the unselective inhibition of multiple MMP family members could interfere not only with tumor-promoting pathways but also with normal physiological processes and protective host responses.

Advances in structural biology also revealed that MMPs possess complex multidomain structures extending beyond the catalytic domain. Accordingly, most secreted MMPs contain additional structural elements, including the pro-peptide domain, the hemopexin-like domain, and specific domains, such as collagen-binding or fibronectin-like domains. Initially, these protein regions received relatively little attention because they were not directly involved in catalysis. However, the elucidation of full-length MMP structures demonstrated that these domains contribute to several biological functions at the molecular level, such as substrate recognition, enzyme activation, protein-protein interactions, cellular localization, and signal transduction. Although the exploitation of these non-catalytic domains belongs to more recent generations of inhibitor design, their identification represented a conceptual shift from the limitations encountered with catalytic-site inhibitors [38].

One notable exception to the general clinical failure of MMPIs is doxycycline hyclate, marketed as Periostat, which remains the only FDA-approved MMP inhibitor, although for the treatment of periodontitis rather than cancer. Unlike synthetic hydroxamate inhibitors, the precise molecular mechanism by which doxycycline suppresses MMP activity has not been fully elucidated. The putative mechanisms include the interaction with catalytic metal ions, modulation of MMP gene expression, and indirect anti-inflammatory effects. In periodontal disease, its clinical efficacy is believed to derive from its ability to accumulate on tooth surfaces, thus allowing for the sustained local inhibition of MMP activity while minimizing the systemic exposure.

The overall experience accumulated during three decades of MMPI development enabled an in-depth knowledge of MMP biology. In this context, MMPI played a key role as tool compounds and molecular probes to elucidate the pathological pathways and molecular axes involving various MMP subtypes in the context of different tumors. Concerning the understanding of MMP as cancer targets, nowadays these enzymes are recognized not only for their capability to degrade the extracellular matrix and to promote tumor invasion, but for their role as regulators involved in various biological processes, such as inflammation, innate immunity, tissue repair, angiogenesis, and maintenance of tissue homeostasis. Another feature that has been elucidated in the context of the biological role of MMPs is the protective or homeostatic functions under specific physiological and pathological conditions, specifically for gelatinases, MMP-3, MMP-8, and MMP-12. This is an important observation in line with the clinical evidence that the indiscriminate inhibition of the entire MMP family is unlikely to provide therapeutic benefit and may instead interfere with essential biological processes. Consequently, the principal lesson derived from the early generations of MMP inhibitors is that the successful achievement of MMP-based therapeutics requires the selective modulation of pathogenic MMP activities while preserving the physiological functions of enzymes that contribute to normal tissue homeostasis [39].

4. Classical MMP inhibitors as investigational drugs

In the 1990s, following the results of cancer research showing that malignant tumors use matrix metalloproteinases as a kind of "molecular scissors" to destroy the extracellular matrix, allowing tumor cells to invade surrounding tissue, enter the bloodstream, and metastasize [40], big pharmaceutical companies began a global race to develop molecules capable of blocking these enzymes, resulting to a series of lead compounds that were studied in clinical trials (Fig. 4 and Table 1).

As discussed, the first generation of inhibitors was developed based on peptidomimetics that aimed to mimic the structure of collagen. Among these compounds, batimastat (shown in Fig. 4), reached clinical phase I in 1994 and continued until phase III, which it did not pass due to poor oral availability and therefore did not enter the market [41]. Similarly, marimastat (Fig. 4) was designed specifically to address the shortcomings of the

Table 1 Representative clinical MMPI drug candidates.

Inhibitor	Class	Selectivity	Indication	Study phase
Batimastat	Peptidomimetic	Broad spectrum	Malignant ascites	Failed in phase III
Marimastat	Peptidomimetic	Broad spectrum	Breast cancer, small- and non-small cell lung cancers	Failed in phase III
CGS 27023A	Small molecule	Broad spectrum	Advanced solid cancer	Failed in phase I
Prinomastat	Small molecule	Broad spectrum	Non-small cell lung cancer, glioblastoma, prostate cancer	Failed in phase III
Tanomastat	Small molecule	MMP-2, MMP-3, MMP-9	Ovarian cancer, pancreatic adenocarcinoma, non-small cell lung cancer, rheumatoid arthritis, organ transplant rejection	Failed in phase III
Doxycycline	Tetracycline derivative	Broad spectrum	Approved for periodontal diseases, multiple sclerosis, type II diabetes	Approved

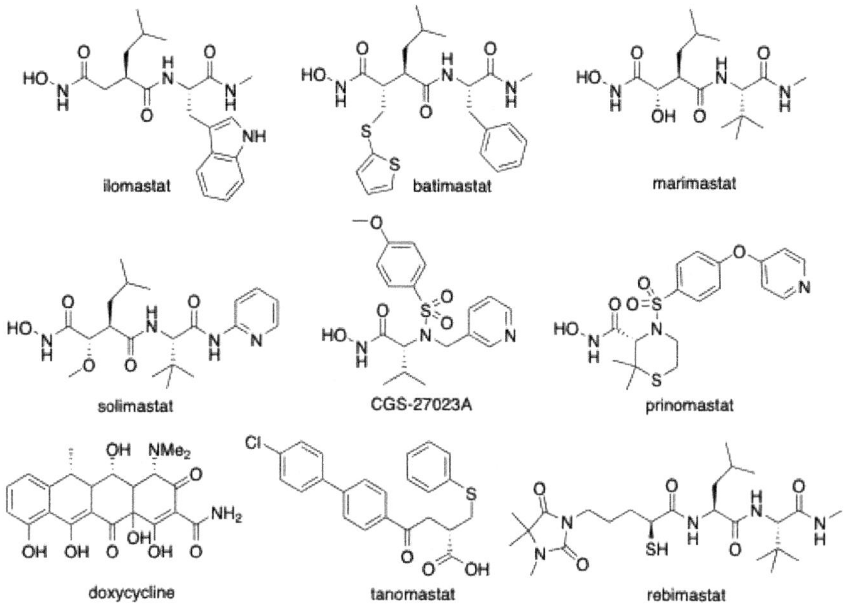

Fig. 4 Structure of representative MMP inhibitors that entered clinical trials.

analogous batimastat, which, however, failed to even pass clinical phase I due to its poor therapeutic efficacy. To improve oral availability and selectivity, additional non-peptidomimetic inhibitors based on the 3D conformation of the active site were developed in subsequent years. For example, tanomastat (Fig. 4) was developed but never reached market, as phase III trials of this drug were terminated due to poor results obtained on pancreatic and lung cancer cells. Over the last 20 years, numerous experiments have been carried out to try to resolve these issues. The problem of lack of selectivity has always been the real obstacle to the development of inhibitors and remains the greatest challenge in research today. Major MMPIs as tool compounds or clinical candidates are herein summarized.

5. Ilomastat (GM6001)

Glycomed Inc. was historically recognized for developing and supplying potent synthetic matrix metalloproteinase inhibitors, with GM6001 being the most important, also known as ilomastat or galardin. This molecule is generally considered as a standard MMPI, establishing the canonical binding mode consisting of a bidentate coordination of the hydroxamate to the catalytic Zn ion, and the occupation of the S1′ pocket that drove later inhibitor design. In this view, ilomastat has become a widely used laboratory tool for studying metalloproteinases [42]. Specifically, it serves as the reference control compound used by scientists to understand the processes by which MMPs destroy tissue. It is employed in experimental models to study pulmonary fibrosis, chronic inflammation in cystic fibrosis, radiation damage, and the mechanisms of cell migration involved in metastasis.

It was developed in the early 1990s, as a peptidomimetic belonging to the hydroxamic acid class. Its chemical structure mimics the epitope of collagen binding sites, and the chemical moiety key for inhibition is the hydroxamate group. Ilomastat is a broad-spectrum inhibitor, as it blocks almost all major metalloproteinases, including MMP-1, MMP-2, MMP-3, MMP-8, and MMP-9, and exhibits very poor oral bioavailability, which severely limited its development as a systemic cancer drug. Also, ilomastat demonstrated a remarkable ability to inhibit bacterial metalloproteinases, too, such as the elastase produced by *Pseudomonas aeruginosa*.

Due to absorption issues and the failure of its therapeutic class in oncology, researchers have partially redirected ilomastat toward topical and local applications. Specifically, it has been studied for the treatment of corneal ulcers and chemical eye injuries. Due to its ability to inhibit cutaneous collagenase (MMP-1), the molecule has been evaluated for both the prevention of photo-induced skin aging and the treatment of hypertrophic scars.

6. Batimastat

Batimastat is one of the earliest and most extensively investigated matrix metalloproteinase inhibitors in preclinical cancer research. Numerous studies demonstrated its antitumor and antiangiogenic properties across a variety of experimental models. Batimastat was developed by the British company British Biotech as a synthetic derivative of hydroxamic acid. Its discovery followed the general strategy that was successful for the design of ACE inhibitors, incorporating a scissile bond surrogate into a specific pseudopeptide substrate. In this context, the development of batimastat focused on *C*-terminal analogues containing residues capable of interacting with corresponding MMP subsites and containing a hydroxamate moiety as a scissile bond replacement. Designed to mimic natural peptides, it interacts with the zinc atom in the active site of matrix metalloproteinases (MMP-1, −2, −7, −9), thereby deactivating them. The broad-spectrum inhibition activity of batimastat against MMPs is evinced by high activity towards

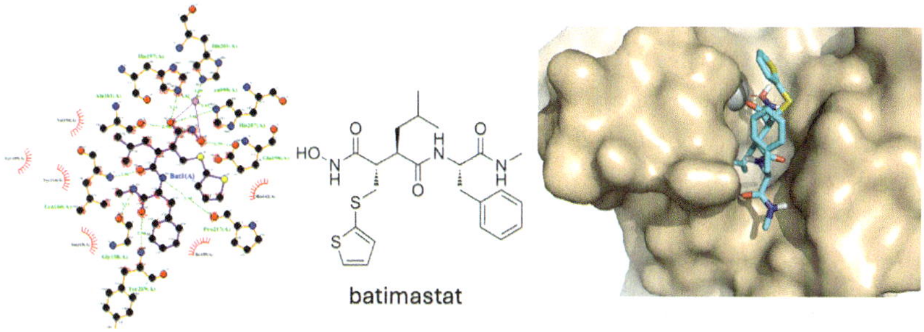

Fig. 5 Left: Structure and 2D binding mode of batimastat (BB94) and MMP8 with LigPlot+; center: structure of batimastat; right: close-up of the X-ray crystallographic structure of the compound in the active site of MMP8 (PDB: 1MMB).

fibroblast collagenase, neutrophil collagenase and gelatinases, all in the low nM range).

Crystallographic data [43] of batimastat bound to MMP8 show the canonical binding mode of peptidomimetic hydroxamates characterized by the strong chelating interaction of the hydroxamic acid group with the zinc ion and by the interaction of a non-polar residue, here the isobutyl side-chain of a leucine isostere, with the deep hydrophobic S1' subsite of the enzyme (Fig. 5).

Concerning the preclinical evidence of MMP inhibition *in vivo*, batimastat caused tumor burden reduction in human ovarian carcinoma xenograft. An elegant demonstration of the specific MMP inhibition resulting to such *in vivo* effects was achieved through the inactive diastereomer of batimastat that had no effect on tumor biology. Also, in animal models and *in vitro* studies, batimastat demonstrated remarkable capabilities, as it drastically reduced tumor vascularization, particularly in colon and breast cancer in mice, thanks of an anti-angiogenic effect, and almost completely blocked the ability of tumor cells to invade the extracellular matrix and metastasize. These findings provided strong evidence for its ability to interfere with tumor-associated blood vessel formation. Nevertheless, although monotherapy delayed tumor progression and extended lifespan, all treated animals eventually did not survive to the disease. Interestingly, the concomitant administration of batimastat with cisplatin enhanced the antitumor efficacy of chemotherapy by further suppressing tumor growth and limiting the intraperitoneal spread. These observations suggest that MMP inhibitors may serve as valuable adjuvants to conventional cytotoxic therapies, either by enhancing their effectiveness or by improving treatment responses.

Batimastat was the first MMP inhibitor to enter phase I clinical evaluation. Despite its enormous therapeutic potential, batimastat had very low water solubility as a critical chemical shortcoming [44], thus impairing its oral bioavailability. To circumvent this issue, British Biotech was forced to formulate it as a suspension for local administration. Initial results from phase I and II trials, which began around 1994, appeared promising regarding local efficacy, reducing the accumulation of tumor-related fluid and the need for frequent and painful drainage procedures [45]. However, serious obstacles arose, as repeated injection of the suspension into the abdominal cavity caused severe local inflammation and chemical peritonitis in many patients. In 1995, during the transition from laboratory production to industrial-scale manufacturing for phase III, unexpected pain in treated patients forced a temporary suspension of the trials. Thus, it was realized that an anticancer drug requiring frequent, painful injections into the abdomen or chest would never have a sustainable commercial future. In this context, shortcomings of batimastat inspired the modification of the molecular structure by introducing a hydroxyl group, resulting to marimastat, which retained the same mechanism of action but displayed improved oral bioavailability required for its formulation in capsules. Consequently, the development of batimastat was gradually abandoned in favor of its successor as a more practical therapeutic alternative. All development projects involving batimastat for any medical indication, including subsequent attempts to test it for different biomedical applications like stroke, were permanently terminated in 2002, and, despite initial encouraging data, the inhibitor never reached the pharmaceutical market.

7. Marimastat (BB-2516)

As above discussed, British Biotech designed marimastat (also known as BB-2516) as a synthetic broad-spectrum matrix metalloproteinase inhibitor structurally derived from batimastat and the first orally bioavailable MMPI. At the molecular level, the oral bioavailability was achieved through the improvement of the polar character of the compound with respect to batimastat, by introducing a polar hydroxyl group. The crystallographic structure of marimastat bound to the Zn-metalloprotease aggrecanase-2 [46] highlights the binding mode of this non-selective, broad spectrum metalloprotease inhibitor based on the hydroxamic acid, interacting with the zinc ion with the hydroxamic group and addressing the S1' subsite of the enzyme with a hydrophobic residue, here the isobutyl side-chain of a

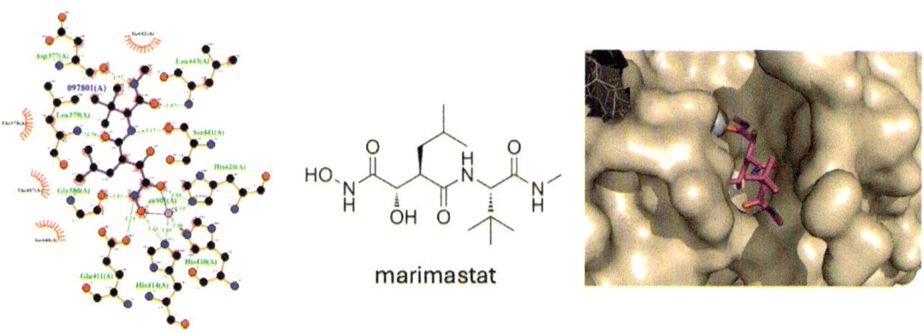

Fig. 6 Left: Structure and 2D binding mode of marimastat (BB-2516) and Zn-metalloprotease ADAMTS-5 with LigPlot+; center: structure of marimastat; right: close-up of the X-ray crystallographic structure of the compound in the active site of ADAMTS-5 (PDB: 3HY7).

leucine isostere, similarly to batimastat (Fig. 6). The S1' pocket, also referred to as the specificity pocket, is the recognition site for the different substrates for metalloproteinases. Marimastat has a very short P1' moiety, hence it can be recognized by most of the metalloproteinases.

Preclinical studies identified several toxic effects associated with this compound, including gastrointestinal ulceration, body weight reduction, hemorrhagic events, fibrosis, inflammatory reactions in joints and skeletal muscle necrosis. Nevertheless, clinical evaluations indicated an acceptable safety profile in patients with advanced malignancies, reporting musculoskeletal symptoms as the principal adverse effect and dose-limiting toxicity. Thus, the clinical development of marimastat progressed through multiple stages, applying this experimental drug both as monotherapy and in combination with conventional cytotoxic agents, resulting to promising outcomes. Early phase I and II trials showed that the drug could reduce tumor markers in some patients. This success produced great speculative excitement, with media and investors addressing marimastat as the future "miracle pill" against cancer.

In 2000, to secure final approval, marimastat underwent massive phase III clinical trials involving thousands of patients with advanced-stage cancers, including pancreatic, lung, breast, and brain cancer. Nevertheless, the drug did not significantly prolong patient survival as compared to a placebo or standard therapies. Also, many patients developed musculoskeletal syndrome characterized by a combination of severe joint pain, stiffness and inflammation. In 2002, following the failure of yet another trial for glioblastoma, British Biotech permanently terminated the development of marimastat.

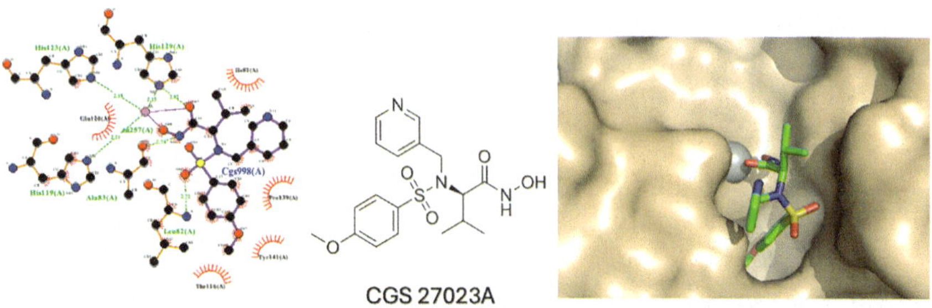

Fig. 7 Left: Structure and 2D binding mode of CGS-27023A (MMI270) and macrophage elastase MMP-12 with LigPlot+; center: structure of CGS-27023A; right: close-up of the X-ray crystallographic structure of the compound in the active site of MMP12 (PDB: 1JIZ).

8. CGS-27023A (MMI270)

In the laboratories of Ciba-Geigy, which merged with Sandoz in 1996 to form Novartis, the non-peptide hydroxamate compound designated CGS-27023A was synthesized [17,47]. This molecule is a non-peptidic, broad-spectrum matrix metalloproteinase inhibitor based on the structure of valine-linked hydroxamate derivatives, with the aim of creating a molecule with optimal metabolic stability to be orally bioavailable. Similarly to batimastat and marimastat, it belongs to the first-generation of MMPIs studied as potential therapeutic agents for the treatment of cancer.

This compound can interact with the zinc ion in the catalytic site of a wide range of MMPs, particularly MMP-2, MMP-3, MMP-9, and stromelysin-1, and can inhibit MMPs proteolytic activity at nM concentrations *in vitro*. The crystallographic structure of CGS-27023A bound to the MMP-12 (Fig. 7) highlights the binding mode of this metalloprotease inhibitor interacting with the zinc ion with the hydroxamic group and addressing the S1' subsite of the enzyme with the arylsulfonyl group. Valine side-chain and the pyridyl moiety further interact with subsites flanking the zinc ion catalytic core.

This MMPI showed anticancer activity by preventing tumor growth in experimental tumors and showed antiangiogenic properties in experimental models. Based on promising preclinical results, Novartis renamed the molecule MMI270 and initiated the first human clinical trials (phase I and II). Specifically, in early 2000, the drug entered phase I clinical trials in humans to evaluate its pharmacokinetics and tolerability in patients with advanced solid tumors [48]. Then, it was tested to evaluate its ability to inhibit metastasis [49]. The drug was also tested alone and in combination with traditional chemotherapy on patients with advanced solid tumors, such as colorectal cancer and non-small cell lung cancer (NSCLC) [48]. However, musculoskeletal syndrome as a severe side effect emerged during human trials that had not been anticipated in animal trials. This occurred because the drug, despite its potency, was not selective enough. Therefore, by indiscriminately blocking different MMPs, it altered the normal remodeling of connective tissues and joints in the human body. Thus, due to joint toxicity and the lack of clear tumor volume regression in advanced-stage patients, clinical development of MMI270 was prematurely terminated and abandoned in the early 2000s, in analogy to almost all MMPIs of that generation, such as marimastat or prinomastat.

9. Tanomastat (BAY12-9566)

Tanomastat, also named BAY12-9566, is a synthetic, non-peptide MMP inhibitor characterized by a biphenyl scaffold and a carboxylate group that chelates zinc at the enzyme active site. The compound exhibits remarkable substrate selectivity, effectively inhibiting MMP-2, MMP-3, and MMP-9 while showing little or no activity against MMP-1. Tanomastat was developed by Bayer researchers using a different approach, and specifically as a non-peptidic biphenyl derivative [50].

The molecules displayed excellent oral bioavailability and preclinical success due to the absence of joint pain. Specifically, in animal models, tanomastat yielded extraordinary results, demonstrating to possesses anti-invasive, anti-metastatic, and anti-angiogenic effects without exhibiting any systemic toxicity. Taking advantage of its oral bioavailability, it was evaluated in several experimental cancer models, including human colon carcinoma, human breast carcinoma, murine melanoma, and Lewis lung carcinoma.

Clinical investigations indicated that the drug was generally well tolerated. Phase I studies involving both healthy volunteers and patients with cancer reported no severe toxicities, as it did not cause the painful musculoskeletal issues associated with its predecessors and was highly tolerated by patients. This prompted Bayer to immediately launch tanomastat in large-scale phase III clinical trials worldwide involving patients with various types of cancer, such as advanced pancreatic carcinoma, small-cell lung cancer, non-small-cell lung cancer and ovarian cancer. Nevertheless, in 1999, Bayer received interim analysis data and made the decision to immediately terminate all tanomastat clinical trials due to efficacy and safety concerns. The failure of tanomastat marked the end of the "gold rush" for first- and second-generation metalloproteinase inhibitors. The scientific community ultimately learned that unselective inhibition produced the blocking of protective MMPs, thus impairing key natural defenses, and that the drug was ideal for preventing the very early stages of metastasis acting as a "brake," but it proved entirely useless or even harmful when administered to patients with advanced or metastatic tumors.

10. Prinomastat (AG-3340)

Prinomastat (also known as AG3340) is an orally active, non-peptidic matrix metalloproteinase inhibitor belonging to the hydroxamic acid class, representing an attempt to create a targeted matrix metalloproteinase inhibitor designed specifically to avoid the toxicological failures seen with early compounds like marimastat [51]. Specifically, the molecule was designed to specifically target only those MMPs responsible for matrix degradation and tumor angiogenesis, such as gelatinases (MMP-2 and MMP-9), MMP-13 and MMP-14, and avoiding binding to MMP-1, to stop cancer progression without causing joint toxicity.

The structure of this molecule is characterized by a thiomorpholine scaffold functionalized by a hydroxamate zinc-binding group and a pyridyloxyphenyl sulfonamido group that enabled the improvement of selectivity toward gelatinases and membrane-type MMPs while displaying reduced affinity for MMP-1 and MMP-7 of about three orders of magnitude, demonstrating that substantial improvements in selectivity could be achieved through structure-guided design. This resulted also to a pharmacokinetic profile and therapeutic index superior to those of first-generation inhibitors.

In animal models, it demonstrated the ability to inhibit tumor growth and to significantly reduce the number of spontaneous metastases by preventing cancer cells from migrating. Specifically, it proved to suppress tumor development, reduce metastatic spread, and improve survival outcomes in mice bearing a variety of human tumor xenografts, including colon, lung, breast, prostate, and glioma models. Another significant characteristic was its lipophilic nature, which allowed it to cross the blood-brain barrier, making it potentially useful against brain tumors, too.

Encouraged by laboratory data and good initial tolerability in phase I trials [36], Pfizer launched massive international phase III clinical trials combining prinomastat with standard chemotherapy [52]. However, in 2000, following preliminary disappointing data, Pfizer decided to immediately terminate the two main phase III trials, namely advanced non-small cell lung cancer, when tested in combination with chemotherapy, as the drug showed no improvement in survival or time to disease progression compared to patients treated with placebo alone. Furthermore, despite its targeted design, prolonged use over several months revealed that the drug still caused severe side effects such as arthralgia, muscle stiffness, and joint swelling, thus necessitating frequent dosage reductions or treatment suspensions.

Like other MMPI that terminated clinical studies without entering the drug market, prinomastat now exists solely in scientific databases and is sold exclusively as a biochemical reagent for *in vitro* laboratory research.

11. Periostat (doxycycline)

Periostat represents the only clinical success to emerge from the major research efforts regarding matrix metalloproteinase inhibitors in the 1990s. While molecules originally developed for oncology failed due to toxicity, periostat found a successful alternative path, becoming the first and only MMP-inhibiting drug approved by the FDA for chronic systemic use [53]. Periostat was not identified as a new synthetic chemical entity, but through the repurposing of doxycycline, a well-known broad-spectrum antibiotic belonging to the tetracycline family, discovered in the 1960s. The application of doxycycline in this context emerged from early discovery in the 1980s by researchers at Stony Brook University that tetracyclines could block the destruction of connective tissue caused by collagenase enzymes (MMPs) at very low molecular concentrations and independently of their bactericidal action.

The major obstacle to using an antibiotic over long periods is the risk of destroying gut flora and developing dangerous antibiotic resistance. To circumvent this problem, researchers formulated capsules containing five- to tenfold reduced dosage with respect to conventional antimicrobial treatments, thus impairing antibacterial activity and maintaining the ability to effectively block MMP-1, MMP-8, and MMP-9, although its precise mechanism of action on MMPs activity remains unclear.

Developed commercially by CollaGenex Pharmaceuticals, Periostat (20 mg doxycycline hyclate) received historic FDA approval on September 30, 1998. It was approved as an adjunct therapy to deep cleaning for the treatment of chronic adult periodontitis [54]. Instead of targeting plaque bacteria, Periostat treated periodontitis by modulating the host's immune response, specifically preventing the patient's white blood cells from producing excessive metalloproteinases, thus protecting the bone and tooth-supporting tissues from resorption and destruction.

Overall, doxycycline, an antibiotic from the mid-20th century, remains the only FDA-approved MMPi. Beyond periodontitis, doxycycline has demonstrated beneficial effects in small clinical studies involving multiple sclerosis and type II diabetes mellitus. In patients with multiple sclerosis, treatment in combination with interferon-β reduced circulating MMP-9 levels, decreased lesion activity, and limited monocyte migration while maintaining an acceptable safety profile. Although these findings remain preliminary and require validation in larger clinical trials, they show that the selective modulation of MMP activity can have a therapeutic potential when appropriate disease contexts and pharmacological strategies are met.

12. Conclusion

MMP inhibitors have been studied over the last three decades as therapeutic agents due to the pivotal roles of matrix metalloproteinases in tumor progression, resulting in the development of several generations of inhibitors and candidate cancer drugs. Starting from first-generation inhibitors that were designed as peptide-based inhibitors, hydroxamate-based peptidomimetics such as batimastat and marimastat were among the first investigational new drugs entering clinical trials, although their clinical success was limited by poor bioavailability and selectivity, resulting in severe side effects. Second-generation MMPIs consisted of alternative zinc-binding groups and non-peptide structures, although they again failed in clinical trials due to low oral bioavailability and safety concerns. Specifically, the failure of prinomastat and tanomastat, which were conceived as improved molecules starting from previous clinical experience, reconsidered the MMPs as a valid therapeutic strategy for the treatment of cancer. Doxycycline remains the only FDA-approved MMP inhibitor.

Nevertheless, given the prominent role of this class of enzymes in the context of cancer and other major pathologies, research is still active in view of developing more selective and safe molecules capable of interacting solely with the overexpressed MMP subtype. Advances in structural biology highlighting the structural differences among MMPs is expected to facilitate the development of more selective inhibitors with improved efficacy and reduced toxicity and enable new MMPI as candidate cancer drugs.

References

[1] Z. Werb, ECM and cell surface proteolysis: regulating cellular ecology, Cell 91 (4) (1997) 439–442.

[2] H. Birkedal-Hansen, Proteolytic remodeling of extracellular matrix, Curr. Opin. Cell Biol. 7 (5) (1995) 728–735.
[3] L.J. McCawley, L.M. Matrisian, Matrix metalloproteinases: multifunctional contributors to tumor progression, Mol. Med. Today 6 (4) (2000) 149–156.
[4] G. Tu, W. Xu, H. Huang, S. Li, Progress in the development of matrix metalloproteinase inhibitors, Curr. Med. Chem. 15 (14) (2008) 1388–1395.
[5] J.M. Whitelock, A.D. Murdoch, R.V. Iozzo, P.A. Underwood, The degradation of human endothelial cell-derived perlecan and release of bound basic fibroblast growth factor by stromelysin, collagenase, plasmin, and heparanases, J. Biol. Chem. 271 (1996) 10079–10086.
[6] A.F. Chambers, L.M. Matrisian, Changing views of the role of matrix metalloproteinases in metastasis, J. Natl. Cancer Inst. 89 (1997) 1260–1270.
[7] Z.S. Galis, J.J. Khatri, Matrix metalloproteinases in vascular remodeling and atherogenesis: the good, the bad, and the ugly, Circ. Res. 90 (3) (2002) 251–262.
[8] R. Visse, H. Nagase, Matrix metalloproteinases and tissue inhibitors of metalloproteinases: structure, function, and biochemistry, Circ. Res. 92 (8) (2003) 827–839.
[9] C. Johnson, Z.S. Galis, Matrix metalloproteinase-2 and -9 differentially regulate smooth muscle cell migration and cell-mediated collagen organization, Arter. Thromb. Vasc. Biol. 24 (1) (2004) 54–60.
[10] Y.E. Liu, M. Wang, J. Greene, J. Su, S. Ullrich, H. Li, S. Sheng, P. Alexander, Q.A. Sang, Y.E. Shi, Preparation and characterization of recombinant tissue inhibitor of metalloproteinase 4 (TIMP-4), J. Biol. Chem. 272 (33) (1997) 20479–20483.
[11] J.D. Raffetto, R.A. Khalil, Matrix metalloproteinases and their inhibitors in vascular remodeling and vascular disease, Biochem. Pharmacol. 75 (2) (2008) 346–359.
[12] C. Tallant, A. Marrero, F.X. Gomis-Rüth, Matrix metalloproteinases: fold and function of their catalytic domains, Biochim. Biophys. Acta 1803 (1) (2010) 20–28.
[13] A. Varghese, S.O. Waheed, K. Gorantla, I. DiCastri, C. LaRouche, B. Kaski, G.B. Fields, T.G. Karabencheva-Christova, Catalytic mechanism of collagen hydrolysis by zinc(II)-dependent matrix metalloproteinase-1, J. Phys. Chem. B 127 (45) (2023) 9697–9709.
[14] L. Aureli, M. Gioia, I. Cerbara, S. Monaco, G.F. Fasciglione, S. Marini, P. Ascenzi, A. Topai, M. Coletta, Structural bases for substrate and inhibitor recognition by matrix metalloproteinases, Curr. Med. Chem. 15 (22) (2008) 2192–2222.
[15] A. Gimeno, R. Beltrán-Debón, M. Mulero, G. Pujadas, S. Garcia-Vallvé, Understanding the variability of the S1′ pocket to improve matrix metalloproteinase inhibitor selectivity profiles, Drug Discov. Today 25 (1) (2020) 38–57.
[16] J. Hu, P.E. Van den Steen, Q.X. Sang, G. Opdenakker, Matrix metalloproteinase inhibitors as therapy for inflammatory and vascular diseases, Nat. Rev. Drug Discov. 6 (6) (2007) 480–498.
[17] K. Li, F.R. Tay, C.K.Y. Yiu, The past, present and future perspectives of matrix metalloproteinase inhibitors, Pharmacol. Ther. 207 (2020) 107465.
[18] R.E. Vandenbroucke, C. Libert, Is there new hope for therapeutic matrix metalloproteinase inhibition? Nat. Rev. Drug Discov. 13 (12) (2014) 904–927.
[19] B.W. Matthews, Structural basis of the action of thermolysin and related zinc peptidases, Acc. Chem. Res. 21 (1988) 333–340.
[20] G.B. Fields, The rebirth of matrix metalloproteinase inhibitors: moving beyond the dogma, Cells 8 (9) (2019) 984.
[21] B. De, M.G. Natchus, M. Cheng, S. Pikul, N.G. Almstead, Y.O. Taiwo, C.E. Snider, L. Chen, B. Barnett, F. Gu, M. Dowty, The next generation of MMP inhibitors. Design and synthesis, Ann. N. Y. Acad. Sci. 878 (1999) 40–60.
[22] M. Whittaker, C.D. Floyd, P. Brown, A.J. Gearing, Design and therapeutic application of matrix metalloproteinase inhibitors, Chem. Rev. 99 (9) (1999) 2735–2776.

[23] J.W. Skiles, N.C. Gonnella, A.Y. Jeng, The design, structure, and therapeutic application of matrix metalloproteinase inhibitors, Curr. Med. Chem. 8 (4) (2001) 425–474.
[24] J.A. Jacobsen, J.L. Major Jourden, M.T. Miller, S.M. Cohen, To bind zinc or not to bind zinc: an examination of innovative approaches to improved metalloproteinase inhibition, Biochim. Biophys. Acta 1803 (1) (2010) 72–94.
[25] B. Pirard, Insight into the structural determinants for selective inhibition of matrix metalloproteinases, Drug Discov. Today 12 (15–16) (2007) 640–646.
[26] W. Bode, C. Fernandez-Catalan, H. Tschesche, F. Grams, H. Nagase, K. Maskos, Structural properties of matrix metalloproteinases, Cell Mol. Life Sci. 55 (4) (1999) 639–652.
[27] S.P. Gupta, Quantitative structure-activity relationship studies on zinc-containing metalloproteinase inhibitors, Chem. Rev. 107 (7) (2007) 3042–3087.
[28] J.F. Fisher, S. Mobashery, Recent advances in MMP inhibitor design, Cancer Metastasis Rev. 25 (1) (2006) 115–136.
[29] H.E. Van Wart, H. Birkedal-Hansen, The cysteine switch: a principle of regulation of metalloproteinase activity with potential applicability to the entire matrix metalloproteinase gene family, Proc. Natl. Acad. Sci. U. S. A. 87 (14) (1990) 5578–5582.
[30] B.G. Rao, Recent developments in the design of specific Matrix Metalloproteinase inhibitors aided by structural and computational studies, Curr. Pharm. Des. 11 (3) (2005) 295–322.
[31] L.M. Coussens, B. Fingleton, L.M. Matrisian, Matrix metalloproteinase inhibitors and cancer: trials and tribulations, Science 295 (5564) (2002) 2387–2392.
[32] A.O. Stewart, P.A. Bhatia, J.G. Martin, J.B. Summers, K.E. Rodriques, M.B. Martin, J.H. Holms, J.L. Moore, R.A. Craig, T. Kolasa, J.D. Ratajczyk, H. Mazdiyasni, F.A. Kerdesky, S.L. DeNinno, R.G. Maki, J.B. Bouska, P.R. Young, C. Lanni, R.L. Bell, G.W. Carter, C.D. Brooks, Structure-activity relationships of N-hydroxyurea 5-lipoxygenase inhibitors, J. Med. Chem. 40 (13) (1997) 1955–1968.
[33] L. Devel, B. Czarny, F. Beau, D. Georgiadis, E. Stura, V. Dive, Third generation of matrix metalloprotease inhibitors: gain in selectivity by targeting the depth of the S1′ cavity, Biochimie 92 (11) (2010) 1501–1508.
[34] M.B. Onaran, A.B. Comeau, C.T. Seto, Squaric acid-based peptidic inhibitors of matrix metalloprotease-1, J. Org. Chem. 70 (26) (2005) 10792–10802.
[35] D.R. Shalinsky, J. Brekken, H. Zou, C.D. McDermott, P. Forsyth, D. Edwards, S. Margosiak, S. Bender, G. Truitt, A. Wood, N.M. Varki, K. Appelt, Broad antitumor and antiangiogenic activities of AG3340, a potent and selective MMP inhibitor undergoing advanced oncology clinical trials, Ann. N. Y. Acad. Sci. 878 (1999) 236–270.
[36] K.R. Hande, M. Collier, L. Paradiso, J. Stuart-Smith, M. Dixon, N. Clendeninn, G. Yeun, D. Alberti, K. Binger, G. Wilding, Phase I and pharmacokinetic study of prinomastat, a matrix metalloprotease inhibitor, Clin. Cancer Res. 10 (3) (2004) 909–915.
[37] C.M. Overall, O. Kleifeld, Towards third generation matrix metalloproteinase inhibitors for cancer therapy, Br. J. Cancer 94 (7) (2006) 941–946.
[38] N. Sela-Passwell, G. Rosenblum, T. Shoham, I. Sagi, Structural and functional bases for allosteric control of MMP activities: can it pave the path for selective inhibition? Biochim. Biophys. Acta 1803 (1) (2010) 29–38.
[39] E. Nuti, T. Tuccinardi, A. Rossello, Matrix metalloproteinase inhibitors: new challenges in the era of post broad-spectrum inhibitors, Curr. Pharm. Des. 13 (20) (2007) 2087–2100.
[40] Y. Shi, X. Ma, G. Fang, X. Tian, C. Ge, Matrix metalloproteinase inhibitors (MMPIs) as attractive therapeutic targets: recent progress and current challenges, NanoImpact 21 (2021) 100293.
[41] J.A. Low, M.D. Johnson, E.A. Bone, R.B. Dickson, The matrix metalloproteinase inhibitor batimastat (BB-94) retards human breast cancer solid tumor growth but not ascites formation in nude mice, Clin. Cancer Res. 2 (7) (1996) 1207–1214.

[42] D. Grobelny, L. Poncz, R.E. Galardy, Inhibition of human skin fibroblast collagenase, thermolysin, and Pseudomonas aeruginosa elastase by peptide hydroxamic acids, Biochemistry 31 (31) (1992) 7152–7154.
[43] F. Grams, M. Crimmin, L. Hinnes, P. Huxley, M. Pieper, H. Tschesche, W. Bode, Structure determination and analysis of human neutrophil collagenase complexed with a hydroxamate inhibitor, Biochemistry 34 (43) (1995) 14012–14020.
[44] S. Wojtowicz-Praga, J. Low, J. Marshall, E. Ness, R. Dickson, J. Barter, M. Sale, P. McCann, J. Moore, A. Cole, M.J. Hawkins, Phase I trial of a novel matrix metalloproteinase inhibitor batimastat (BB-94) in patients with advanced cancer, Invest. New Drugs 14 (2) (1996) 193–202.
[45] S.L. Parsons, S.A. Watson, R.J. Steele, Phase I/II trial of batimastat, a matrix metalloproteinase inhibitor, in patients with malignant ascites, Eur. J. Surg. Oncol. 23 (6) (1997) 526–531.
[46] M.D. Tortorella, A.G. Tomasselli, K.J. Mathis, M.E. Schnute, S.S. Woodard, G. Munie, J.M. Williams, N. Caspers, A.J. Wittwer, A.M. Malfait, H.S. Shieh, Structural and inhibition analysis reveals the mechanism of selectivity of a series of aggrecanase inhibitors, J. Biol. Chem. 284 (36) (2009) 24185–24191.
[47] L.J. MacPherson, E.K. Bayburt, M.P. Capparelli, B.J. Carroll, R. Goldstein, M.R. Justice, L. Zhu, S. Hu, R.A. Melton, L. Fryer, R.L. Goldberg, J.R. Doughty, S. Spirito, V. Blancuzzi, D. Wilson, E.M. O'Byrne, V. Ganu, D.T Parker, Discovery of CGS 27023A, a non-peptidic, potent, and orally active stromelysin inhibitor that blocks cartilage degradation in rabbits, J. Med. Chem. 40 (16) (1997) 2525–2532.
[48] N.C. Levitt, F.A. Eskens, K.J. O'Byrne, D.J. Propper, L.J. Denis, S.J. Owen, L. Choi, J.A. Foekens, S. Wilner, J.M. Wood, M. Nakajima, D.C. Talbot, W.P. Steward, A.L. Harris, J. Verweij, Phase I and pharmacological study of the oral matrix metalloproteinase inhibitor, MMI270 (CGS27023A), in patients with advanced solid cancer, Clin. Cancer Res. 7 (7) (2001) 1912–1922.
[49] T. Kasaoka, H. Nishiyama, M. Okada, M. Nakajima, Matrix metalloproteinase inhibitor, MMI270 (CGS27023A) inhibited hematogenic metastasis of B16 melanoma cells in both experimental and spontaneous metastasis models, Clin. Exp. Metastasis 25 (7) (2008) 827–834.
[50] C. Gatto, M. Rieppi, P. Borsotti, S. Innocenti, R. Ceruti, T. Drudis, E. Scanziani, A.M. Casazza, G. Taraboletti, R. Giavazzi, BAY 12-9566, a novel inhibitor of matrix metalloproteinases with antiangiogenic activity, Clin. Cancer Res. 5 (11) (1999) 3603–3607.
[51] R. Scatena, Prinomastat, a hydroxamate-based matrix metalloproteinase inhibitor. A novel pharmacological approach for tissue remodelling-related diseases, Expert Opin. Investig. Drugs 9 (9) (2000) 2159–2165.
[52] D. Bissett, K.J. O'Byrne, J. von Pawel, U. Gatzemeier, A. Price, M. Nicolson, R. Mercier, E. Mazabel, C. Penning, M.H. Zhang, M.A. Collier, F.A. Shepherd, Phase III study of matrix metalloproteinase inhibitor prinomastat in non-small-cell lung cancer, J. Clin. Oncol. 23 (4) (2005) 842–849.
[53] J. Caton, M.E. Ryan, Clinical studies on the management of periodontal diseases utilizing subantimicrobial dose doxycycline (SDD), Pharmacol. Res. 63 (2) (2011) 114–120.
[54] P.M. Preshaw, A.F. Hefti, M.J. Novak, B.S. Michalowicz, B.L. Pihlstrom, R. Schoor, C.L. Trummel, J. Dean, T.E. Van Dyke, C.B. Walker, M.H. Bradshaw, Subantimicrobial dose doxycycline enhances the efficacy of scaling and root planing in chronic periodontitis: a multicenter trial, J. Periodontol. 75 (8) (2004) 1068–1076.

CHAPTER FOUR

Matrix metalloproteinase-13 selective inhibitors

Simone Giovannuzzi*, Claudiu T. Supuran
NEUROFARBA Department, Pharmaceutical and Nutraceutical Section, University of Florence, Firenze, Italy
*Corresponding author. e-mail addresses: simone.giovannuzzi@unifi.it, claudiu.supuran@unifi.it.

Contents

Abstract

Matrix metalloproteinase-13 (MMP13) is the primary protease responsible for the irreversible degradation of type II collagen in the extracellular matrix (ECM), playing a pivotal role in the pathogenesis of osteoarthritis (OA), rheumatoid arthritis (RA), and various malignant cancers. While early clinical efforts with broad-spectrum zinc-chelating inhibitors were abandoned due to dose-limiting musculoskeletal syndrome (MSS), recent research has focused on achieving high isoform selectivity by targeting unique structural features such as the deep S1′ specificity pocket and the MMP13-specific S1″ side pocket. This chapter reviews a wide set of advanced inhibitor scaffolds, including selective sulfonamide hydroxamates, hydantoins such as **AZD6605**, and non-zinc-binding templates like quinazoline-2-carboxamides and pyrimidine-dicarboxamides. Furthermore, we discuss innovative strategies such as the use of allosteric inhibitors, exosite-binding agents targeting the hemopexin domain, and selectively deuterated molecules (e.g., **AQU-019**) to improve metabolic stability. Finally, the chapter highlights the clinical potential of neutralizing antibodies (e.g., **14D10**), targeted polypharmacology for dual MMP10/13 inhibition, and the development of selective PET radiotracers for non-invasive monitoring of disease progression and atherosclerotic plaque instability.

Keywords: Antibodies, Matrix metalloproteinase, MMP13, Non-zinc binding inhibitors, Osteoarthritis, Rheumatoid arthritis, Selective inhibition, Tumor, Zinc-binding inhibitors

Fungal and Protozoan Carbonic Anhydrases, Volume 60
ISSN 1874-6047
https://doi.org/10.1016/bs.enz.2026.03.001

1. Introduction

Matrix metalloproteinases (MMPs) constitute a group of zinc-dependent proteases that degrade a wide range of extracellular matrix (ECM) components [1–3]. They are generally divided into several subclasses, such as collagenases (MMP1, -8, -13, -18), gelatinases (MMP2, -9), membrane-anchored (MMP14, -15, -16, -17, -24, -25), and other members of the family (MMP7, -12, -19, -20, -23, -26, -28) [1–4]. Structurally, MMPs share a modular architecture composed of a conserved signal peptide, a propeptide region, and a catalytic domain. With the exception of MMP-7, -23, and -26, they also feature a proline-rich linker (hinge region) and a C-terminal hemopexin-like domain [5]. As illustrated in Fig. 1, the catalytic domain has a spherical shape and a characteristic fold containing three α-helices, five β-sheets, and eight connecting loops [1–4]. This domain includes a catalytic Zn(II) ion coordinated by three histidine residues, along with an additional zinc ion and three calcium ions that contribute to structural stability. Substrate recognition by MMPs is governed by a network of subsites, or pockets, that interact with complementary substrate residues [6]. These six pockets surround the catalytic zinc ion on both sides, S1, S2, and S3 on one side and their primed counterparts S1', S2', and S3' on the other, creating the basis for substrate and inhibitor specificity. Among these regions, the S1' pocket shows the greatest variation in amino acid composition and depth across the MMP family [7–9]. Depending on the enzyme, the S1' pocket may be shallow (as in MMP1 and -7), of intermediate depth (MMP2, -8, and -9), or deep (MMP3, -11, -12, -13, and -14) [10]. Notably, MMP13 possesses a large and hydrophobic S1' site formed in part by a flexible "S1' specificity loop" (also known as the Ω-loop, residues 245–253). This structural feature is thought to play a key role in determining the selective binding of MMP13 inhibitors [11]. It is a long, open channel whose contours are defined by L218 residue, located on one side of the pocket, and by the conformation of the S1′ specificity loop [11]. The S1′ pocket in MMP13 is significantly larger than in other MMPs due to the presence of a Leucine and a Proline residue within the active site, and it can accommodate large groups; therefore, this region is considered to be an important structural basis for designing selective MMP13 inhibitors. In addition, MMP13 contains negatively charged residues, such as Aspartate or Glutamate, located just before the third Histidine residue (H232) of the catalytic Zn(II)-binding site. The presence of a negatively charged residue at position 231 may influence the polarization of the Zn(II)-bound water

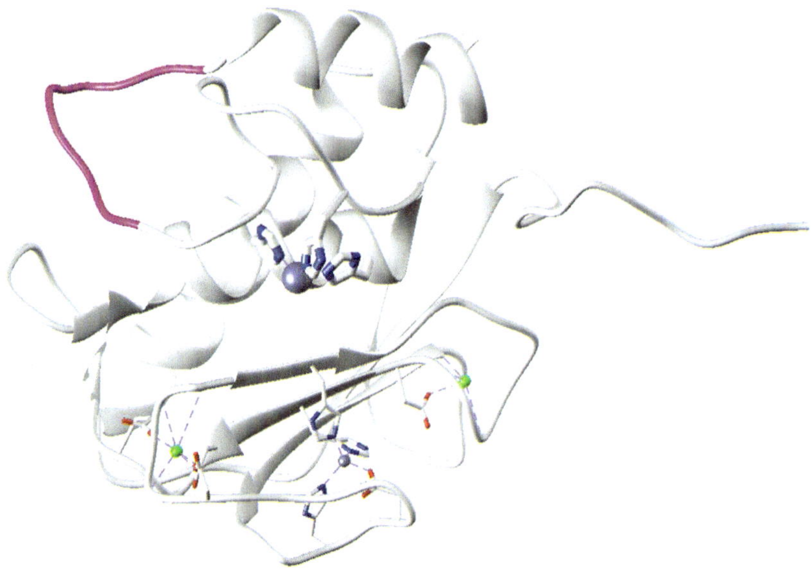

Fig. 1 Structure of MMP13 (PDB: 3KRY). The catalytic domain of MMP13 is represented by three histidine residues, the catalytic Zn(II) ion in dark grey, and the Ω-loop in magenta; the structural Zn(II) and Ca(II) ions are shown in grey and green, respectively.

molecule within these enzymes, increasing its nucleophilic character. Indeed, this would certainly explain the high proteolytic efficiency of MMP13, which cleaves interstitial collagens I–III into the typical ¾ and ¼ fragments, and preferentially hydrolyzes type II collagen over types I, five times faster, and III, six times faster [12,13].

MMP13 is the principal catabolic factor involved in cartilage degradation due to its distinctive ability to cleave type II collagen, which is the main protein component of this tissue. Cartilage breakdown products stimulate type A synoviocytes to release inflammatory cytokines such as TNF-α, IL-1β, and IL-6, as well as additional MMPs, particularly MMP13, thereby triggering and/or accelerating the progression of osteoarticular diseases, the most emblematic of which is osteoarthritis (OA) [14–16]. Wernicke *et al.* further confirmed the expression of MMP13 also in the synovial membrane of patients with rheumatoid arthritis, expanding the potential pathological role of the degradative MMP13 enzyme to rheumatoid arthritis [17]. To confirm, Billinghurst *et al.* demonstrated that RS-102,481, a small-molecule preferential inhibitor of MMP13, was able to potently inhibit the release of type II collagen degradation peptides from human osteoarthritis articular cartilage explants [18].

Nonetheless, MMP13 has also been found highly expressed in numerous types of cancers, as shown by Johansson *et al.*, who detected high levels of MMP13 mRNA in cell lines derived from squamous cell carcinomas of the head and neck [19]. These authors suggested that the catalytic activity of MMP13 may play a role in the high invasion capacity of such cancer cells. Indeed, degradation of the basement membrane is thought to be an early and critical event influencing tumor invasion and metastasis. Consistent with a role for MMP13 proteolytic activity in invasive malignant growth, Airola *et al.* reported high levels of MMP13 mRNA only in malignant squamous epithelium of the skin, while premalignant and benign tumors were mostly negative for MMP13 [20,21]. Similarly, in melanoma, elevated MMP13 expression correlated with aggressive tumor phenotypes, particularly nodular melanomas, and its expression varied based on BRAF mutation status, suggesting a role in invasive progression [22]. Moreover, in breast cancer, MMP13 was upregulated by an integrin-mediated pathway, significantly contributing to stromal invasion in ductal carcinoma in situ, thereby facilitating the transition to invasive breast cancer [23]. MMP13 also plays a significant role in bone-related malignancies, particularly multiple myeloma, in which it contributes to osteolytic lesions [24].

Given that MMP13 plays a crucial role in grave pathologies, such as OA, RA and tumors progression and metastasis, it has become a hot spot that researchers are paying attention to. The scientific community might be able to improve treatments if it can detect or synthesize a species that blocks a specific step in the transcriptional regulation pathway of MMP13, inhibits its synthesis or its activity. In this context, there are some natural compounds such as resveratrol [25], curcumin [26], and epigallocatechin-3-gallate [27], that showed protective effects against matrix degradation and inflammation in OA-affected chondrocytes by indirectly inhibiting the expression of MMP13. However, these natural compounds still have potentially toxic effects on other human tissues because they do not work only on MMP13 expression [25–27].

2. MMP13 selective inhibitors

In recent years, the development of MMP13 inhibitors has benefited from major advances in small-molecule drug discovery technologies. Small molecules continue to dominate therapeutic development because they are easy to manufacture, stable during storage and transport, capable of crossing cell membranes, and generally have low immunogenicity. Since the efficiency of drug discovery depends heavily on screening strategies, improvements in these methods directly affect the quality of lead compounds and the overall feasibility of progressing candidates toward clinical use [28]. Alongside classical strategies, such as modifying known bioactive molecules or employing high-throughput screening (HTS), a broad suite of computational tools has become integral to modern drug design. Computer-aided drug design (CADD) now encompasses structure-based approaches (e.g., molecular docking and dynamic simulations), ligand-based techniques (such as QSAR analysis and pharmacophore modelling), and fragment-based design workflows. More recently, machine-learning and artificial-intelligence methods, including deep learning models, graph-based neural networks, chemical language models, and reinforcement learning, have introduced new avenues for generating and optimizing candidate structures [29,30]. These technological developments have been directly applied to MMP13 inhibitor discovery. Examples include the use of ligand–protein fingerprinting tools like FLAP and natural-product-derived fragment libraries (NPDF) for de novo lead generation, in addition to HTS and computational screening pipelines [31,32]. Collectively, these strategies have accelerated the identification of selective MMP13 inhibitors, which generally fall into two main classes: zinc-chelating and non-zinc-chelating compounds.

2.1 Zinc-chelating inhibitors

Zinc-chelating inhibitors (ZCI), such as those incorporating hydroxamic acid, target the Zn(II) ion in the catalytic site of MMP13, which is the engine of the enzyme. Such inhibitors strongly impair its catalytic activity and generally achieve nanomolar or sub-nanomolar IC_{50} values. However, the catalytic site, including the zinc-binding motif, is highly conserved across all 23 human MMPs. Because ZCIs are so "sticky" to the zinc, they often inhibit other MMPs, such as MMP1, -2 or -14, that the body needs for healthy functions. This lack of selectivity historically led to Musculoskeletal Syndrome (MSS).

First ZCIs were produced by the Becker and Barta's research group between 2000 and 2010 [33–38]. Initially, they described a novel series of aryl hydroxamic acid derivatives as ZCIs that inhibit MMP2 and-13 but spare MMP1 [33]. The series, labelled in Fig. 2 as Series **A**, was conceived *de novo*, upon consideration of the bond distances and geometry of the active

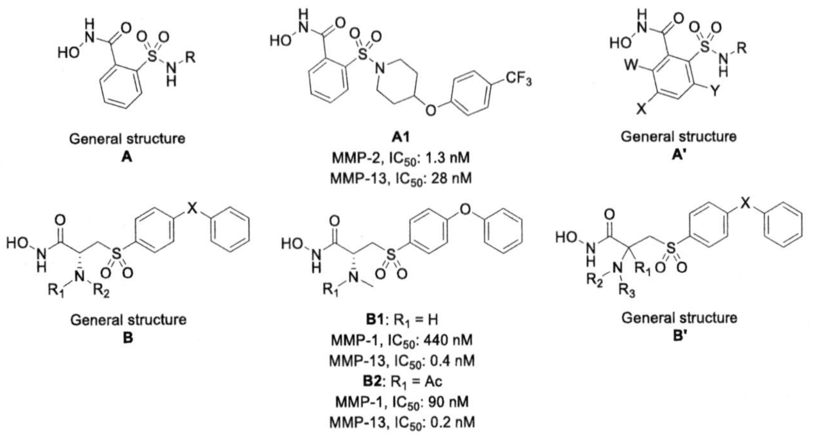

Fig. 2 Representation of general structures **A, A′, B,** and **B′** with a few relative compounds, **A1, B1** and **B2** [33–36].

site of these enzymes to maintain a pivotal hydrogen bond with residue L160 and to install a relatively long substituent into loop P1', to favour selectivity against MMP1, as MMP1 has a relatively shallower pocket [33]. It consists of a phenyl hydroxamic-based scaffold including different amines as a linker, such as piperazines, piperidines, and pyrrolidines, connected to an aryl backbone. Such derivatives were assayed in vitro for their inhibition of MMP1, -2 and -13, displaying high selectivity and nanomolar IC_{50} values against both isozymes 2 and 13 over isozyme 1, which was not inhibited at a concentration of 10 μM [33]. In 2001, Barta *et al.* treated these compounds in rats and discovered that one of the compounds (**A1**) was orally available and achieved a $t_{1/2}$ of 0.8h. Consequently, to improve its half-life in rats, the authors produced a new series by increasing steric hindrance around the potentially metabolizable hydroxamate moiety [34], introducing halogens, methoxy groups or a pegylated chain in positions W, X, and Y (Fig. 2, General structure **A'**). Definitely, they confirmed the high potency and selectivity for MMP2 and -13 over MMP1 and enhanced pharmacokinetic properties in rats [34]. In the same year, Becker *et al.* proposed a series of α-amino-β-sulphone hydroxamates (Fig. 2, general structure **B**) and α-alkyl-α-amino-β-sulphone hydroxamates (Fig. 2, general structure **B'**) to inhibit MMP13 [35,36]. Both series were further assessed for their inhibitory activity against several MMP isozymes displaying a notable potency against MMP13 but even an increased activity on isozyme 1, showing IC_{50}s ranging from 90 nM to 10,000 nM depending on the nature of substituents

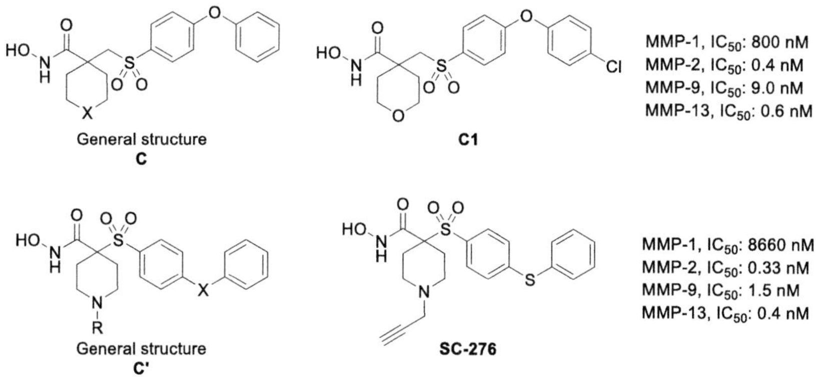

Fig. 3 Representation of general structures **C** and **C'** with a few relative compounds **C1** and **SC-276** [37].

$R_{1\text{-}2}$ [35,36]. Furthermore, derivatives **B1** and **B2** showed good absorption when administered orally in the rat [35]. In 2005, Becker *et al.* explored a series of α-Piperidine-β-sulfone hydroxamate derivatives (Fig. 3, general structure C) and α-Piperidine-α-sulfone hydroxamate derivatives (Fig. 3, general structure C') as potent MMP2, -9 and -13 inhibitors [37]. The α-Piperidine-β-sulfone hydroxamate series was first designed and synthesized with the intent to enhance their pharmacokinetic properties with respect to compounds related to structures **C** and **C'**. However, despite their strong capability exhibited in MMP2 and -13 inhibition, the studied pharmacokinetic properties, such as C_{max} and C_{6h}, were not exhaustive [37]. Indeed, compound **C1** (Fig. 3), which exhibited the most potent activity against targeted MMPs, showed low values of C_{max} and C_{6h}, 1372 ng/mL and 537, respectively. Given such behavior, the authors enclosed the α-Piperidine-α-sulfone hydroxamate series in this study and noted a general and notable enhancement in pharmacokinetic properties, as reported for compound **SC-276** (Fig. 3), showing comparable activity against targeted MMPs and a higher value of C_{max}, 13,630 ng/mL [37]. **C'1** was further advanced through antitumor and antiangiogenesis assays demonstrating excellent antitumor activity vs the MX-1 breast tumor in mice when dosed orally as monotherapy or in combination with paclitaxel [37].

In 2010, α-Sulfone-α-piperidine and α-tetrahydropyranyl hydroxamates were explored by Becker *et al.* as potent inhibitors of MMP2,-9, and -13, with oral efficacy in inhibiting tumor growth in mice and left-ventricular hypertrophy in rats and in the bovine cartilage degradation *ex vivo* explant system [38]. Based on the general structure **C'** and its best candidate

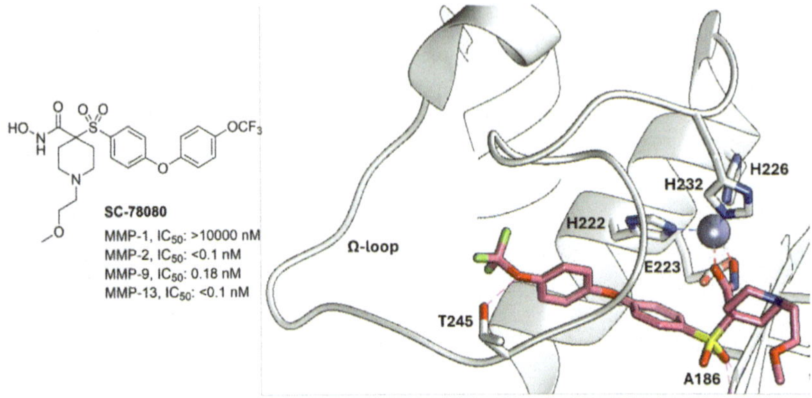

Fig. 4 Complex of **SC-78,080** bound at the MMP-13 active site (PDB: 3KRY); **SC-78,080** in pink, hydrogen bonds as dashed lines in magenta, backbone in white, and Zn(II) in dark grey [38].

SC-276, the authors expanded the series including other substituents to improve pharmacokinetic parameters and *in vitro/in vivo* efficacy. Among all compounds with subnanomolar activity against MMP2 and -13, **SC-78,080** stood out as one of the best candidates for further studies, and it was cocrystallized with MMP13 (Fig. 4), showing that the two oxygen atoms of the hydroxamate coordinate to the metal, and an additional interaction is formed with the side chain of Glu223. Definitely, **SC-78,080** is a 100 pM inhibitor of MMP13 and the gelatinases (MMP2 and MMP9), with >100,000-fold selectivity over MMP1 for MMP2 and MMP13 and >50,000-fold selectivity for MMP9 over MMP1; moreover, the compound has good oral bioavailability (52–82 percent) in the mouse, rat, dog, and cynomolgus monkey [38]. It is rapidly absorbed after oral dosing, with a T_{max} ranging from 10 min (mouse) to 4 h (rat) and the $t_{1/2}$ ranges from 4 to 6 h. Conversely, **SC-78,080** caused less fibroplasia in rats [38].

In 2010, Kolodziej *et al.* also described a novel series of N-aryl isonipecotamide α-sulfone hydroxamate derivatives combining solution-phase and resin-bound library technologies, as outlined in Scheme 1 using a modified Wang resin, to afford compounds belonging to general structure **D** [39]. This series was further assessed for its capability in MMP13 inhibition showing a high potency and a remarkable selectivity on isozyme 13 over isozyme 2. In fact, such compounds displayed selectivity indices MMP2/MMP13 (SIs 2/13) values ranging from 34 to 1659 [39]. Despite the high potency and selectivity achieved with compounds belonging to the general structure

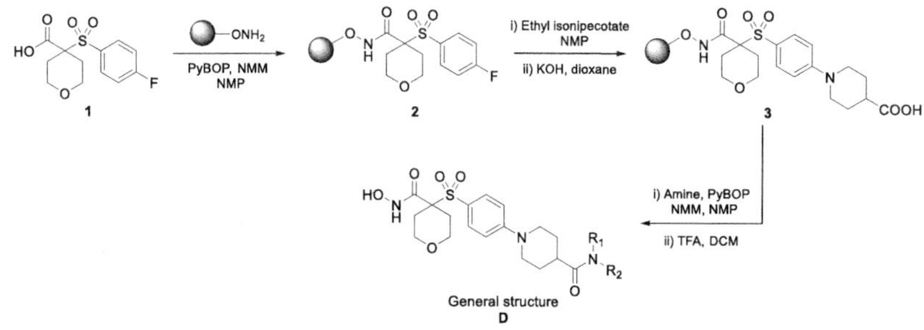

Scheme 1 Synthesis of isonipecotamide sulfone hydroxamates [39].

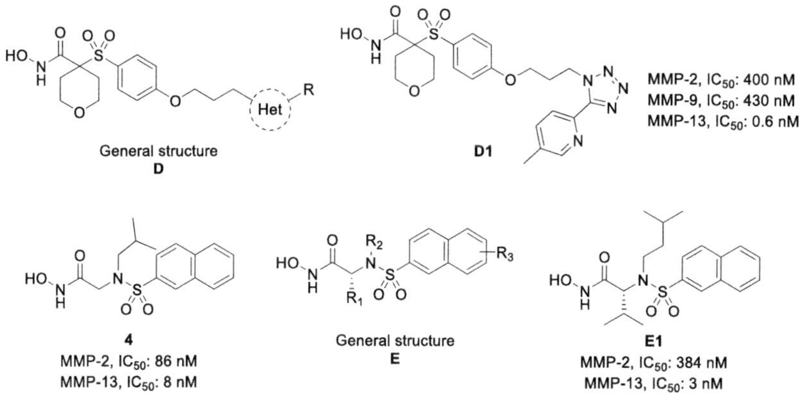

Fig. 5 Representation of general structures **D** and **E** and compounds **D1, 4** and **E1** [40,41].

D, such compounds displayed short half-life in rats, presumably due to hydrolytic cleavage of the amide bond. Therefore, Barta *et al.* replaced the amide moiety with non-hydrolyzable bioisosteric heterocycles, such as tetrazole, oxadiazole and imidazole, in an effort to improve half-life (Fig. 5, general structure **D**) [40]. A large set of compounds was synthesized and further evaluated for their inhibitory potency against a panel of MMPs and for their pharmacokinetic features in rats [40]. Among these compounds, derivative **D1** (Fig. 5) stood out as the most promising lead with a K_I of 0.6 nM against MMP13 and showed more than 400-fold selectivity versus MMP8 and 600-fold selectivity versus MMP2; moreover, no activity was detected against MMP1 and -14. Furthermore, pharmacokinetic parameters in rats were also ameliorated given that compound **D1** exhibited a 4.8 h of half-life and a C_{max} of 3000 ng/mL [40]. Successively, Tommasi *et al.* analyzed the vast SAR established from previous approaches towards

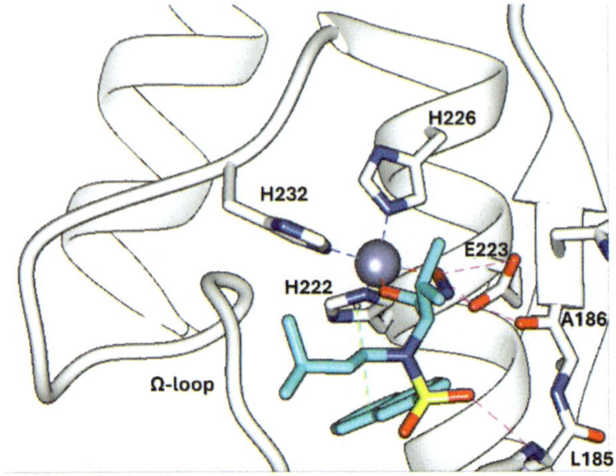

Fig. 6 Complex of **E1** bound at the MMP13 active site (PDB: 3ZXH); **E1** in Cyan, hydrogen bonds as magenta dashed lines, π-π stacking as green dashed lines, backbone in white, and Zn(II) in dark grey [41].

MMP13 inhibitors and noticed that the 2-naphthyl sulfonamide-substituted inhibitor **4** exhibited modest selectivity (10-fold) on MMP13 over isozyme 2 [41]. Consequently, they reported the optimization of compound **4** to discover inhibitors with higher selectivity, substituting the isopropyl moiety with different groups and adding several aliphatic moieties in position α to hydroxamic acid (Fig. 5, general structure **E**). This strategy led to derivative **E1** that was cocrystallized in the MMP13 active site (Fig. 6) to elucidate the observed selectivity [41]. Compound **E1** was further evaluated for its rat pharmacokinetic profile via cassette dosing at 5 mg/kg through oral administration (po) and 1 mg/kg through intravenous (iv), revealing that it had moderate in vivo clearance (2.4 L/h/kg), $t_{1/2}$ of 5.7 h, and marginal oral bioavailability ($F = 23.8$ percent) with oral exposure of 1.5 μM h [41].

In 2012, Santamaria *et al.* described the development of first-in-class dual MMP12/MMP13 inhibitors designed to treat RA and OA while avoiding the selectivity issues of earlier MMPIs [42]. By optimizing an N–O-isopropyl sulfonamide hydroxamate scaffold (general structure **F**, Fig. 7), the authors identified compounds that exhibit slow-binding kinetics, leading to prolonged target residence time. In this context, derivative **F1** (Fig. 7) emerged as a lead, showing low-nanomolar potency against both enzymes, favorable binding kinetics, and strong selectivity that spares key protective

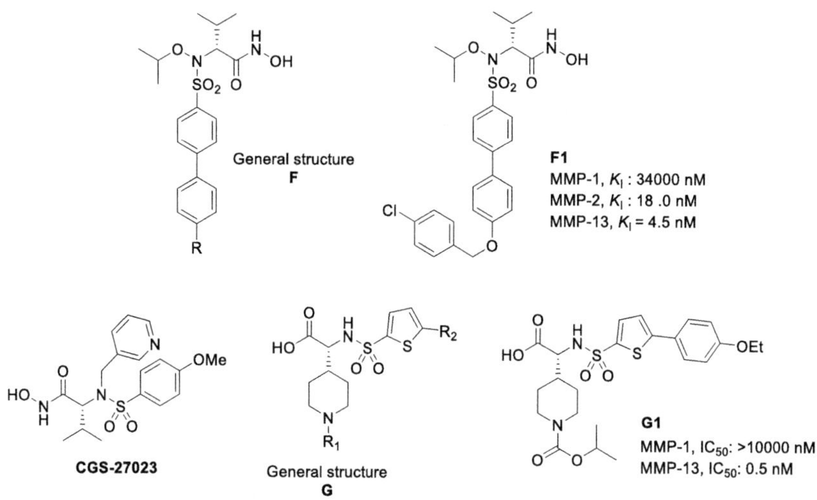

Fig. 7 Representation of general structures **F** and **G** and compounds **CGS-27,023, F1** and **G1** [42,43].

MMPs, including MMP8. Structural studies confirmed deep engagement of the S1′ pocket as the basis for potency and selectivity [42].

Following a different drug-design strategy, Monovich *et al.* hypothesized that selective inhibition of MMP13 versus other metalloproteinases could be achieved with alternate zinc-binding moiety [43]. Thus, a primary goal of this research was to replace the hydroxamic acid moiety with a carboxylic acid to reduce the potential for pan-MMP and off-target inhibition. Such direct replacement of the hydroxamic acid in commercial **CGS-27,023** (Fig. 7) with a carboxylic acid moiety resulted in a 50-fold loss of potency versus MMP-13 (IC_{50}s from 1.9 nM to 89 nM), reflecting the relatively weaker affinity of the carboxylic acid for zinc [42]. Consequently, to achieve low-nanomolar inhibitors, modifications of the side chains were performed to give a large set of novel derivatives having a general structure **G** as reported in Fig. 6 [43]. Among them, derivative **G1** (Fig. 7) resulted in being a sub-nanomolar inhibitor of MMP13 (IC_{50} value of 0.5 nM and K_I of 0.19 nM) avoiding cross-reactivity with MMP1 (IC_{50} over 10,000 nM). Furthermore, in a rat model of MMP13-induced cartilage degradation, compound **G1** significantly reduced proteoglycan release following oral dosing at 30 mg/kg (75 percent inhibition) and at 10 mg/kg (40 percent inhibition) [43]. In 2016, Nara *et al.* reported a structure-guided strategy to develop highly selective MMP13 inhibitors for osteoarthritis, by designing a

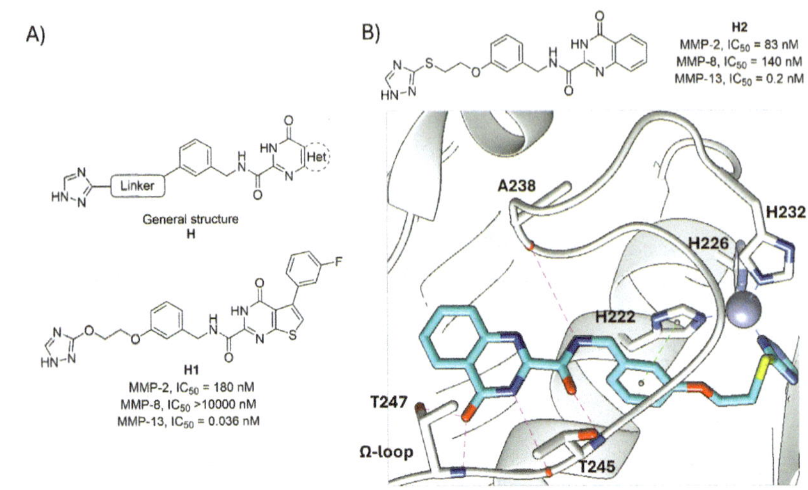

Fig. 8 (A) General structure **H** and compound **H1**. (B) Complex of **H2** in MMP13 active site (PDB: 5B5P); **H2** in Cyan, hydrogen bonds as magenta dashed lines, π-π stacking as green dashed lines, backbone in white, and Zn(II) in dark grey [44].

hybrid scaffold that combines deep S1′ pocket engagement with a novel triazole-based zinc-binding group, the authors achieved sub-nanomolar potency and exceptional selectivity [44]. They produced a large series of compounds having general structure **H** (Fig. 8A) and then assessed their inhibitory profile against a panel of MMPs. In this context, **H1** exhibited an IC_{50} of 0.036 nM and over 1500-fold selectivity against a broad panel of MMPs (MMP-1, -2, -3, -7, -8, -9, -10, and -14) and ADAM17/TACE [44]. Further studies demonstrated strong oral bioavailability and effective protection against cytokine-induced cartilage degradation in a bovine nasal cartilage (BNC) assay for compound **H1**. inhibiting collagen release by 70.8 percent at 1 μM [44].

The study by De Savi *et al.* (2013) reports the discovery of **AZD6605** (Fig. 9A), a potent and selective MMP13 inhibitor developed for the treatment of OA [45]. The key innovation by the AstraZeneca team was the replacement of the classical hydroxamate zinc-binding group with an imidazolidinedione (hydantoin) moiety, ameliorating inhibitory potency and metabolic stability. First, medicinal chemistry efforts focused on optimizing two main series, with aryl piperidines and piperazines (General structures **I** and **I'**, Fig. 9A). Among them, **AZD6605** emerged as the lead candidate, combining high potency (IC_{5n} of 4.7 nM), excellent selectivity, and favorable

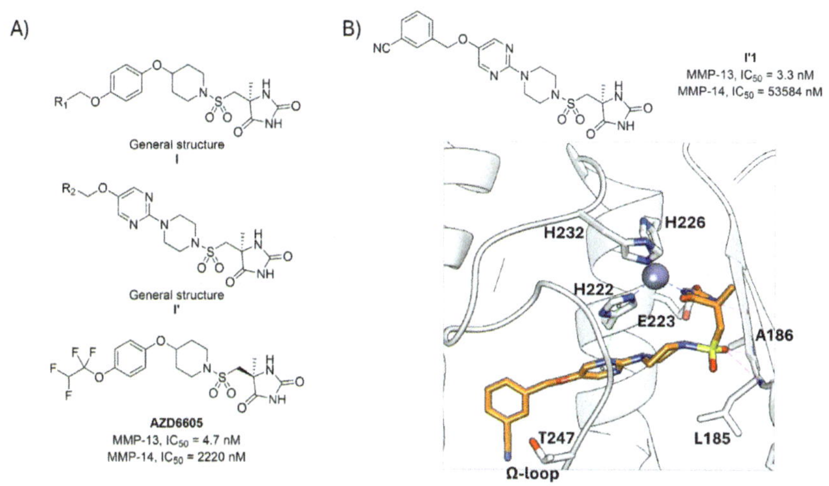

Fig. 9 A) General structures **I** and **I'** and compound **AZD6605**. B) Complex of **I'1** in MMP13 active site (PDB: 4JPA); **I'1** in orange, hydrogen bonds as magenta dashed lines, π-π stacking as green dashed lines, backbone in white, and Zn(II) in dark grey [45].

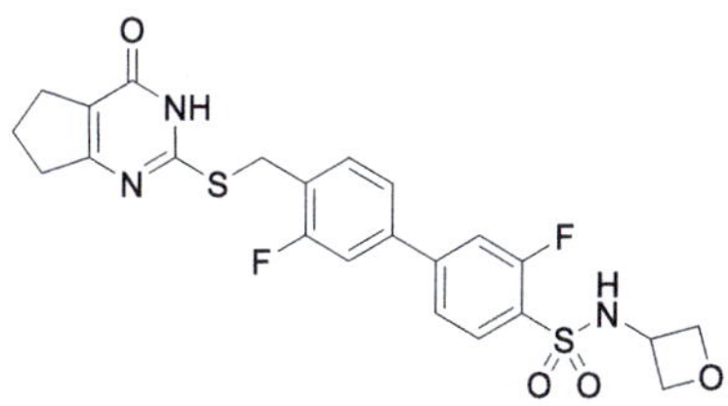

5
MMP-13, IC_{50} = 42 nM

Fig. 10 Structure of compound **5** [46].

pharmacokinetics [45]. One of its analogues, here labelled **I'1**, was also cocrystallized with MMP13 to reveal that the hydantoin group coordinates the catalytic Zn(II) in a monodentate mode (Fig. 9B). Successively, in vivo validation demonstrated a strong chondroprotective effect of **AZD6605**, with complete prevention of tibial cartilage volume loss in a spontaneous guinea pig OA model, as assessed by MRI. It also showed outstanding selectivity (up to >2000-fold over MMP1 and MMP7), no evidence of musculoskeletal toxicity in 28-day rat studies, and a pharmacokinetic profile characterized by low clearance and a long predicted human half-life, supporting once-daily oral dosing [45].

In 2022, Fuerst *et al.* describe the optimization of compound **5** (Fig. 10) as a highly selective MMP13 inhibitor for OA therapy [46]. Starting from a non–zinc-chelating lead with limited selectivity, the authors performed a focused SAR study near the catalytic zinc site, introducing small heterocycles to improve metabolic stability and pharmacokinetics. Incorporation of an oxetane ring proved crucial, yielding compound **5** with an excellent balance between potency and selectivity. Although moderately potent (IC_{5n} = 42 nM), compound **5** showed outstanding selectivity, sparing other MMP isoforms, fully inhibited type II collagen degradation in vitro, and reached therapeutically relevant concentrations in synovial fluid *in vivo*. Docking studies suggest a weak, non-classical interaction between the oxetane oxygen and zinc, explaining both selectivity and reduced systemic toxicity [46].

While traditional inhibitors rely on a ZBG to achieve potency, this approach often leads to pan-MMP inhibitors and consequently poor selectivity across the MMP family due to the highly conserved nature of the active site. In fact, in the search for highly selective MMP13 inhibitors, moving away from the catalytic zinc ion has become a transformative strategy.

2.2 Non-zinc-chelating inhibitors

In order to bypass limitations related to ZBG inhibitors, recent medicinal chemistry efforts have focused on non-zinc chelating inhibitors. These ligands achieve high affinity and exceptional selectivity by targeting the deep, variable S1' hydrophobic pocket or distinct allosteric sites unique to MMP13. By avoiding direct interaction with the zinc ion, these inhibitors minimize cross-reactivity with other metalloenzymes, offering a more precise pharmacological profile for the treatment of osteoarthritis.

Some of the first non-zinc-chelating inhibitors were described by Johnson *et al.* [47]. The authors screened their compound library in search of MMP13 inhibitors and discovered two novel derivatives, here labelled as compounds **6** and **7** (Fig. 11), with remarkable inhibition against MMP13, K_Is of 30 nM and 0.67 nM, respectively, but without any detectable activity against other MMPs, K_Is over 100,000 nM. The non-competitive mechanism was confirmed by X-ray crystal structures of both compounds with MMP13 (Fig. 11), which bind within the S_1'-specificity loop of the protein and extend past this pocket out toward solvent [47]. Transition to *ex vivo* models confirmed efficacy in a more complex tissue context using bovine nasal cartilage explants stimulated with inflammatory cytokines (IL-1β and oncostatin M). In this system, Compound **7** significantly inhibited hydroxyproline release, a direct marker of collagen degradation, demonstrating that selective MMP13 inhibition alone is sufficient to protect the cartilage matrix despite the presence of other cytokine-induced proteases

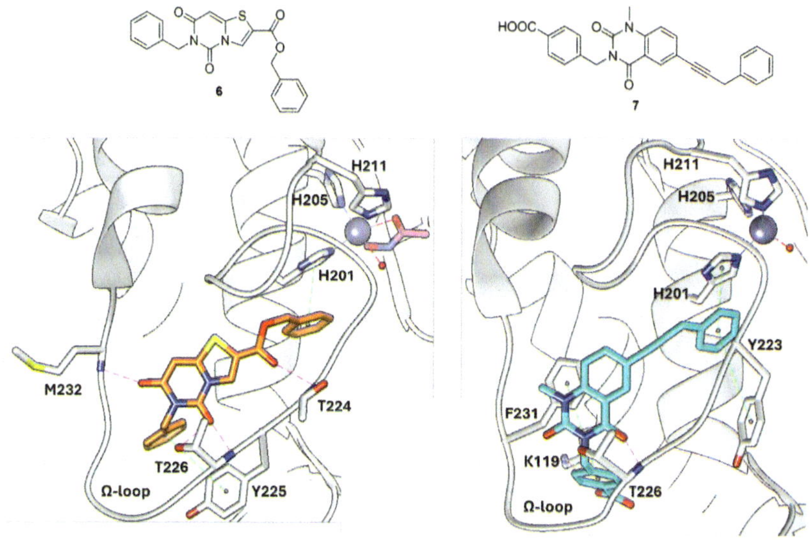

Fig. 11 Structures of compounds **6** and **7** and their complex in the MMP13 active site. Compound **6** (PDB: 2OW9) in orange and Compound **7** (PDB: 2OZR) in cyan; H_2O as a red sphere, hydrogen bonds as magenta dashed lines, π-π stacking as green dashed lines, π-cation as yellow dashed line, backbone in white, 1-acetohydroxamicacid in pink (ligand added to prevent autolysis bind the catalytic Zn(II) in a bidentate mode), and Zn(II) in dark grey [47].

[47]. These findings provided the rationale for further in vivo studies, where oral administration reduced synovial fluid TIINE levels in a rat model of acute cartilage damage following intra-articular injection of exogenous MMP13. Long-term therapeutic efficacy was subsequently confirmed in a chronic surgical rabbit model of OA, in which twice-daily oral treatment markedly reduced cartilage lesion areas (up to 68 percent on tibial plateaus and 51 percent on femoral condyles). Importantly, safety was demonstrated in a rat fibroplasia model designed to mimic musculoskeletal syndrome associated with earlier MMP inhibitors, as the selective MMP13 inhibitor caused no histological or clinical alterations, highlighting that extreme selectivity and a non-chelating mechanism are critical for halting cartilage degradation without compromising joint health [47]. Successively, based on this work, Barani *et al.* proposed a different series of non-competitive MMP13 selective inhibitors (structures not shown) and assessed the best candidate in normal bovine articular cartilage and human OA cartilage cultures that were stimulated with IL-1 and OSM to degrade collagen.

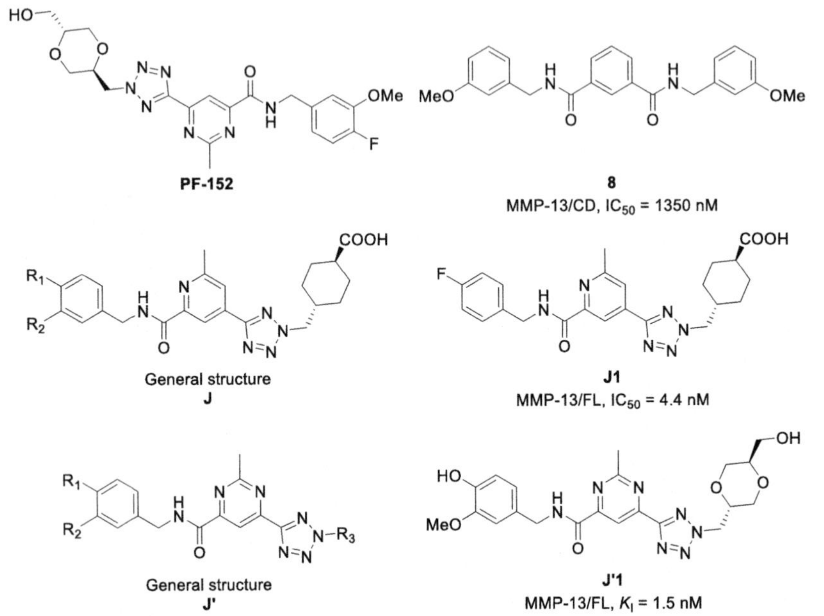

Fig. 12 Structure of compounds **PF-152, 8, J1** and **J'1** [49–51].

Collectively, such data showed that MMP13 inhibition is sufficient to inhibit collagen degradation in a physiologic system [48].

Based on Schnute *et al.* achievements about tetrazole-based MMP13 inhibitors [49] the study reported by Settle *et al.* represents a landmark contribution to disease-modifying osteoarthritis drug research by introducing **PF-152** (Fig. 12), a highly selective MMP13 inhibitor [49]. Unlike earlier broad-spectrum MMP inhibitors associated with musculoskeletal toxicity, **PF-152** is a non-hydroxamate compound that binds the S1′ pocket of MMP13 without chelating the catalytic Zn(II), achieving more than 3000-fold selectivity over other metalloproteinases. Further in vitro assays confirmed the high potency of **PF-152** (IC_{50} = 1.5 nM against human MMP13) and its ability to inhibit collagen degradation. *Ex vivo* experiments using cytokine-stimulated human cartilage explants demonstrated dose-dependent reductions in collagen type II (TIINE) and aggrecan fragments (ARGN, AGEG) [49]. These findings supported progression to *in vivo* evaluation, where an oral treatment with **PF-152** significantly reduced macroscopic cartilage lesions (by approximately 51–55 percent) and preserved tibial plateau histological integrity, particularly in peripheral regions.

Consistently, urinary TIINE levels returned to pre-surgical baseline at all tested doses, while synovial aggrecan biomarkers were reduced by up to 44 percent [49]. Successively, the same group from Pfizer described the discovery and optimization of a new class of MMP13 inhibitors based on a (pyridin-4-yl)-2H-tetrazole scaffold [50]. A central innovation of this work was the development of non–zinc-chelating inhibitors. Unlike classical hydroxamate-based compounds, these molecules selectively occupy the S1′ pocket of MMP13. As part of the original HTS which led to the discovery of compound **7**, phenyl dicarboxamide **8** was also identified as a modest inhibitor of MMP13, IC_{50} of 1.35 μM [50]. Optimization began from a modest screening hit and focused on isosteric replacement (General structure **J**, Fig. 12), introducing a tetrazole ring and a pyridine core, giving derivative **J1** (Fig. 12) as the lead compound. It showed high potency (K_I of 4.4 nM) against full-length MMP13 isozyme and exceptional selectivity over a broad panel of other MMP and ADAMTS enzymes. It also exhibited 100 percent oral bioavailability in rats, low clearance, and robust in vivo efficacy. In a rat medial meniscal tear model, **J1** significantly reduced the collagen degradation biomarker TIINE and provided measurable protection against tibial cartilage damage [50]. In 2016, the study by Ruminski *et al.* from Pfizer represented a major advance in the development of MMP13 inhibitors for OA, addressing a critical limitation of earlier candidates: renal toxicity [51]. The authors hypothesized that these acidic compounds were substrates of the human organic anion transporter hOAT3, leading to active uptake and accumulation in renal cells. Through extensive SAR optimization, the team introduced several structural innovations to improve pharmacokinetics and safety, leading to General structure **J'** (Fig. 12) [51]. These efforts led to the identification of compound **J'1**, which combined high potency (K_I of 1.5 nM) with exceptional selectivity, over 10,000-fold over MMP3 over 14 MMPs. Successively, in a canine OA model, **J'1** reduced levels of the cartilage degradation biomarker TIINE, confirming in vivo efficacy, and no renal toxicity in a 14-day monkey study at doses up to 1000 mg/kg was detected, achieving exposures far exceeding those that caused nephrotoxicity with earlier acidic inhibitors [51].

Panico *et al.* investigated 2-benzo[d]isothiazolyl-imino-5-benzylidene-4-thiazolidinone derivatives as potential MMPIs and evaluated their antidegenerative activity on human chondrocyte cultures stimulated by IL-1β [52]. Among these short series, having a general structure **K** (Fig. 13), compound **K1** (Fig. 13) displayed a nanomolar inhibition of MMP13, K_I of 0.036 μM, and was further in vitro assayed on chondrocytes at different concentrations

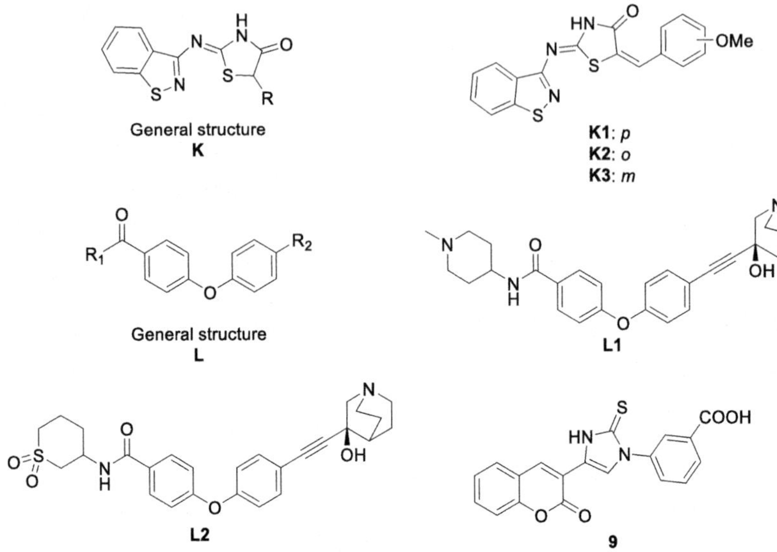

Fig. 13 Structures of compound **K1-K3, L1, L2** and **9** [51–56].

from 10 μg/mL to 0.01 μg/mL, giving the best performance at the higher concentration [52]. Considering the promising data, compound **K1** was used as a lead compound for a successive development proposed by Vicini *et al.* [53]. Two methoxybenzylidene positional isomers **K2** and **K3** (Fig. 13) were synthesized to investigate structure-activity relationships (SARs). **K1** showed the highest enzymatic potency (IC_{50} of 36 nM), with strong and selective inhibition of MMP13 and significant restoration of glycosaminoglycan levels in IL-1β–stimulated human chondrocytes. The other isomers, **K2** and **K3**, did not display inhibition of MMP13 but retained cartilage-protective effects [53].

In 2012, De Savi *et al.* screened over 19,000 different compounds to identify a promising candidate as an MMP13 inhibitor for successive development [54]. The authors confirmed that derivative **6** bound the MMP13 active site by using nuclear magnetic resonance (NMR) chemical shift perturbation mapping with ^{15}N-labelled MMP13 (180μM) on a 600 MHz spectrometer, and consequently produced a large set of compounds having general structure **L** (Fig. 13). This series exhibited a better solubility and metabolic stability than other previously described in the literature, showing good half-life and moderate clearance in rats [54]. The same authors described an optimization campaign aimed at developing selective

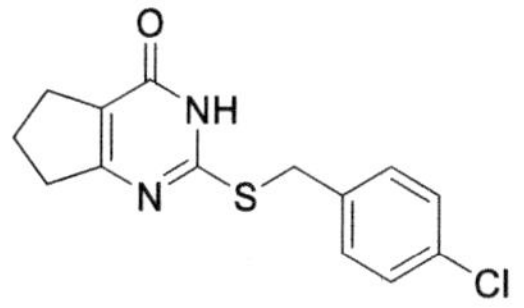

10
MMP-1, IC_{50} > 40 μM
MMP-8, IC_{50} > 40 μM
MMP-13, IC_{50} = 3.8 μM

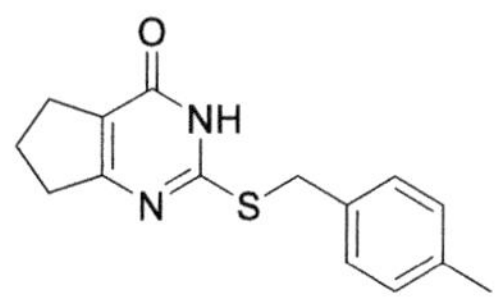

11
MMP-1, IC_{50} > 40 μM
MMP-8, IC_{50} > 40 μM
MMP-13, IC_{50} = 0.8 μM

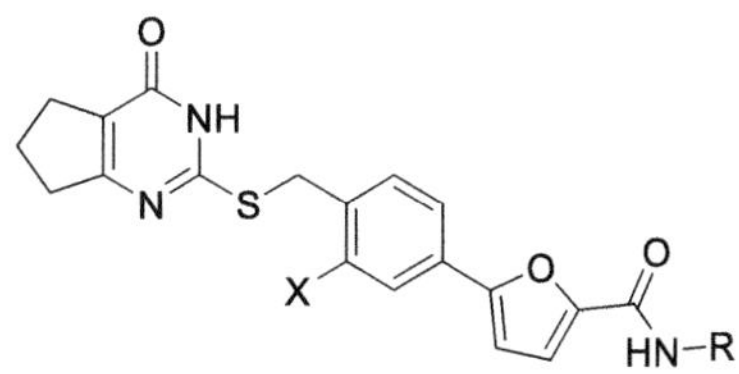

General structure
M

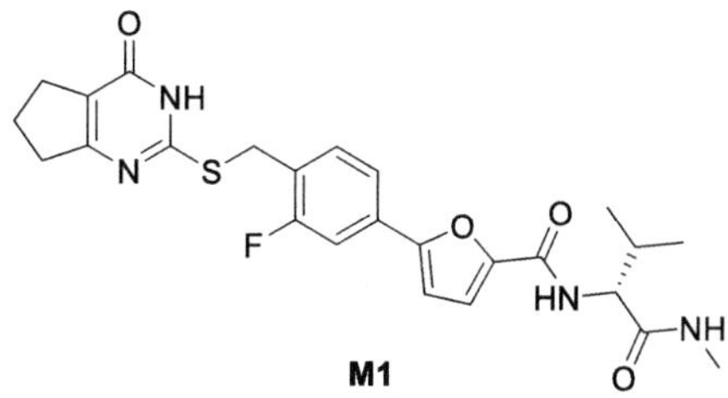

M1
MMP-1, IC_{50} > 5000 nM
MMP-2, IC_{50} > 5000 nM
MMP-13, IC_{50} = 2.7 nM

Fig. 14 Structures of **10, 11** and **M1** [57,58].

MMP13 inhibitors starting from compounds previously published [55]. The most significant results were achieved with the identification of two lead candidates (General structure **L**, Fig. 13), compounds **L1** (a methyl-piperazine derivative) and **L2** (a diastereomer of the pyranic sulfone), which demonstrated exceptional selectivity on MMP13 (>10,000-fold) over other MMPs and ADAMTS proteases, high solubility and very low in vitro metabolic clearance in both human microsomes and hepatocytes, no inhibition of cytochrome P450 enzymes and no interaction with the hERG channel, minimizing cardiovascular and drug-drug interaction risks [55]. A different technique to identify novel MMP13 selective inhibitory leads was applied by La Pietra *et al.* in 2012 [56]. Using virtual screening on the Life Chemicals database, a collection of approximately 7769 drug-like compounds, followed by enzymatic validation, five new lead compounds were identified. Among them, the chromenone scaffold emerged as the most promising one (compound **9** in Fig. 13), and an early structure-based optimization significantly improved its potency. However, these compounds resulted in less active inhibitors than classical hydroxamates [56].

Adopting a similar *in silico* approach based on HTS, Roth *et al.* identified compound **10** (Fig. 14), a low-micromolar inhibitor with a small

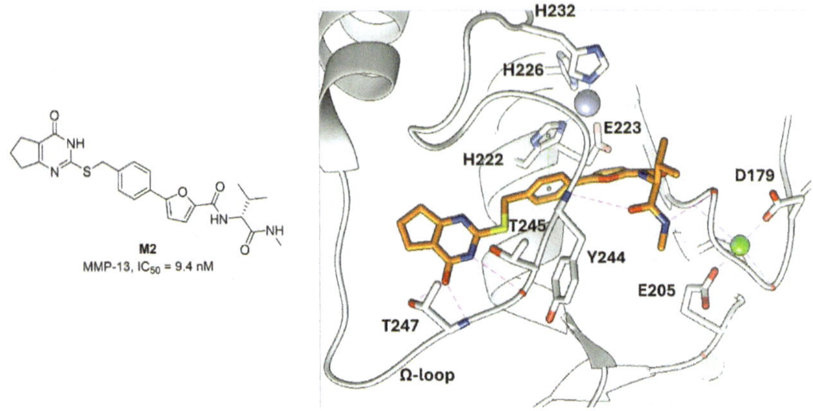

Fig. 15 Complex of **M2** bound at the MMP-13 active site (PDB: 5UWL); **M2** in orange, hydrogen bonds as magenta dashed lines, π-π stacking as green dashed lines, backbone in white, Zn(II) in dark grey, and Ca(II) in green [58].

molecular scaffold (2-(arylmethylthio)-cyclopentapyrimidin-4-one) that already showed indications of a mechanism of action not based on zinc chelation [57]. It was further optimized using a combination of cheminformatics and targeted chemical synthesis, leading to novel derivatives with improved potency and selectivity, particularly compound **11** (Fig. 14), which showed a fivefold increase in inhibitory activity and no off-target inhibition of MMP1 or -8. Next structural analyses suggested binding to a previously unexplored site on MMP13 rather than the canonical S1′ pocket, highlighting a novel strategy for achieving selective cartilage-protective inhibition [57]. In 2017, Choi *et al.* moved on to another step when they proposed a refined structure-guided design strategy divided into three steps [58]. Briefly, by using compound **10** merged with chelating ligands as structural anchors and then replacing them with polar, solvent-oriented groups, they achieved potent inhibition while dramatically improving selectivity. This approach preserved key spatial interactions while avoiding the liabilities associated with metal chelation [58]. Among several compounds having general structure **M** (Fig. 14), derivative **M1** (Fig. 14) showed nanomolar activity, strong protection against collagen degradation, and minimal off-target effects across a broad protease panel. Analogue **M2** was then cocrystallized in the MMP13 active site (Fig. 15) [58].

In 2012, Gege *et al.* described a highly selective MMP13 inhibitor designed for intra-articular administration [59]. Based on pyrimidine-based

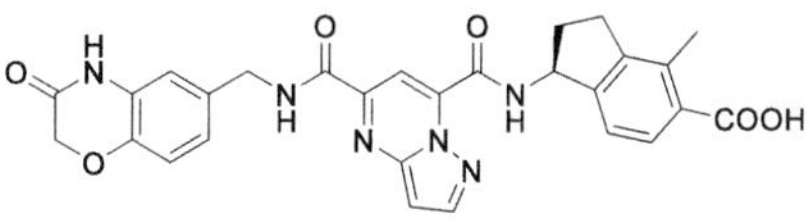

12

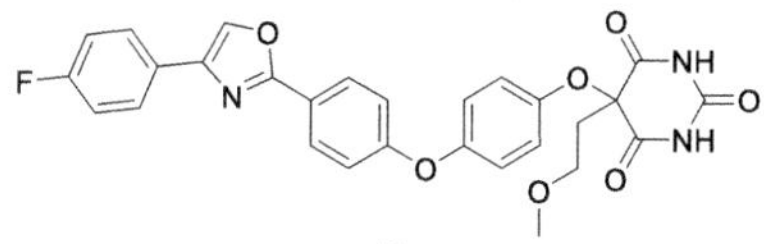

13

Fig. 16 Structures of compounds **12** and **13** [59,60].

drugs reported in the literature, such as **PF-152** (Fig. 12), the authors designed compound **12** (Fig. 16), showing sub-picomolar potency against human MMP13 (IC_{50} of 0.03 nM) and unprecedented selectivity margins over other MMP isozymes. Further cartilage explant studies demonstrate robust, dose-dependent inhibition of collagen degradation, confirming functional target engagement in a tissue context. These findings are further reinforced in an *ex vivo* rat model, where a single intra-articular dose provides complete protection against collagen loss for several weeks. Importantly, the pharmacokinetic behavior of **12** strongly supports its suitability for local therapy: rapid clearance from systemic circulation is coupled with prolonged retention within the cartilage matrix at therapeutically effective concentrations [59].

In the same year, Shah *et al.* demonstrated that selective inhibition of MMP13 using the highly specific pyrimidinetrione-based inhibitor **13** (Fig. 16, synthesized at Pfizer) can slow breast tumor growth and significantly reduce tumor-induced bone destruction without inducing musculoskeletal toxicity [60]. Using both xenograft (human MDA-MB-231 cells) and syngeneic (murine 4T1.2 cells) cancer mouse models, the study documented that preventive treatment with compound **13** significantly slowed primary tumor growth and, in parallel, delayed the onset and reduced the severity of osteolytic metastatic lesions induced by tumor cells in an intracardiac metastasis model. However, such treatments did not reduce metastatic burden in soft tissues such as the lungs and brain, nor did they improve overall survival in advanced disease models [60].

In 2014, Nara *et al.* described the development of a novel class of quinazoline-2-carboxamide inhibitors of MMP13 that achieve exceptional potency and selectivity [61]. Using a structure-based drug design approach, the authors optimized a screening hit by exploiting MMP13-specific binding features, discovering a moderately potent (IC_{50} of 12 nM) quinazoline-2-carboxamide compound, **14** (Fig. 17A). It was cocrystallized in the MMP13 active site (Data not shown) revealing that the quinazoline core occupies the deep S1′ pocket, forming stabilizing hydrogen-bond interactions with the

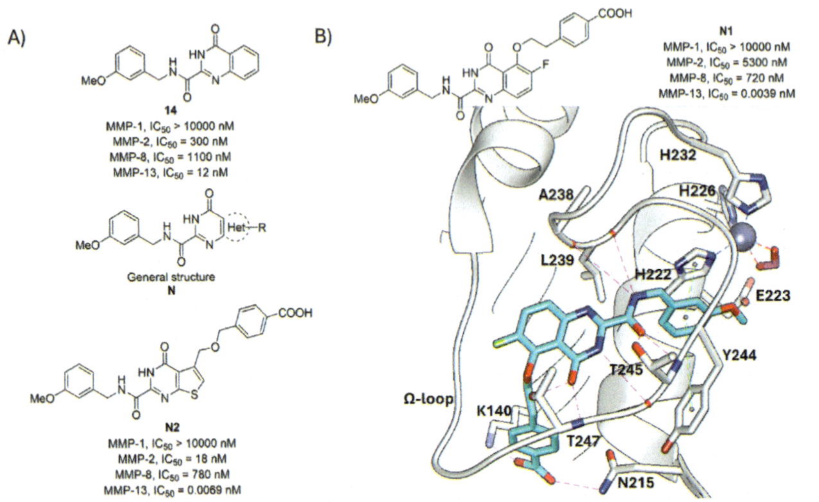

Fig. 17 **(A)** General structure **N** and compounds **14** and **N1**. **(B)** Complex of **N1** in MMP13 active site (PDB: 3WV1); **N1** in cyan, formic acid in pink, hydrogen bonds as magenta dashed lines, π-π stacking as green dashed lines, backbone in white, and Zn(II) in dark grey [61,62].

protein backbone, while extension into the adjacent S1″ pocket, a region unique to MMP13, proved critical for selectivity [61]. Consequently, the authors produced a large series of quinazoline-based derivatives (General structure **N**, Fig. 17A) finding in compound **N1** (Fig. 17B) the most potent inhibitor. Compound **N1** exhibited picomolar potency (IC5_n of 3.9 pM) and over 41,000-fold selectivity over other MMPs and ADAM17/TACE, with no overt toxicity in short-term rat studies. To improve pharmacokinetic properties, it was converted in a sodium salt and further assessed in rat osteoarthritis models, where it significantly reduced cartilage degradation [61]. Also, derivative **N1** was cocrystallized in the MMP13 active site showing a strong interaction with K140 residue at the bottom of the S1″ pocket (Fig. 17B) [61]. Successively, the same research group reported the optimization of compound **14** modifying the linker region, the central core, and the substituent directed into the S1″ pocket [62]. Among the synthesized derivatives, **N2** (Fig. 17A) emerged as the lead candidate, displaying picomolar potency (IC5_n of 6.9 pM) and extraordinary selectivity (>2600-fold over several MMPs and ADAM17/TACE). It effectively inhibited cytokine-induced collagen degradation in bovine cartilage explants and showed good oral bioavailability across multiple species. Importantly, repeated-dose

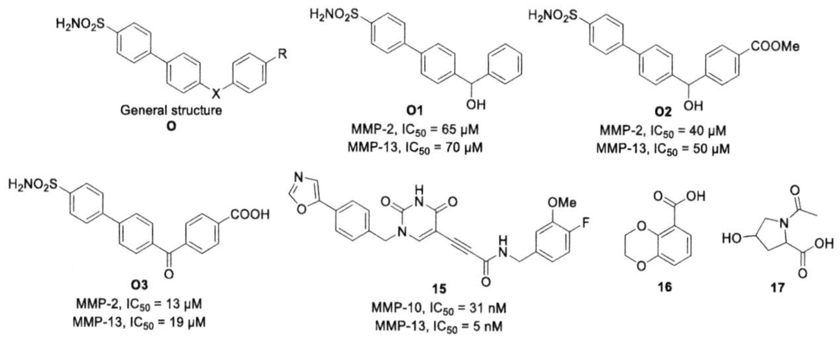

Fig. 18 Structures of compounds **O1–O3**, and **15–17** [63,64,68].

toxicity studies in rats demonstrated a high safety margin (NOAEL 60 mg/kg/day) with no musculoskeletal side effects [62].

In 2015, the study by Santamaria *et al.* repurposed a series of 4,4′-biphenylsulfonamides (General structure **O**, Fig. 18) as selective, non–zinc-binding inhibitors of MMP2 and -13 [63]. Originally developed as carbonic anhydrase (CA) inhibitors, such derivatives were repurposed to target MMPs involved in tumor invasion and angiogenesis, enabling dual CA/MMP inhibition for potential anticancer applications. Among them, three compounds **O1–O3** emerged as leads, exhibiting micromolar potency against MMP2 and -13 and clear selectivity over other MMP isozymes. Compounds **O1** and **O3** acted as non-competitive inhibitors, whereas compound **O2** displayed competitive behavior likely due to steric overlap with the substrate-binding groove [63]. A few years later, a multitargeting strategy was also implemented by Senn *et al.*, who reported the discovery of Compound **15** (Fig. 18), as a highly potent and selective non-hydroxamate dual inhibitor of MMP10 and -13 [64]. The authors utilized structure-based design to evolve a dual inhibitor starting from a previous lead that had high affinity for MMP13 but poor affinity for MMP10. Such derivative showed outstanding potency (IC_{5n} = 5 nM against MMP13 and 31 nM against MMP10) and exceptional selectivity, displaying minimal activity against other MMPs and even discriminating between closely related stromelysins [64]. The same multitargeting approach was further studied by El Ashry *et al.* in 2020, who implemented a sophisticated synthetic strategy, based on photoirradiation-mediated Dimroth rearrangement, to achieve selective MMP10/13 pharmacological agents [65], and by Morcos *et al.* in 2023, via s-triazine-based MMP10/13 inhibitors armed with electrophilic warheads for concomitant ferroptosis induction [66,67].

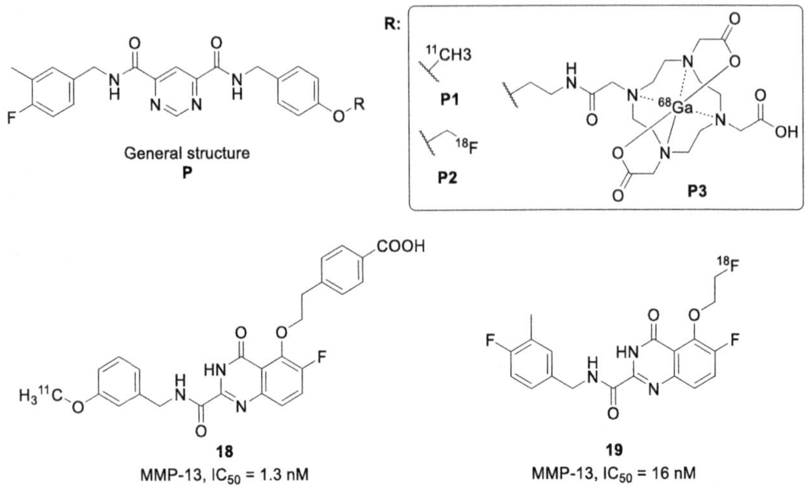

Fig. 19 Structures of radiolabeled tracers **P1–P3, 18** and **19** [69,70].

Next, Kothapalli *et al.* proposed a selective strategy to inhibit MMP13 by targeting the enzyme's C-terminal hemopexin domain rather than the catalytic zinc site [68]. Through virtual screening and experimental validation, two compounds, **16** (2,3-dihydro-1,4-benzodioxine-5-carboxylic acid) and **17** (1-acetyl-4-hydroxypyrrolidine-2-carboxylic acid), were identified as selective MMP13 inhibitors (Fig. 18). Both showed micromolar binding affinity ($K_d = \sim 1.3–1.4\ \mu M$), selectively inhibited MMP13 without affecting MMP2 or -9, and reduced invasion of aggressive breast and colon cancer cell lines. Among the tested compounds, **16** demonstrated the strongest anti-invasive effects [68].

A novel approach was proposed by Hugenberg *et al.*, whose developed the first radiolabeled tracers for selective, noninvasive PET imaging of MMP13 activity [69]. To ensure high specificity and avoid the toxicity of zinc-binding broad-spectrum inhibitors, the researchers designed inhibitors based on an N,N′-bis(benzyl)pyrimidine-4,6-dicarboxamide scaffold **P** (also reported by Schnute et al. and Ruminski et al. [50,51]) that binds selectively to the MMP13 specificity loop. The scaffold's flexibility enabled radiolabeling with clinically relevant isotopes: carbon-11 (**P1**), fluorine-18 (**P2**), and gallium-68 (**P3**), producing tracers suitable for different PET imaging settings (Fig. 19). Importantly, radiolabeling did not reduce biological activity. Indeed, several compounds showed very high potency (IC_{5n} values as low as 0.2 nM) and strong selectivity, with minimal inhibition of other MMPs. Further in vivo studies in mice demonstrated favorable pharmacokinetics:

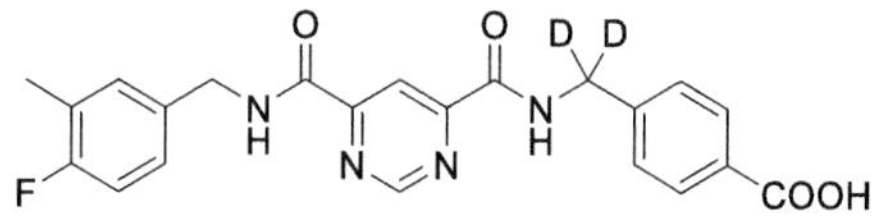

AQU-019
MMP-2, IC_{50} >100000 nM
MMP-8, IC_{50} >100000 nM
MMP-13, IC_{50} = 4.8 nM

Fig. 20 Structure of **AQU-019** [71].

lipophilic ^{11}C and ^{18}F tracers were cleared hepatobiliarily with rapid blood washout, while the more hydrophilic ^{68}Ga-DOTA tracer showed mixed renal and hepatic clearance and longer circulation times. The tracers displayed moderate metabolic stability and no significant off-target accumulation or defluorination [69]. Radio tracers were also produced by Buchler *et al.* in 2023 based on a quinazoline-2-carboxamide scaffold to image MMP13 activity in atherosclerotic plaques [70]. To achieve high selectivity and avoid off-target effects, the tracers were designed using structure-based drug design to bind outside the catalytic zinc site, instead targeting the S1′ and unique S1″ pockets. This U-shaped binding mode, particularly interactions with K140 and N215 residues, provided strong affinity and specificity for MMP13. After synthesizing several candidates that were radiolabeled for in vivo evaluation. The potent carboxylate **18** showed unfavorable kidney and liver accumulation, prompting the development of ether derivatives. Among them, **19** emerged as the best tracer, maintaining good potency (IC_{5n} = 16 nM), high selectivity, low myocardial uptake, and improved metabolic stability. *Ex vivo* studies confirmed specific accumulation of **19** in lipid-rich atherosclerotic plaques, with reduced signal upon competitive blocking [70].

The study by Bendele *et al.* (2021) details the development of **AQU-019** (Fig. 20), a novel and highly selective allosteric inhibitor of MMP-13 [71]. A major innovation was selective deuteration at the benzyl position, which enhanced metabolic stability by reducing cytochrome P450-mediated degradation without introducing chirality. This modification improved pharmacokinetics, increasing total drug exposure by 68 percent in rats. Biochemically, **AQU-019** showed strong potency (IC_{5n} of 4.8 nM for MMP13) and exceptional selectivity, with negligible inhibition of other MMPs. In the MIA rat model of osteoarthritis, intra-articular administration significantly protected cartilage, reducing proteoglycan and cell loss and preserving joint structure-effects superior to the comparator steroid treatment [71].

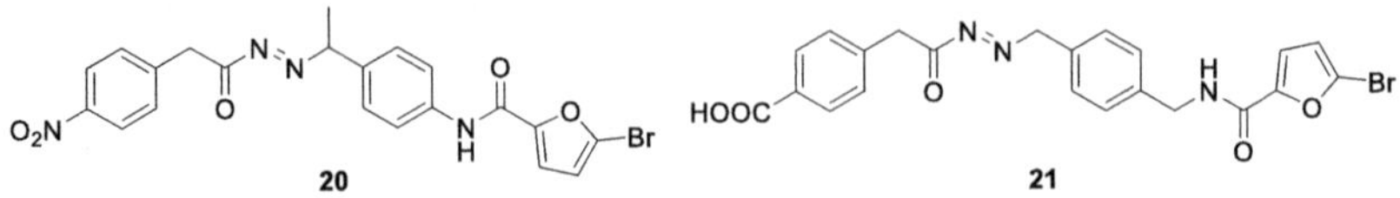

20
MMP-1, IC_{50} = 91 μM
MMP-2, IC_{50} = 99 μM
MMP-13, IC_{50} = 14.6 μM

21
MMP-1, IC_{50} = 177 μM
MMP-2, IC_{50} = 3.6 μM
MMP-13, IC_{50} = 1.8 μM

Fig. 21 Structures of **20** and **21** [72].

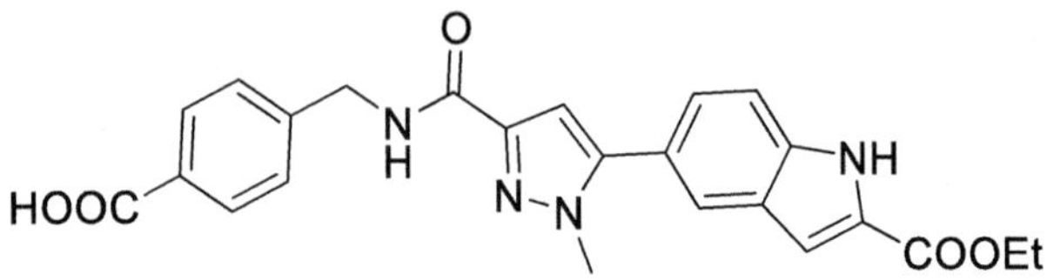

BI-4394
MMP-13, IC_{50} = 1 nM

Fig. 22 Structure of **BI-4394** [73].

In 2023, Cuffaro *et al.* reported the discovery of a novel class of non-zinc-binding MMP13 inhibitors based on an N-acyl hydrazone scaffold [72]. Using a multi-step virtual screening workflow, combining molecular weight filtering, shape similarity, structure-based docking, and SAR criteria, the researchers screened over 212,000 compounds and identified compound **20** (Fig. 21) as an initial hit ($IC_{5n} = 14.6\ \mu M$) with promising selectivity. Subsequent chemical optimization produced compound **21** (Fig. 21), which introduced a carboxylic acid and methylene spacer to enhance electrostatic interaction with K140 in the S1″ pocket, improving potency eightfold ($IC_{5n} = 1.8\ \mu M$) and achieving ~100-fold selectivity over related MMPs. Further docking studies confirmed selective binding within the S1″ pocket, and ADMET predictions indicated good drug-like properties and absorption [72].

In 2024, Roy *et al.* (2024) presented a breakthrough in OA treatment through the development of an enzyme-responsive injectable hydrogel designed for the "on-demand" release of a highly selective MMP13 blocker known as **BI-4394** (Fig. 22) [73]. By encapsulating this potent inhibitor in a biocompatible delivery system, the study successfully prevented cartilage degradation and improved mobility in animal models, offering a more effective and patient-friendly alternative to frequent joint injections. To improve drug retention in the joint, the inhibitor was encapsulated in a biocompatible triglycerol monostearate (TG-18) hydrogel. The system releases the drug in response to esterases and MMPs present in inflamed joints, enabling proportional, "on-demand" delivery. Its shear-thinning properties allow easy injection. In a rat osteoarthritis model, a single hydrogel dose preserved cartilage structure, reduced bone erosion, and restored mobility more effectively than saline or free drug injections [73].

2.3 Antibodies

In addition to small-molecule inhibitors, antibody-based strategies have emerged as a highly selective approach to targeting MMP13. By recognizing unique structural epitopes outside the conserved catalytic zinc site, monoclonal antibodies can selectively neutralize the active enzyme while minimizing off-target effects that previously limited broad-spectrum MMP inhibitors.

Naito *et al.* (2012) developed **14D10**, a monoclonal neutralizing antibody that selectively targets the active form of MMP13, in context of treating OA [74]. Unlike earlier small-molecule zinc-chelating inhibitors that caused off-target effects and musculoskeletal toxicity due to poor selectivity, **14D10** was generated against a unique peptide sequence in the S1′ specificity loop of MMP13, ensuring high specificity. The antibody binds exclusively to the active enzyme (not the latent pro-form or related MMPs), showing very high affinity ($K_d = 0.33$ nM) and effectively blocking type II collagen cleavage. It acts through a noncompetitive mechanism, likely disrupting zinc coordination and enzyme activity rather than competing with substrates. Beyond therapeutic potential, **14D10** enabled the development of a selective ELISA to quantify active MMP13 levels, revealing that only a small fraction (~15 percent) of total enzyme is active in chondrocyte cultures. Overall, **14D10** provides both a precise research tool and a promising strategy for targeted diagnosis and treatment of osteoarthritis by specifically inhibiting active MMP13 [74].

3. Conclusions

The development of selective MMP13 inhibitors has shifted from early broad-spectrum zinc chelation, often associated with poor selectivity, musculoskeletal syndrome, and off-target toxicity, to Structure-Based Drug Design strategies that exploited unique features of the MMP13's binding site. By targeting the less conserved S1′ and S1″ pockets rather than the catalytic zinc, modern inhibitors achieve high specificity through defined molecular anchoring interactions with T245 and T247 and salt-bridge formation with residue K140, resulting in picomolar potency and exceptional isoform selectivity [44,48,51,55]. At the same time, safety and pharmacokinetic challenges have been addressed through chemical optimization [47,51,62,73]. Replacing nephrotoxic carboxylic acids with non-acidic scaffolds, alongside innovations such as dioxane or oxetane rings and selective deuteration, has improved oral bioavailability, metabolic stability, and systemic tolerability [46,48,51,71]. Beyond traditional small molecules, emerging approaches include neutralizing antibodies, dual-target inhibitors, and enzyme-responsive hydrogels for localized drug delivery, ameliorated therapeutic applications while minimizing systemic exposure [64–67,73,74]. Complementary advances in selective PET radiotracers also enable non-invasive monitoring of tissue remodeling, linking diagnostics with treatment [69,70].

Overall, the integration of selective scaffolds, safer chemistries, targeted delivery, and real-time imaging brings MMP13 inhibition closer to clinically viable disease-modifying therapies, particularly for osteoarthritis. It is also demonstrated by the number of patented small-molecule inhibitors that were documented in the recent past [75,76].

References

[1] M. Whittaker, C.D. Floyd, P. Brown, A.J. Gearing, Design and therapeutic application of matrix metalloproteinase inhibitors, Chem. Rev. 99 (1999) 2735–2776.

[2] J.W. Skiles, N.C. Gonnella, A.Y. Jeng, The design, structure, and therapeutic application of matrix metalloproteinase inhibitors, Curr. Med. Chem. 8 (2001) 425–474.

[3] J.W. Skiles, N.C. Gonnella, A.Y. Jeng, The design, structure, and clinical update of small molecular weight matrix metalloproteinase inhibitors, Curr. Med. Chem. 11 (2004) 2911–2977.

[4] G. Tu, W. Xu, H. Huang, S. Li, Progress in the development of matrix metalloproteinase inhibitors, Curr. Med. Chem. 15 (2008) 1388–1395.

[5] A. Dufour, N.S. Sampson, S. Zucker, J. Cao, Role of the hemopexin domain of matrix metalloproteinases in cell migration, J. Cell. Physiol. 217 (2008) 643–651.

[6] H. Laronha, J. Caldeira, Structure and function of human matrix metalloproteinases, Cells 9 (2020) 1076.

[7] L. Aureli, M. Gioia, I. Cerbara, S. Monaco, G.F. Fasciglione, S. Marini, P. Ascenzi, A. Topai, M. Coletta, Structural bases for substrate and inhibitor recognition by matrix metalloproteinases, Curr. Med. Chem. 15 (2008) 2192–2222.

[8] K. Maskos, W. Bode, Structural basis of matrix metalloproteinases and tissue inhibitors of metalloproteinases, Mol. Biotechnol. 25 (2003) 241–266.

[9] C.A. Kontogiorgis, P. Papaioannou, D.J. Hadjipavlou-Litina, Matrix metalloproteinase inhibitors: a review on pharmacophore mapping and (Q)SARs results, Curr. Med. Chem. 12 (2005) 339–355.

[10] H.I. Park, Y. Jin, D.R. Hurst, C.A. Monroe, S. Lee, M.A. Schwartz, Q.X. Sang, The intermediate S1′ pocket of the endometase/matrilysin-2 active site revealed by enzyme inhibition kinetic studies, protein sequence analyses, and homology modeling, J. Biol. Chem. 278 (2003) 51646–51653.
[11] B. Lovejoy, A.R. Welch, S. Carr, C. Luong, C. Broka, R.T. Hendricks, J.A. Campbell, K.A. Walker, R. Martin, H. Van Wart, M.F. Browner, Crystal structures of MMP-1 and -13 reveal the structural basis for selectivity of collagenase inhibitors, Nat. Struct. Biol. 6 (1999) 217–221.
[12] V. Knäuper, C. López-Otin, B. Smith, G. Knight, G. Murphy, Biochemical characterization of human collagenase-3, J. Biol. Chem. 271 (1996) 1544–1550.
[13] J.M. Howes, D. Bihan, D.A. Slatter, S.W. Hamaia, L.C. Packman, V. Knauper, R. Visse, R.W. Farndale, The recognition of collagen and triple-helical toolkit peptides by MMP-13: sequence specificity for binding and cleavage, J. Biol. Chem. 289 (2014) 24091–24101.
[14] Q. Hu, M. Ecker, Overview of MMP-13 as a promising target for the treatment of osteoarthritis, Int. J. Mol. Sci. 22 (2021) 1742.
[15] E.E. Mehana, A.F. Khafaga, S.S. El-Blehi, The role of matrix metalloproteinases in osteoarthritis pathogenesis: an updated review, Life Sci. 234 (2019) 116786.
[16] H. Li, D. Wang, Y. Yuan, J. Min, New insights on the MMP-13 regulatory network in the pathogenesis of early osteoarthritis, Arthritis Res. Ther. 19 (2017) 248.
[17] D. Wernicke, C. Seyfert, B. Hinzmann, E. Gromnica-Ihle, Cloning of collagenase 3 from the synovial membrane and its expression in rheumatoid arthritis and osteoarthritis, J. Rheumatol. 23 (1996) 590–595.
[18] R.C. Billinghurst, W. Wu, M. Ionescu, A. Reiner, L. Dahlberg, J. Chen, H. van Wart, A.R. Poole, Comparison of the degradation of type II collagen and proteoglycan in nasal and articular cartilages induced by interleukin-1 and the selective inhibition of type II collagen cleavage by collagenase, Arthritis Rheum. 43 (2000) 664–672.
[19] N. Johansson, K. Airola, R. Grénman, A.L. Kariniemi, U. Saarialho-Kere, V.M. Kähäri, Expression of collagenase-3 (matrix metalloproteinase-13) in squamous cell carcinomas of the head and neck, Am. J. Pathol. 151 (1997) 499–508.
[20] K. Airola, N. Johansson, A.L. Kariniemi, V.M. Kähäri, U.K. Saarialho-Kere, Human collagenase-3 is expressed in malignant squamous epithelium of the skin, J. Invest. Dermatol. 109 (1997) 225–231.
[21] J.J. Li, A.R. Johnson, Selective MMP13 inhibitors, Med. Res. Rev. 31 (2011) 863–894.
[22] G. Zamolo, M. Grahovac, G. Žauhar, D. Vučinić, L. Kovač, N. Brajenić, B. Grahovac, Matrix metalloproteinases MMP-1, MMP-2, and MMP-13 are overexpressed in primary nodular melanoma, J. Cutan. Pathol. 47 (2020) 139–145.
[23] S.V. Gibson, E. Tomas Bort, L. Rodríguez-Fernández, M.D. Allen, J.J. Gomm, I. Goulding, U. Auf dem Keller, A. Agnoletto, C. Brisken, B. Peck, A.J. Cameron, J.F. Marshall, J.L. Jones, E.P. Carter, R.P. Grose, TGFβ-mediated MMP13 secretion drives myoepithelial cell dependent breast cancer progression, NPJ Breast Cancer 9 (2023) 9.
[24] C.H. Lo, G. Shay, J.J. McGuire, T. Li, K.H. Shain, J.Y. Choi, R. Fuerst, W.R. Roush, A.M. Knapinska, G.B. Fields, C.C. Lynch, Host-derived matrix metalloproteinase-13 activity promotes multiple myeloma-induced osteolysis and reduces overall survival, Cancer Res. 81 (2021) 2415–2428.
[25] H. Gu, Y. Jiao, X. Yu, X. Li, W. Wang, L. Ding, L. Liu, Resveratrol inhibits the IL-1 -induced expression ofMMP-13 and IL-6 in human articular chondrocytes viaTLR4/MyD88-dependent and-independent signaling cascades, Int. J. Mol. Med. 39 (2017) 734–740.
[26] Q. Yang, S. Wu, X. Mao, W. Wang, H. Tai, Inhibition effect of curcumin on TNF- and MMP-13 expression induced by advanced glycation end products in chondrocytes, Pharmacology 91 (2013) 77–85.

[27] Z. Rasheed, A.N. Anbazhagan, N. Akhtar, S. Ramamurthy, F.R. Voss, T.M. Haqqi, Green tea polyphenol epigallocatechin-3-gallate inhibits advanced glycation end product-induced expression of tumor necrosis factor- and matrix metalloproteinase-13 in human chondrocytes, Arthritis Res. Ther. 11 (2009) R71.
[28] Y. Wang, H. Jeon, 3D cell cultures toward quantitative high-throughput drug screening, Trends Pharmacol. Sci. 43 (2022) 569–581.
[29] V.D. Mouchlis, A. Afantitis, A. Serra, M. Fratello, A.G. Papadiamantis, V. Aidinis, I. Lynch, D. Greco, G. Melagraki, Advances in de novo drug design: from conventional to machine learning methods, Int. J. Mol. Sci. 22 (2021) 1676.
[30] K. Atz, L. Cotos, C. Isert, M. Håkansson, D. Focht, M. Hilleke, D.F. Nippa, M. Iff, J. Ledergerber, C.C.G. Schiebroek, V. Romeo, J.A. Hiss, D. Merk, P. Schneider, B. Kuhn, U. Grether, G. Schneider, Prospective de novo drug design with deep interactome learning, Nat. Commun. 15 (2024) 3408.
[31] R. Kothapalli, A.M. Khan, Basappa, A. Gopalsamy, Y.S. Chong, L. Annamalai, Chem-informatics-based drug design approach for identification of inhibitors targeting the characteristic residues of MMP-13 hemopexin domain, PLoS One 5 (2010) e12494.
[32] A. Di Pizio, M. Agamennone, P. Tortorella, Non-zinc-binding inhibitors of MMP-13: GRID-based approaches to rationalize the binding process, Curr. Top. Med. Chem. 16 (2016) 449–459.
[33] T.E. Barta, D.P. Becker, L.J. Bedell, G.A. De Crescenzo, J.J. McDonald, G.E. Munie, S. Rao, H.S. Shieh, R. Stegeman, A.M. Stevens, C.I. Villamil, Synthesis and activity of selective MMP inhibitors with an aryl backbone, Bioorg. Med. Chem. Lett. 10 (2000) 2815–2817.
[34] T.E. Barta, D.P. Becker, L.J. Bedell, G.A. De Crescenzo, J.J. McDonald, P. Mehta, G.E. Munie, C.I. Villamil, Selective, orally active MMP inhibitors with an aryl backbone, Bioorg. Med. Chem. Lett. 11 (2001) 2481–2483.
[35] D.P. Becker, T.E. Barta, L. Bedell, G. De Crescenzo, J. Freskos, D.P. Getman, S.L. Hockerman, M. Li, P. Mehta, B. Mischke, G.E. Munie, C. Swearingen, C.I. Villamil, alpha-Amino-beta-sulphone hydroxamates as potent MMP-13 inhibitors that spare MMP-1, Bioorg. Med. Chem. Lett. 11 (2001) 2719–2722.
[36] D.P. Becker, G. De Crescenzo, J. Freskos, D.P. Getman, S.L. Hockerman, M. Li, P. Mehta, G.E. Munie, C. Swearingen, alpha-Alkyl-alpha-amino-beta-sulphone hydroxamates as potent MMP inhibitors that spare MMP-1, Bioorg. Med. Chem. Lett. 11 (2001) 2723–2725.
[37] D.P. Becker, C.I. Villamil, T.E. Barta, L.J. Bedell, T.L. Boehm, G.A. De Crescenzo, J.N. Freskos, D.P. Getman, S. Hockerman, R. Heintz, S.C. Howard, M.H. Li, J.J. McDonald, C.P. Carron, C.L. Funckes-Shippy, P.P. Mehta, G.E. Munie, C.A. Swearingen, Synthesis and structure-activity relationships of beta- and alpha-piperidine sulfone hydroxamic acid matrix metalloproteinase inhibitors with oral antitumor efficacy, J. Med. Chem. 48 (2005) 6713–6730.
[38] D.P. Becker, T.E. Barta, L.J. Bedell, T.L. Boehm, B.R. Bond, J. Carroll, C.P. Carron, G.A. De Crescenzo, A.M. Easton, J.N. Freskos, C.L. Funckes-Shippy, M. Heron, S. Hockerman, C.P. Howard, J.R. Kiefer, M.H. Li, K.J. Mathis, J.J. McDonald, P.P. Mehta, G.E. Munie, T. Sunyer, C.A. Swearingen, C.I. Villamil, D. Welsch, J.M. Williams, Y. Yu, J. Yao, Orally active MMP-1 sparing α-tetrahydropyranyl and α-piperidinyl sulfone matrix metalloproteinase (MMP) inhibitors with efficacy in cancer, arthritis, and cardiovascular disease, J. Med. Chem. 53 (2010) 6653–6680.
[39] S.A. Kolodziej, S.L. Hockerman, G.A. De Crescenzo, J.J. McDonald, D.A. Mischke, G.E. Munie, T.R. Fletcher, N. Stehle, C. Swearingen, D.P. Becker, MMP-13 selective isonipecotamide alpha-sulfone hydroxamates, Bioorg. Med. Chem. Lett. 20 (2010) 3561–3564.

[40] T.E. Barta, D.P. Becker, L.J. Bedell, A.M. Easton, S.L. Hockerman, J. Kiefer, G.E. Munie, K.J. Mathis, M.H. Li, J.G. Rico, C.I. Villamil, J.M. Williams, MMP-13 selective α-sulfone hydroxamates: a survey of P1′ heterocyclic amide isosteres, Bioorg. Med. Chem. Lett. 21 (2011) 2820–2822.
[41] R.A. Tommasi, S. Weiler, L.W. McQuire, O. Rogel, M. Chambers, K. Clark, J. Doughty, J. Fang, V. Ganu, J. Grob, R. Goldberg, R. Goldstein, S. Lavoie, R. Kulathila, W. Macchia, R. Melton, C. Springer, M. Walker, J. Zhang, L. Zhu, M. Shultz, Potent and selective 2-naphthylsulfonamide substituted hydroxamic acid inhibitors of matrix metalloproteinase-13, Bioorg. Med. Chem. Lett. 21 (2011) 6440–6445.
[42] S. Santamaria, E. Nuti, G. Cercignani, L. Marinelli, V. La Pietra, E. Novellino, A. Rossello, N-O-isopropyl sulfonamido-based hydroxamates: kinetic characterisation of a series of MMP-12/MMP-13 dual target inhibitors, Biochem. Pharmacol. 84 (2012) 813–820.
[43] L.G. Monovich, R.A. Tommasi, R.A. Fujimoto, V. Blancuzzi, K. Clark, W.D. Cornell, R. Doti, J. Doughty, J. Fang, D. Farley, J. Fitt, V. Ganu, R. Goldberg, R. Goldstein, S. Lavoie, R. Kulathila, W. Macchia, D.T. Parker, R. Melton, E. O'Byrne, G. Pastor, T. Pellas, E. Quadros, N. Reel, D.M. Roland, Y. Sakane, H. Singh, J. Skiles, J. Somers, K. Toscano, A. Wigg, S. Zhou, L. Zhu, W.C. Shieh, S. Xue, L.W. McQuire, Discovery of potent, selective, and orally active carboxylic acid based inhibitors of matrix metalloproteinase-13, J. Med. Chem. 52 (2009) 3523–3538.
[44] H. Nara, A. Kaieda, K. Sato, T. Naito, H. Mototani, H. Oki, Y. Yamamoto, H. Kuno, T. Santou, N. Kanzaki, J. Terauchi, O. Uchikawa, M. Kori, Discovery of novel, highly potent, and selective matrix metalloproteinase (MMP)-13 inhibitors with a 1,2,4-triazol-3-yl moiety as a zinc binding group using a structure-based design approach, J. Med. Chem. 60 (2017) 608–626.
[45] C. De Savi, D. Waterson, A. Pape, S. Lamont, E. Hadley, M. Mills, K.M. Page, J. Bowyer, R.A. Maciewicz, Hydantoin based inhibitors of MMP13—discovery of AZD6605, Bioorg. Med. Chem. Lett. 23 (2013) 4705–4712.
[46] R. Fuerst, J.Y. Choi, A.M. Knapinska, M.D. Cameron, C. Ruiz, A. Delmas, M.S. Sundrud, G.B. Fields, W.R. Roush, Development of a putative Zn^{2+}-chelating but highly selective MMP-13 inhibitor, Bioorg. Med. Chem. Lett. 76 (2022) 129014.
[47] A.R. Johnson, A.G. Pavlovsky, D.F. Ortwine, F. Prior, C.F. Man, D.A. Bornemeier, C.A. Banotai, W.T. Mueller, P. McConnell, C. Yan, V. Baragi, C. Lesch, W.H. Roark, M. Wilson, K. Datta, R. Guzman, H.K. Han, R.D. Dyer, Discovery and characterization of a novel inhibitor of matrix metalloprotease-13 that reduces cartilage damage in vivo without joint fibroplasia side effects, J. Biol. Chem. 282 (2007) 27781–27791.
[48] V.M. Baragi, G. Becher, A.M. Bendele, R. Biesinger, H. Bluhm, J. Boer, H. Deng, R. Dodd, M. Essers, T. Feuerstein, B.M. Gallagher Jr., C. Gege, M. Hochgürtel, M. Hofmann, A. Jaworski, L. Jin, A. Kiely, B. Korniski, H. Kroth, D. Nix, B. Nolte, D. Piecha, T.S. Powers, F. Richter, M. Schneider, C. Steeneck, I. Sucholeiki, A. Taveras, A. Timmermann, J. Van Veldhuizen, J. Weik, X. Wu, B. Xia, A new class of potent matrix metalloproteinase 13 inhibitors for potential treatment of osteoarthritis: evidence of histologic and clinical efficacy without musculoskeletal toxicity in rat models, Arthritis Rheum. 60 (2009) 2008–2018.
[49] M.E. Schnute, P.M. O'Brien, J. Nahra, M. Morris, W. Howard Roark, C.E. Hanau, P.G. Ruminski, J.A. Scholten, T.R. Fletcher, B.C. Hamper, J.N. Carroll, W.C. Patt, H.S. Shieh, B. Collins, A.G. Pavlovsky, K.E. Palmquist, K.W. Aston, J. Hitchcock, M.D. Rogers, J. McDonald, A.R. Johnson, G.E. Munie, A.J. Wittwer, C.F. Man, S.L. Settle, O. Nemirovskiy, A. Agawal, R.D. Dyer, T. Sunyer, Discovery of (pyridin-4-yl)-2H-tetrazole as a novel scaffold to identify highly selective matrix metalloproteinase-13 inhibitors for the treatment of osteoarthritis, Bioorg. Med. Chem. Lett. 20 (2010) 576–580.

[50] S. Settle, L. Vickery, O. Nemirovskiy, T. Vidmar, A. Bendele, D. Messing, P. Ruminski, M. Schnute, T. Sunyer, Cartilage degradation biomarkers predict efficacy of a novel, highly selective matrix metalloproteinase 13 inhibitor in a dog model of osteoarthritis: confirmation by multivariate analysis that modulation of type II collagen and aggrecan degradation peptides parallels pathologic changes, Arthritis Rheum. 62 (2010) 3006–3015.

[51] P.G. Ruminski, M. Massa, J. Strohbach, C.E. Hanau, M. Schmidt, J.A. Scholten, T.R. Fletcher, B.C. Hamper, J.N. Carroll, H.S. Shieh, N. Caspers, B. Collins, M. Grapperhaus, K.E. Palmquist, J. Collins, J.E. Baldus, J. Hitchcock, H.P. Kleine, M.D. Rogers, J. McDonald, G.E. Munie, D.M. Messing, S. Portolan, L.O. Whiteley, T. Sunyer, M.E. Schnute, Discovery of N-(4-Fluoro-3-methoxybenzyl)-6-(2-(((2S,5R)-5-(hydroxymethyl)-1,4-dioxan-2-yl) methyl)-2H-tetrazol-5-yl)-2-methylpyrimidine-4-carboxamide, a highly selective and orally bioavailable matrix metalloproteinase-13 inhibitor for the potential treatment of osteoarthritis, J. Med. Chem. 59 (2016) 313–327.

[52] A.M. Panico, P. Vicini, A. Geronikaki, M. Incerti, V. Cardile, L. Crascì, R. Messina, S. Ronsisvalle, Heteroarylimino-4-thiazolidinones as inhibitors of cartilage degradation, Bioorg. Chem. 39 (2011) 48–52.

[53] P. Vicini, L. Crascì, M. Incerti, S. Ronsisvalle, V. Cardile, A.M. Panico, Benzisothiazolyliminothiazolidin-4-ones with chondroprotective properties: searching for potent and selective inhibitors of MMP-13, ChemMedChem 6 (2011) 1199–1202.

[54] C. De Savi, A.D. Morley, A. Ting, I. Nash, K. Karabelas, C.M. Wood, M. James, S.J. Norris, G. Karoutchi, N. Rankine, G. Hamlin, P.A. Macfaul, D. Ryan, S.V. Baker, D. Hargreaves, S. Gerhardt, Selective non zinc binding inhibitors of MMP13, Bioorg. Med. Chem. Lett. 21 (2011) 4215–4219.

[55] C. De Savi, A.D. Morley, I. Nash, G. Karoutchi, K. Page, A. Ting, S. Gerhardt, Lead optimisation of selective non-zinc binding inhibitors of MMP13. Part 2, Bioorg. Med. Chem. Lett. 22 (2012) 271–277.

[56] V. La Pietra, L. Marinelli, S. Cosconati, S.F. Di Leva, E. Nuti, S. Santamaria, I. Pugliesi, M. Morelli, F. Casalini, A. Rossello, C. La Motta, S. Taliani, R. Visse, H. Nagase, F. da Settimo, E. Novellino, Identification of novel molecular scaffolds for the design of MMP-13 inhibitors: a first round of lead optimization, Eur. J. Med. Chem. 47 (2012) 143–152.

[57] J. Roth, D. Minond, E. Darout, Q. Liu, J. Lauer, P. Hodder, G.B. Fields, W.R. Roush, Identification of novel, exosite-binding matrix metalloproteinase-13 inhibitor scaffolds, Bioorg. Med. Chem. Lett. 21 (2011) 7180–7184.

[58] J.Y. Choi, R. Fuerst, A.M. Knapinska, A.B. Taylor, L. Smith, X. Cao, P.J. Hart, G.B. Fields, W.R. Roush, Structure-based design and synthesis of potent and selective matrix metalloproteinase 13 inhibitors, J. Med. Chem. 60 (2017) 5816–5825.

[59] C. Gege, B. Bao, H. Bluhm, J. Boer, B.M. Gallagher, B. Korniski, T.S. Powers, C. Steeneck, A.G. Taveras, V.M. Baragi, Discovery and evaluation of a non-Zn chelating, selective matrix metalloproteinase 13 (MMP-13) inhibitor for potential intra-articular treatment of osteoarthritis, J. Med. Chem. 55 (2012) 709–716.

[60] M. Shah, D. Huang, T. Blick, A. Connor, L.A. Reiter, J.R. Hardink, C.C. Lynch, M. Waltham, E.W. Thompson, An MMP13-selective inhibitor delays primary tumor growth and the onset of tumor-associated osteolytic lesions in experimental models of breast cancer, PLoS One 7 (2012) e29615.

[61] H. Nara, K. Sato, T. Naito, H. Mototani, H. Oki, Y. Yamamoto, H. Kuno, T. Santou, N. Kanzaki, J. Terauchi, O. Uchikawa, M. Kori, Discovery of novel, highly potent, and selective quinazoline-2-carboxamide-based matrix metalloproteinase (MMP)-13 inhibitors without a zinc binding group using a structure-based design approach, J. Med. Chem. 57 (2014) 8886–8902.

[62] H. Nara, K. Sato, T. Naito, H. Mototani, H. Oki, Y. Yamamoto, H. Kuno, T. Santou, N. Kanzaki, J. Terauchi, O. Uchikawa, M. Kori, Thieno[2,3-d]pyrimidine-2-carboxamides bearing a carboxybenzene group at 5-position: highly potent, selective, and orally available MMP-13 inhibitors interacting with the S1″ binding site, Bioorg. Med. Chem. 22 (2014) 5487–5505.
[63] S. Santamaria, E. Nuti, G. Cercignani, G. La Regina, R. Silvestri, C.T. Supuran, A. Rossello, Kinetic characterization of 4,4′-biphenylsulfonamides as selective non-zinc binding MMP inhibitors, J. Enzyme Inhib. Med. Chem. 30 (2015) 947–954.
[64] N. Senn, M. Ott, J. Lanz, R. Riedl, Targeted polypharmacology: discovery of a highly potent non-hydroxamate dual matrix metalloproteinase (MMP)-10/-13 inhibitor, J. Med. Chem. 60 (2017) 9585–9598.
[65] E.S.H. El Ashry, L.F. Awad, M. Teleb, N.A. Ibrahim, M.M. Abu-Serie, M.N. Abd Al Moaty, Structure-based design and optimization of pyrimidine- and 1,2,4-triazolo[4,3-a]pyrimidine-based matrix metalloproteinase-10/13 inhibitors via Dimroth rearrangement towards targeted polypharmacology, Bioorg. Chem. 96 (2020) 103616.
[66] C.A. Morcos, S.N. Khattab, N.S. Haiba, R.W. Bassily, M.M. Abu-Serie, M. Teleb, Battling colorectal cancer via s-triazine-based MMP-10/13 inhibitors armed with electrophilic warheads for concomitant ferroptosis induction, the first-in-class dual-acting agents, Bioorg. Chem. 141 (2023) 106839.
[67] C.A. Morcos, N.S. Haiba, R.W. Bassily, M.M. Abu-Serie, A.F. El-Yazbi, O.A. Soliman, S.N. Khattab, M. Teleb, Structure optimization and molecular dynamics studies of new tumor-selective s-triazines targeting DNA and MMP-10/13 for halting colorectal and secondary liver cancers, J. Enzyme Inhib. Med. Chem. 39 (2024) 2423174.
[68] R. Kothapalli, K. Sivaraman Siveen, T.Z. Tan, J.P. Thiery, A.P. Kumar, G. Sethi, K. Swaminathan, Functional characterization of selective exosite-binding inhibitors of matrix metalloproteinase-13 (MMP-13)—experimental validation in human breast and colon cancer, Biosci. Biotechnol. Biochem. 80 (2016) 2122–2131.
[69] V. Hugenberg, S. Wagner, K. Kopka, M. Schäfers, R.C. Schuit, A.D. Windhorst, S. Hermann, Radiolabeled selective matrix metalloproteinase 13 (MMP-13) inhibitors: (Radio)syntheses and in vitro and first in vivo evaluation, J. Med. Chem. 60 (2017) 307–321.
[70] A. Buchler, U.S. Ismailani, N. MacMullin, F. Abdirahman, M. Adi, C. Bi, C. Jany, J.W. Keillor, B.H. Rotstein, Quinazoline-2-carboxamides as selective PET radiotracers for matrix metalloproteinase-13 imaging in atherosclerosis, J. Med. Chem. 66 (2023) 6682–6696.
[71] A.M. Bendele, M. Neelagiri, V. Neelagiri, I. Sucholeiki, Development of a selective matrix metalloproteinase 13 (MMP-13) inhibitor for the treatment of osteoarthritis, Eur. J. Med. Chem. 224 (2021) 113666.
[72] D. Cuffaro, A. Gimeno, B.L. Bernardoni, R. Di Leo, G. Pujadas, S. Garcia-Vallvé, S. Nencetti, A. Rossello, E. Nuti, Identification of N-acyl hydrazones as new non-zinc-binding MMP-13 inhibitors by structure-based virtual screening studies and chemical optimization, Int. J. Mol. Sci. 24 (2023) 11098.
[73] H.S. Roy, P. Murugesan, C. Kulkarni, M. Arora, G.K. Nagar, R. Guha, N. Chattopadhyay, D. Ghosh, On-demand release of a selective MMP-13 blocker from an enzyme-responsive injectable hydrogel protects cartilage from degenerative progression in osteoarthritis, J. Mater. Chem. B. 12 (2024) 5325–5338.
[74] S. Naito, T. Takahashi, J. Onoda, A. Yamauchi, T. Kawai, J. Kishino, S. Yamane, I. Fujii, N. Fukui, Y. Numata, Development of a neutralizing antibody specific for the active form of matrix metalloproteinase-13, Biochemistry 51 (2012) 8877–8884.

[75] N.G. Li, Y.P. Tang, J.A. Duan, New selective inhibitors of MMP-13 for inflammatory diseases: a patent evaluation (W02012151158), Expert Opin. Ther. Pat. 23 (2013) 669–675.
[76] S. Banerjee, S.K. Baidya, N. Adhikari, T. Jha, An updated patent review of matrix metalloproteinase (MMP) inhibitors (2021-present), Expert Opin. Ther. Pat. 33 (2023) 631–649.

CHAPTER FIVE

Tissue inhibitors of metalloproteinases (TIMPs)

Niccolò Paoletti[a,*], Antoine Dufour[b], Claudiu T. Supuran[a]
[a] Neurofarba Department, Pharmaceutical and Nutraceutical Section, University of Florence, Sesto Fiorentino (FI), Italy
[b] Department of Physiology and Pharmacology, Snyder Institute for Chronic Diseases, McCaig Institute for Bone and Joint Health, and Hotchkiss Brain Institute, University of Calgary, Calgary, Canada
*Corresponding author. e-mail address: niccolo.paoletti@unifi.it.

Contents

Abstract

The extracellular matrix (ECM) is a dynamic structural network essential for cellular organization, continuously remodeled by proteases such as matrix metalloproteinases (MMPs). Under physiological conditions, these proteases are strictly controlled through

Fungal and Protozoan Carbonic Anhydrases, Volume 60
ISSN 1874-6047
https://doi.org/10.1016/bs.enz.2026.05.001

gene transcription, zymogen activation, and endogenous tissue inhibitors of metalloproteinases (TIMPs). TIMPs are vital endogenous inhibitors that tightly regulate various metalloproteinases, including MMPs, ADAMs (a disintegrin and metalloproteinase), and ADAMTSs (a disintegrin and metalloproteinase with thrombospondin motifs) to maintain healthy tissue homeostasis. Disrupting this balance between metalloproteases and TIMPs triggers uncontrolled ECM turnover, changes in immune responses, activation of cell surface receptors that can drive pathological conditions such as fibrosis, cancer, autoimmune diseases, arthritis, cardiovascular diseases and many others. Since their initial discovery, four human TIMPs have been identified, each demonstrating unique biological roles and tissue distributions. TIMPs represent an evolutionarily ancient protein lineage found across bacteria, invertebrates, and vertebrates. Despite their critical role in maintaining human health, there is still a significant gap in our understanding of exactly how their dysregulation drives disease progression.

Keywords: ADAMs, ADAMTs, Inhibitor, Matrix metalloproteinases, MMPs, Protease inhibitor, TIMP, Tissue inhibitors of metalloproteinases

1. Introduction

The extracellular matrix (ECM) is a complex, non-cellular 3D macromolecular network that provides necessary structural scaffolding to tissues. The ECM is no longer regarded as a mere static architecture, but rather as a dynamic environment in continuous communication with cellular compartments. Proper assembly is essential in multicellular organisms for the organization of cells and tissues, and for maintaining a microenvironment conducive to cellular function. The principal proteases involved in ECM remodeling and catabolism are matrix metalloproteinases (MMPs) and ADAMTS metalloproteinases (a disintegrin and metalloproteinase with thrombospondin motifs) [1]. Their activity is tightly regulated under physiological conditions at multiple levels, including gene transcription, zymogen activation, and inhibition by endogenous inhibitors [1–3]. Tissue inhibitors of metalloproteinases (TIMPs) are endogenous inhibitors of these proteases and thus play a crucial role in regulating ECM turnover, tissue remodeling, and cellular behavior. Disruption of the finely tuned balance between MMPs and TIMPs has been associated with pathological conditions characterized by uncontrolled ECM turnover, inflammation, fibrosis and aberrant cell growth and migration, including arthritis, cardiovascular diseases, cancer, pulmonary disorders, nephritis, neurological disorders, and tissue ulceration [1]. Despite decades of research, we still have a minimal understanding on how these processes drive disease progression.

The discovery of tissue inhibitors of metalloproteinases dates back to the early 1970s, when inhibitory activity against collagenase (later identified as

TIMP-1) was first detected in cultured human skin fibroblasts, human serum, and bovine tissue extracts [1,4–6]. This inhibitor was purified in 1979 and subsequently shown to inhibit not only collagenases but also a broader range of metalloproteinases, leading to the adoption of the term "tissue inhibitor of metalloproteinases" [7–8]. Interestingly, TIMP-1 was first identified as an erythropoiesis modulator (originally called erythroid-potentiating activity or EPA) in the 1980s [9,10]. Following the identification of TIMP-1, additional family members were discovered. TIMP-2 was identified as a distinct inhibitor with similar activity but different biochemical properties [11–12]. TIMP-3, characterized in the early 1990s, was distinguished by its dominant association with the extracellular matrix [13]. Finally, TIMP-4 was identified in the mid-1990s as the fourth member of the family, exhibiting a more restricted tissue distribution, particularly in cardiovascular tissues [14].

2. General structure of TIMPs

TIMPs constitute an evolutionarily ancient protein lineage whose presence has been documented across the entire animal kingdom, including vertebrates, invertebrates, and bacteria [1,15]. Whereas invertebrates frequently harbor a single TIMP gene (*e.g., Drosophila melanogaster*), the human genome encodes four homologues (TIMP-1, TIMP-2, TIMP-3, and TIMP-4) [16,17]. Notably, their genomic organization is highly conserved: the genes encoding TIMP-1, TIMP-3, and TIMP-4 are embedded within introns of synapsin genes, suggesting an evolutionarily conserved coordination [16].

The four human TIMP isoforms are mature proteins comprising 184–194 amino acid residues, with molecular masses ranging from 22 to 28 kDa (Table 1). Despite exhibiting only moderate sequence identity (37–51 percent), these proteins share a highly conserved structural organization characterized by a distinctive wedge-shaped conformation, divided into two functionally distinct domains (Fig. 1A): an N-terminal domain (~125 residues) and a C-terminal domain (~65 residues). Structural stability is ensured by six conserved disulfide bonds, three per domain, which link distant regions of the polypeptide chain and preserve the integrity of the inhibitory interface [1,17–20]. The N-terminal domain adopts an oligonucleotide/oligosaccharide-binding fold (OB-fold), consisting of a closed, twisted five-stranded β-barrel (Fig. 1B, β-strands sA–sF) arranged in a Greek-key topology. This domain also comprises three α-helices: one located near the N-terminus (hI) and two (hII and hIII) positioned toward the C-terminus, where they contribute to the interface with the C-terminal

Table 1 Summary of the general properties of human TIMPs [1–22].

TIMP	Main Targets	Other MMP inhibition	Unique features	Non-canonical signaling	Tissue distribution	N° of residues	Molecular Weight (kDa)	Glycosylation
TIMP-1	Soluble MMPs	ADAM10	Glycosylated	CD36/integrin $\beta1$	Widely	184	22–22,5	Highly (~30 percent mass)
TIMP-2	MMP-2, MT-MMPs	ADAM12	Dual role	Integrin $\alpha3\beta1$	Widely	194	21	No
TIMP-3	MMPs, ADAMs, ADAMTs	–	ECM-bound	TNFR/VEGFR2	ECM-associated	188	22–27	Partial (C-terminal site)
TIMP-4	Limited MMPs	ADAM17 and 28	Platelet-associated	Limited	Restricted (cardiac, muscular)	194	22	No

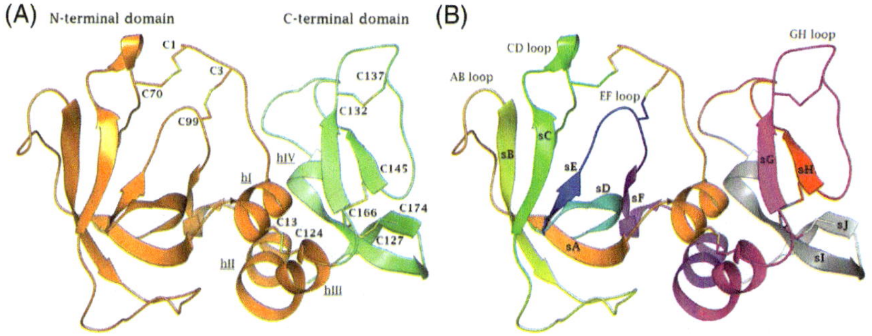

Fig. 1 General structure of TIMPs. A) Domain organization highlighting the N-terminal (orange) and C-terminal (green) domains; the six disulfide bonds and four α-helices (hI–hIV) are shown. B) Distribution and localization of the ten β-strands (sA–sJ), with each section color-coded. The image was generated from the crystal structure of TIMP-1 (PDBID: 1UEA) [21].

domain. The extreme N-terminal segment is highly conserved across all TIMPs, reflecting its essential role in MMP inhibition. Several cysteine residues are critical for maintaining the three-dimensional architecture of this domain: Cys1 and Cys3 form disulfide bonds with Cys70 and Cys99, respectively (numbering based on TIMP-1), while Cys13 is disulfide-linked to Cys124. The C-terminal domain comprises a pair of parallel β-strands (sG and sH) connected by a loop, followed by an α-helix (hIV) and a pair of antiparallel β-strands (sI and sJ) forming a β-hairpin motif. This domain is further stabilized by three additional disulfide bonds involving Cys127, Cys132, and Cys145, which are covalently linked to Cys174, Cys137, and Cys166, respectively. Although the C-terminal domain is less directly involved in catalytic inhibition, it mediates critical protein–protein interactions and receptor binding [1,21,22].

One of the key biochemical differences between TIMP family members is their degree of glycosylation. TIMP-1 is the only isoform that is extensively glycosylated, with two well-established N-linked glycosylation sites at Asn30 and Asn78. These carbohydrate moieties are structurally heterogeneous and account for approximately 30 percent of the total molecular mass of the protein [22]. Regarding its function, removing glycosylation does not appear to significantly affect TIMP-1's inhibitory activity towards MMPs. The Asn30 site is structurally located on the protein surface in a region that does not interfere with metalloproteinase binding. Consistently, deglycosylation does not impair inhibitory activity in vitro. However, glycosylation enhances solubility and stability in vivo [21,22]. In contrast, TIMP-2 and TIMP-4 are non-glycosylated proteins, whereas TIMP-3 exhibits only partial glycosylation, with a potential glycosylation site located in the C-terminal region [1,22].

3. Molecular basis of TIMP-mediated inhibition: Structural determinants of selectivity and specificity

TIMP inhibition follows a conserved "wedge mechanism" that physically occludes the catalytic zinc site. This inhibition occurs with a 1:1 stoichiometry through the insertion of the so-called reactive ridge of the inhibitor. This structural element comprises the N-terminal segment (residues Cys1-Pro5) together with the CD loop (Met66-Cys70), which are covalently constrained by the Cys1-Cys70 disulfide bond. This wedge-like motif occupies approximately 60–75 percent of the enzyme-inhibitor interface within the catalytic cleft, enabling the insertion of the highly conserved N-terminal Cys1 above the catalytic Zn^{2+} ion of the MMP active site (Fig. 2) [21,23]. Cys1 is the principal determinant of inhibitory potency; its α-amino and carbonyl groups coordinate the catalytic Zn^{2+} in a bidentate fashion, displacing the water molecule required for proteolysis and thereby abolishing enzymatic activity [1,20–24]. This interaction is further stabilized by hydrogen bonding with the catalytic glutamate residue (Glu202, MMP-3 numbering). In addition, the N-terminal segment adopts a substrate-like extended conformation within the active-site cleft, occupying the S1–S3′ subsites but in an inverted orientation that precludes proteolytic cleavage [1,20,21]. Within this framework, residue 2 (Thr/Ser, conserved across vertebrate TIMPs) is positioned within the S1′ specificity pocket, a key determinant of MMP selectivity, where it engages in multiple hydrogen bonds with the backbone carbonyl and amide groups of L164, A165, and P221 (Fig. 2B) [25]. Whereas, residues 3–5 further interact with the primed subsites (S1′-S3′), and Ser68 and Val69 contribute by interacting to the non-primed S2 and S3 subsites, respectively (Fig. 2C). Additional binding energy is provided by auxiliary exosite interactions involving the AB loop, the EF loop, and some residues of the C-terminal domain, which flank the catalytic groove and modulate protease recognition [1].

TIMPs are generally broad-spectrum inhibitors with inhibition constants in the low nanomolar range for most MMPs (Table 2). However, subtle structural variations confer marked selectivity toward specific MMP subfamilies (Fig. 3 and Table 3).

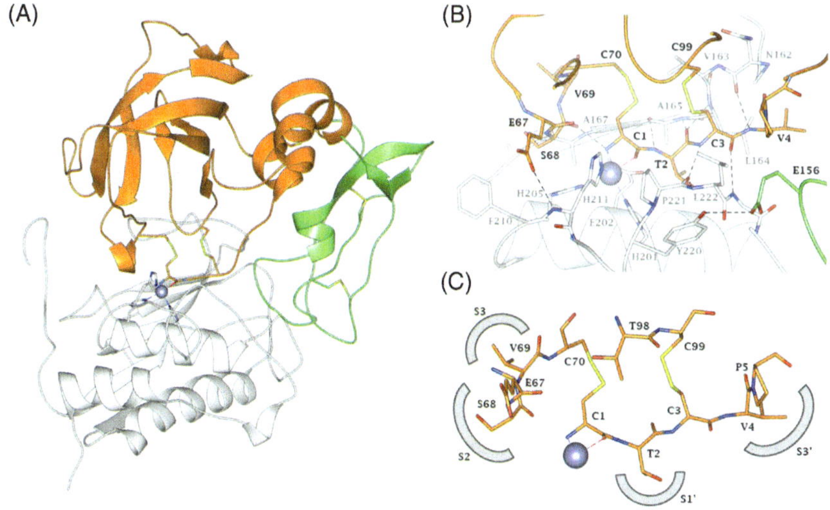

Fig. 2 General binding mode of TIMPs to MMPs. A) Overview of the binding mode of TIMP-1 in complex with MMP-3. The N-terminal and C-terminal domains of TIMP-1 are shown in orange and green, respectively, while MMP-3 is rendered in white. The catalytic zinc ion is depicted as a gray sphere. B) Detailed view of the interaction between the "reactive ridge" of TIMP-1 and MMP-3, maintaining the color scheme described in panel A. Hydrogen bonds are shown as black dashed lines, and residue labels are color-coded for TIMP-1 (black) and MMP-3 (gray). C) Two-dimensional representation of the binding mode, highlighting the S1–S3′ subsites. The image was generated from the crystal structure of the TIMP-1/MMP-3 complex (PDBID: 1UEA) [21].

Table 2 Inhibitory constants (K_I) of TIMPs towards metalloproteinases. Reported K_I values vary depending on experimental conditions, enzyme constructs, and assay systems; values shown represent approximate ranges derived from multiple independent studies [1,20–36].

Target	**K_I nM**			
	TIMP-1	**TIMP-2**	**TIMP-3**	**TIMP-4**
MMP-1	1–3	~1	~1	0.5–3
MMP-2	~1	fM-pM	1–5	pM-nM
MMP-3	1–3	0.5–2	1–10	1–2
MMP-9	nM	nM	nM	nM
MMP-14	Weak(~100)	~1	~1	~1
ADAM10	nM	Weak	nM	Weak
ADAM17 (TACE)	Weak/inactive	Inactive	0.5–2	Weak
ADAMTS-4, −5	Inactive	Weak	0.5–2	Inactive

Table 3 Structural determinants of TIMP selectivity (mutations → effects).

Position (TIMP-1 numbering)	Mutation	Structural effect	Functional consequence	Target specificity
Cys1	Ala or Ser	Loss of Zn coordination	Abolished inhibition	All MMPs
Thr2	Gly or Ala	Reduced S1' pocket interaction	Strong loss of affinity	MMP-1, −2, −3
	Val or Leu or Met	Improved hydrophobic packing	Increased affinity	MMP-3 vs MMP-1
Val4	Ala or Leu	Alters S3' interactions	Modulates binding energy	MMP-dependent
Leu34	Gly	Increased backbone flexibility	Stabilized Tyr35-Tyr153 clasp	MMP-3
Tyr35 (AB loop)	Ala or Phe	Alters exosite interaction	Reduced MT-MMP binding	MMP-14
AB loop (length)	Extension	Increased exosite contacts	Enhanced MT-MMP inhibition	MMP-14
AB loop (chimeric)	Introduction TIMP-3 AB loop		Increase affinity	ADAM17
Ser68 (CD loop)	Ala	Alters ridge orientation	Indirect affinity reduction	Broad
Thr98	Leu	Alters surface complementary	Gain of ADAM inhibition	ADAM-17, MT-MMPs
Gly154	Ala	Induces α-helix formation	Enhances interdomain packing	Broad
Leu34/Gly154	Gly34/Ala143	Stabilized Tyr35-Tyr153 clasp	Increase affinity	MMP-3
Position 4/38/71 (TIMP-2)	Pro4/Asn38/Ile71	Interface reshaping	Increase selectivity	MMP-9 vs MMP-14
Position 38/71 (TIMP-2)	Gln38/Asn71	Interface reshaping	Increase affinity	MT1-MMP

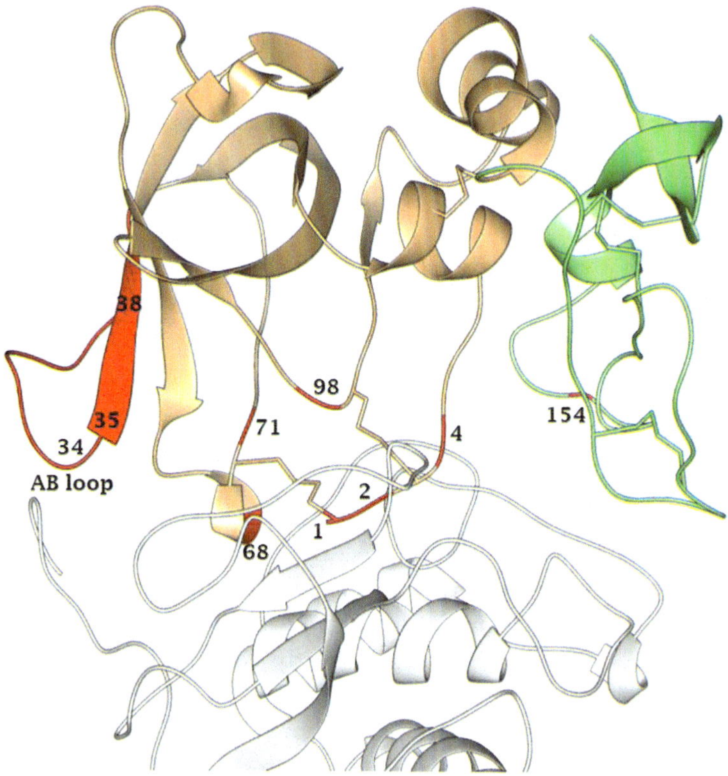

Fig. 3 Graphical representation of the main mutation sites. The N-terminal domain of TIMPs is depicted in orange, the C-terminal domain in green, and mutation sites are highlighted in red. The image was generated from the crystal structure of TIMP-1 (PDBID: 1UEA) [21].

Position 2 mutations (Thr/Ser). Residue 2 directly engages the S1′ pocket of MMPs and represents a critical selectivity determinant. Mutagenesis studies on Thr2 in N-TIMP-1 have shown that substitution with diverse amino acids profoundly modulates K_I values in a target-dependent manner, reflecting differences in S1′ pocket architecture. Small residues such as Gly or Ala lead to a pronounced loss of affinity, particularly toward MMP-1, MMP-2, and MMP-3, underscoring the importance of side-chain complementarity for productive anchoring. Conversely, bulky hydrophobic substitutions (Val, Leu, Ile, Phe) enhance binding to MMP-3, which features a more spacious S1′ pocket, while generally reducing affinity toward the sterically constrained MMP-1; MMP-2 typically displays intermediate behavior. Charged substitutions (Asp, Glu, Lys, Arg) are broadly detrimental across these enzymes, whereas polar residues such as Ser produce more moderate effects. Notably, MMP-9 is comparatively insensitive to the Thr2→Gly substitution, suggesting reduced reliance on S1′ pocket interactions [20,26–28,37,38].

Positions 4 and 68 mutations. Residues Val4 and Ser68 contribute cooperatively to MMP selectivity, albeit with lower energetic impact than position 2. Position 4 engages regions adjacent to the catalytic cleft, including the S3′ subsite, thereby modulating affinity in an enzyme-dependent manner. Substitutions at this position differentially affect MMP-1 and MMP-3 selectivity, with more modest effects on MMP-2, reflecting subtle differences in active-site topology. Ser68, located within the CD loop, does not act as a primary energetic hotspot but instead contributes to proper orientation of the N-terminal inhibitory segment; mutations at this position therefore exert indirect but measurable effects on protease binding without abolishing inhibition [1,2,27,39].

Position 98 mutations. Thr98 represents a key determinant of differential recognition between MMPs and ADAM family proteases. In wild-type TIMP-1, this residue supports a binding profile predominantly restricted to soluble MMPs, with minimal activity against membrane-type MMPs (e.g., MMP-14), MMP-19, and particularly ADAM17. Remarkably, the Thr98→Leu substitution induces a pronounced rewiring of selectivity, conferring potent inhibitory activity toward ADAM17 and several MT-MMPs. This effect arises not from direct active-site engagement but from improved complementarity with non-conserved surface regions of the target proteases. Accordingly, residue 98 can be regarded as a "specificity switch," capable of redirecting the TIMP inhibitory spectrum toward ADAM-preferred targets and highlighting its relevance for engineering ADAM-selective TIMP variants [40,41].

AB loop length and composition. The AB loop is a major determinant of TIMP selectivity, as it mediates exosite interactions with variable surface regions of metalloproteinases. An extended AB loop, as observed in TIMP-2 relative to TIMP-1, enables broader contact surfaces and enhances inhibition of enzymes with accessible peripheral epitopes. TIMP-2 exhibits high affinity toward MMP-2 (including within the MT1-MMP/MMP-2 activation complex) and MMP-14, whereas the shorter AB loop of TIMP-1 correlates with reduced activity toward MT-MMPs and preferential inhibition of soluble MMPs such as MMP-1 and MMP-3. Engineering studies have demonstrated that modifications in AB loop length and sequence can substantially reshape selectivity profiles. AB loop extensions in TIMP-1, or chimeric replacements with sequences derived from TIMP-2 or TIMP-3, enhance inhibition of MMP-14 and other MT-MMPs that are otherwise poorly inhibited by wild-type TIMP-1. Similarly, such modifications can confer activity toward ADAM17, a protease efficiently regulated by TIMP-3 but weakly inhibited by TIMP-1. At the level of point mutations, residues such as Tyr36 within the AB loop contribute to selective recognition, particularly of MMP-14, and influence interactions with gelatinases (MMP-2 and MMP-9), consistent with its role in exosite-driven binding [2,23,29,40,42].

As can be seen from Table 2, three key conclusions can be deduced:

- *The weak inhibitory activity of TIMP-1 toward MT-MMPs.* TIMP-1 exhibits relatively weak inhibitory potency against membrane-type matrix metalloproteinase, displaying a K_I of approximately 100 nM for MMP-14. This renders TIMP-1 a comparatively inefficient inhibitor under physiological conditions when contrasted with other TIMP family members. The structural basis for this reduced affinity lies in specific steric and interaction constraints at the enzyme-inhibitor interface. A key determinant is the shortened AB loop of TIMP-1, which limits exosite engagement with MT-MMP surfaces. By contrast, the longer AB loop of TIMP-2 contains critical residues, including Tyr36, that are optimally positioned to insert into a complementary cavity on MT1-MMP, thereby stabilizing the complex. An additional crucial factor is residue 98 within TIMP-1, where a threonine is present, whereas TIMP-2 harbors a leucine at the corresponding position. Notably, the Thr98→Leu substitution alone is sufficient to markedly enhance TIMP-1 affinity toward MT-MMPs, underscoring the pivotal role of this site as a structural determinant of membrane-type MMP recognition and binding efficiency [1,20,24,29].
- *The extraordinarily high affinity of TIMP-2 for MMP-2.* The K_I value in the femtomolar range, reported between ~0.6 and 6 fM, indicative of near-irreversible complex formation. This exceptional potency arises from cooperative domain interactions: beyond N-terminal engagement of the catalytic site, the C-terminal domain of TIMP-2, especially residues 186–194, forms extensive contacts with the hemopexin-like domain of MMP-2. Consistently, truncation of the hemopexin domain increases K_I to ~275 nM, confirming its central role in stabilizing the high-affinity complex [43,44].
- *The broad spectrum of TIMP-3 and its specificity for ADAM17.* TIMP-3 represents a unique member of the TIMP family, characterized by a markedly broadened inhibitory spectrum encompassing multiple MMPs (*e.g.*, MMP-1, MMP-2, MMP-3, MMP-9, MMP-14), as well as ADAM

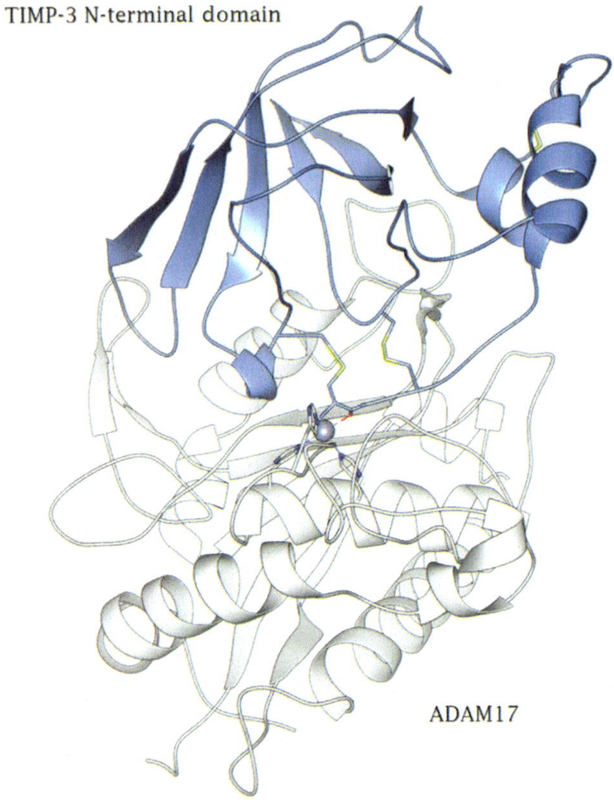

Fig. 4 Binding mode of the N-terminal domain of TIMP-3 with ADAM17. The N-terminal domain of TIMP-3 is shown in blue, while ADAM17 is rendered in white. The image was generated from the crystal structure of the TIMP-3/ADAM17 complex (PDBID: 3CKI) [46].

family members, most notably ADAM17 and ADAM10, and ADAMTS proteases such as ADAMTS-4 and ADAMTS-5, which are key mediators of aggrecan turnover. This expanded activity profile arises from a combination of distinctive structural determinants and its tight association with the ECM. Structurally, TIMP-3 exhibits unique features in surface-exposed loops, including a specialized interaction region located at the C-terminal end of strand sD, which stabilizes complexes with ADAM proteases (Fig. 4). In addition, residues corresponding to position 98 in TIMP-1 are intrinsically optimized in TIMP-3 to favor ADAM engagement, contributing to its elevated affinity for both ADAM and ADAMTS enzymes. A further defining characteristic is its strong ECM binding, mediated through interactions with glycosaminoglycans and proteoglycans. Unlike the soluble TIMP-1 and TIMP-2, TIMP-3 is sequestered within the extracellular matrix, resulting in several functional consequences: (i) increased local inhibitor concentration at sites of pericellular proteolysis, (ii) enhanced control of membrane-associated remodeling enzymes such as MMP-14 and ADAM17, and (iii) effective regulation of shedding events involving receptors and cytokines, as well as ECM degradation processes driven by ADAMTS proteases [1,24,45–48].

4. Non-canonical functions: TIMPs as signaling molecules

The structural architecture of TIMPs provides the foundation for a multifunctionality that extends the mere stoichiometric regulation of MMPs, positioning these proteins as pleiotropic adaptor molecules capable of orchestrating critical cellular signaling pathways through direct interactions with cell surface receptors and ECM components [49,50]. While the N-terminal domain is responsible for inhibiting MMP catalytic activity, the C-terminal domain and distinct surface regions mediate non-canonical functions that are independent of protease inhibition [51].

TIMP-1 is a multi-faceted protein as it additionally exhibits pronounced cytokine-like properties, promoting the growth and differentiation of multiple cell types, including hematopoietic progenitors and germinal center B cells [52]. These mitogenic and cytoprotective effects are mediated through interaction with the tetraspanin receptor CD63, in association with integrin $\beta1$, leading to activation of focal adhesion kinase (FAK), phosphatidylinositol 3-kinase (PI3K), and ERK/MAPK signaling pathways [53–56]. In the context of oncology, these properties confer a dual and paradoxical role to TIMP-1: although it suppresses tumor invasion through MMP inhibition, its ability to promote resistance to apoptosis, induced by chemotherapeutic agents or Fas ligand, supports tumor cell survival and metastatic progression [57–59]. In Burkitt lymphoma cells and B lymphocytes, this anti-apoptotic effect is mediated through modulation of Bcl-2 family members, including upregulation of the anti-apoptotic protein Bcl-xL [50,58,60]. Notably, these functions are independent of protease inhibition, as TIMP-1 variants lacking inhibitory activity, as well as reduced and alkylated forms, retain full growth-promoting capacity, whereas synthetic MMP inhibitors fail to recapitulate these effects [30,58,61,62].

Similarly, TIMP-2 exerts a biphasic control of cell proliferation and angiogenesis via direct interaction with integrin $\alpha3\beta1$ on endothelial cells

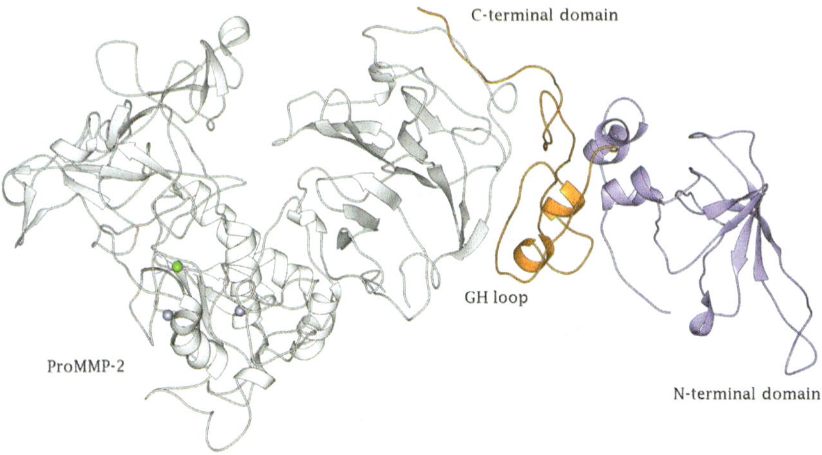

Fig. 5 Binding mode of TIMP-2 with proMMP-2. The N-terminal domain of TIMP-2 is shown in purple, while the C-terminal domain is shown in orange; proMMP-2 is rendered in white. The image was generated from the crystal structure of the TIMP-2/proMMP-2 complex (PDBID: 1GXD) [32].

[63]. This interaction induces SHP-1-mediated dephosphorylation of vascular endothelial growth factor receptor 2 (VEGFR-2), thereby suppressing endothelial proliferation and migration independently of protease inhibition [19,22,64]. In addition, TIMP-2 plays a unique structural and regulatory role in the activation of pro-MMP-2. Acting as a bridging molecule, its C-terminal domain binds the hemopexin domain of pro-MMP-2, specifically through the loop between β-strands sG and sH and the C-terminal tail (Fig. 5), while its N-terminal domain engages membrane-anchored MMP-14. This arrangement facilitates subsequent proteolytic activation by a second, MMP-14 molecule. Once assembled at the cell surface, this ternary complex functions as a receptor for pro-MMP-2, positioning its pro-domain for cleavage by the adjacent active MMP-14. The efficiency of this system is strictly concentration-dependent: at low concentrations, TIMP-2 promotes complex formation and activation, whereas at high concentrations it inhibits activation by saturating all available MMP-14 molecules, thereby preventing the participation of a second activating protease [30,65–73].

In contrast, TIMP-3 is distinguished by its insolubility and sequestration within the ECM through electrostatic interactions with sulfated glycosaminoglycans such as heparin and chondroitin sulfate, establishing it as a critical pericellular regulator [74–76]. Beyond being the only TIMP capable of efficiently inhibiting ADAM10, ADAM12, and ADAM17, TIMP-3 also functions as a direct antagonist of VEGFR-2 by blocking VEGF binding and suppressing angiogenesis [77]. This interaction is specifically mediated by its C-terminal domain. The functional importance of this region is underscored by mutations associated with Sorsby fundus dystrophy (SFD), which introduce additional cysteine residues into the C-terminus, thereby altering TIMP-3 mediated signaling without affecting its MMP inhibitory activity [78–80]. Furthermore, TIMP-3 exerts pro-apoptotic effects in various tumor and vascular smooth muscle cells by stabilizing death receptors at the cell surface, preventing their proteolytic shedding. By inhibiting ADAM17, TIMP-3 blocks the release of extracellular domains of key receptors, including TNFR1, Fas (CD95), and TRAIL-R1. This results in increased retention of functional receptors at the plasma membrane, thereby sensitizing cells to apoptotic signaling [81–83].

Collectively, TIMPs exert a profound influence on tissue differentiation and repair. In the central nervous system, TIMP-1 promotes oligodendrocyte differentiation and myelination, whereas TIMP-2 and TIMP-3 regulate neurogenesis and myogenesis through modulation of integrin $\beta 1$ expression and the availability of ECM-associated growth factors. This intricate network of receptor-mediated and physical interactions underscores that TIMPs function not merely as passive inhibitors of proteolysis, but as dynamic regulators that integrate environmental cues with cellular responses to maintain tissue homeostasis or, under pathological conditions, drive disease progression [30,49,51].

5. Physiopathological role of TIMPs

The functional integration of TIMPs within physiological systems extends far beyond their canonical role as stoichiometric inhibitors of MMPs. Rather, TIMPs function as multi-functional pleiotropic regulators at the interface of extracellular proteolysis, receptor-mediated signaling, and cell fate determination, orchestrating complex biological processes across development, tissue homeostasis, and disease progression [1,52]. A central emerging paradigm is that TIMPs act as regulators of ECM-dependent signaling platforms. Over its structural role, the ECM serves as a reservoir for growth factors and a determinant of tissue stiffness, thereby governing mechanotransduction pathways such as YAP/TAZ signaling. By modulating ECM turnover and composition, TIMPs indirectly regulate these signaling axes, influencing cell proliferation, differentiation, and migration [84].

5.1 Development and tissue homeostasis

During embryogenesis and morphogenesis, TIMPs regulate the balance between matrix remodeling and cellular quiescence. TIMP-3 has been implicated in the regulation of trophoblast invasion during blastocyst implantation through modulation of MMP-9 activity, thereby ensuring tightly controlled ECM turnover [85]. Consistent with this role, genetic studies in model organisms such as *Drosophila* and *Xenopus* demonstrate that loss of TIMP function results in profound developmental abnormalities, underscoring the essential requirement for tightly regulated pericellular proteolysis [86–88]. In bone physiology, TIMPs modulate osteoclast and osteoblast activity through coordinated regulation of multiple molecular targets involved in matrix turnover and cell signaling. Their inhibitory activity toward MMPs such as MMP-9, MMP-13, and MT1-MMP directly limits degradation of the organic bone matrix, thereby restraining osteoclast-mediated resorption. In parallel, TIMP-3–mediated inhibition of ADAM proteases, particularly ADAM17, reduces the shedding of pro-inflammatory and osteoclastogenic mediators, indirectly attenuating RANKL-dependent osteoclast differentiation. Beyond protease inhibition, TIMPs also influence osteoclast survival and function through modulation of integrin-mediated signaling pathways and by regulating the bioavailability of ECM-sequestered growth factors, including TGF-β and IGF-1. Collectively, these mechanisms highlight the role of TIMPs as integrated regulators of bone remodeling [86,89–91].

5.2 Cardiovascular and vascular biology

Cardiovascular homeostasis critically depends on TIMP-mediated regulation of ECM integrity and protease activity. In murine models, TIMP-1 deficiency leads to left ventricular dilation, impaired myocardial architecture, and increased susceptibility to aneurysm formation under hemodynamic stress [92–94]. Conversely, TIMP-3 knockout mouse (*Timp*3$^{-/-}$) results in progressive dilated cardiomyopathy characterized by cardiomyocyte hypertrophy and contractile dysfunction, driven in part by excessive ADAM17-mediated shedding of TNF-α and dysregulated TNF/TGF-β signaling [95,96]. In angiogenesis, TIMPs regulate endothelial cell behavior through coordinated modulation of both proteolytic and non-proteolytic targets. Their inhibitory activity toward MMPs, including MMP-2, MMP-9, and MT1-MMP, limits basement membrane degradation and the release of ECM-bound pro-angiogenic factors, thereby restraining endothelial cell migration and sprouting. In parallel, TIMP-3-mediated inhibition of ADAM proteases, particularly ADAM17 and ADAM10, reduces the shedding of EGFR ligands and inflammatory mediators, attenuating pro-angiogenic signaling cascades. Moreover, TIMP-2 exerts a direct signaling function through binding to integrin $\alpha3\beta1$ on endothelial cells, triggering SHP-1-dependent dephosphorylation of VEGFR-2 and suppression of VEGF-driven proliferation. TIMP-3 further reinforces this effect by directly antagonizing VEGFR-2 signaling. Additionally, by modulating ECM turnover and composition, TIMPs regulate the bioavailability of key growth factors such as VEGF and FGF-2, as well as the structural integrity of the basement membrane. Collectively, these mechanisms position TIMPs as multifaceted inhibitors of angiogenesis, integrating extracellular matrix remodeling with receptor-mediated signaling to tightly control vascular growth [76,97,98].

In vascular pathology, TIMPs also regulate the development and stability of atherosclerotic plaques. TIMP-3 has emerged as a key determinant of plaque stability owing to its dual ability to inhibit both MMPs and ADAM family sheddases. Loss or downregulation of TIMP-3 leads to increased proteolytic activity, including that of MMP-9, resulting in enhanced degradation of collagen and elastin within the fibrous cap and increased plaque vulnerability [99]. In parallel, unchecked ADAM17 activity amplifies inflammatory signaling through increased shedding of TNF-α and EGFR ligands. Indeed, experimental models demonstrate accelerated atherogenesis, increased macrophage infiltration, and necrotic core expansion in the absence of TIMP-3 [100–102].

5.3 Nervous system and aging

In the central nervous system, TIMPs regulate synaptic plasticity, neurogenesis, and neuronal survival. TIMP-1 is rapidly upregulated in response to neuronal activity and is required for proper synaptic remodeling and long-term memory formation [103,104]. In contrast, TIMP-2, whose circulating levels decline with aging, enhances cognitive function and synaptic plasticity, thereby linking TIMP biology to age-associated neurodegeneration [105,106]. Both TIMP-2 and TIMP-3 are enriched in neurogenic niches, where they regulate neural progenitor differentiation through modulation of integrin $\beta1$ signaling and ECM-associated growth factor availability. Under pathological conditions, TIMP functions are highly context-dependent: TIMP-1 generally exerts neuroprotective effects by attenuating glutamate-induced excitotoxicity, whereas TIMP-3 can promote apoptosis and blood-brain barrier disruption under ischemic conditions through enhanced death receptor signaling and ECM degradation [51,107–109].

5.4 Cancer

In cancer, TIMPs exhibit a well-recognized functional duality. While they suppress tumor invasion by inhibiting MMP-mediated ECM degradation, they also modulate tumor cell survival through non-proteolytic signaling pathways. A central mechanistic axis underlying these effects is the regulation of ADAM-dependent growth factor shedding and subsequent activation of receptor tyrosine kinase signaling, particularly the EGFR pathway. ADAM proteases, especially ADAM17 and ADAM10, mediate the proteolytic release of membrane-bound precursors of EGFR ligands, including TGF-α, HB-EGF, Amphiregulin, and Epiregulin. The release of these ligands enables autocrine and paracrine activation of EGFR, triggering downstream signaling cascades such as MAPK/ERK, PI3K/Akt, and STAT3, which collectively promote tumor cell proliferation, survival, and invasiveness. In this context, TIMP-3 functions as a critical tumor suppressor by potently inhibiting ADAM10 and ADAM17, thereby limiting EGFR ligand shedding and attenuating receptor activation. This effect is further complemented by its ability to stabilize death receptors at the cell surface, including TNFR1, Fas, and TRAIL-R1, thereby enhancing apoptotic sensitivity. Loss of TIMP-3 expression, frequently observed in cancer through promoter hypermethylation, results in unchecked ADAM activity, increased availability of EGFR ligands, and amplification of pro-tumorigenic signaling, often accompanied by enhanced inflammatory responses mediated by increased shedding of cytokines such as TNF-α [51,82,110–113]. In contrast, TIMP-1 exerts pro-survival and pro-tumorigenic effects through non-proteolytic mechanisms. By engaging the CD63/Integrin β1 complex, TIMP-1 activates intracellular signaling pathways including PI3K/Akt and MAPK, thereby promoting cell survival and conferring resistance to apoptosis induced by chemotherapeutic agents or death ligands. Elevated circulating levels of TIMP-1 correlate with poor prognosis across multiple malignancies and may contribute to pre-metastatic niche formation through modulation of stromal and growth factor signaling pathways [30,54,57,58,113,114]. Parallel to these effects, TIMP-mediated inhibition of MMPs, including MMP-2 and MMP-9, regulates tumor invasion and metastasis by controlling ECM degradation and the release of matrix-bound growth factors. However, in the absence of effective ADAM inhibition, suppression of MMP activity alone is insufficient to counterbalance EGFR-driven tumor progression. Collectively, these findings establish TIMPs as central regulators of the protease-growth factor signaling interface in cancer, with opposing functional outcomes determined by the balance between ADAM inhibition, MMP regulation, and receptor-mediated signaling [115–117].

5.5 Inflammation and immunity

TIMPs are key regulators of both innate and adaptive immune responses. By controlling MMP and ADAM activity, they modulate the bioavailability of cytokines, chemokines, and their receptors, thereby shaping inflammatory signaling cascades. TIMP-3 functions as a critical checkpoint of inflammatory homeostasis by limiting ADAM17-mediated shedding of TNF-α, IL-6 receptor, and other mediators. In parallel, TIMP-1 influences leukocyte survival and activation through CD63-dependent signaling, while TIMPs collectively regulate macrophage polarization and leukocyte trafficking by controlling ECM degradation and chemokine gradients [30,54,118,119].

5.6 Metabolism

Beyond their roles in tissue remodeling and inflammation, TIMPs also regulate metabolic homeostasis. TIMP-1 is strongly associated with obesity and insulin resistance, where it promotes ECM accumulation, adipose tissue fibrosis, and chronic low-grade inflammation [120]. In contrast, TIMP-3 exerts protective metabolic effects by limiting ADAM17-mediated inflammatory signaling, thereby improving insulin sensitivity and restraining metabolic dysfunction [121].

5.7 Musculoskeletal and fibrotic diseases

In musculoskeletal disorders such as osteoarthritis, TIMP-3 plays a central protective role by inhibiting ADAMTS-4 and ADAMTS-5, which are responsible for cartilage degradation [122]. In fibrotic diseases affecting the liver, lung, and kidney, sustained upregulation of TIMP-1 contributes to pathological ECM accumulation by inhibiting collagen degradation and promoting myofibroblast survival through CD63-dependent signaling and TGF-β pathway activation [123].

5.8 TIMP-4 in diseases

Although comparatively less studied, TIMP-4 represents a functionally distinct member of the TIMP family, characterized by a more restricted tissue distribution and a less pleiotropic profile. Predominantly expressed in cardiovascular and muscular tissues, TIMP-4 contributes to local ECM homeostasis and remodeling, particularly in response to injury. It has been implicated in post-infarction cardiac remodeling and is uniquely enriched in platelets, where it may regulate aggregation and thrombus formation through interactions with platelet-derived MMPs. Its more specialized functional profile and limited involvement in non-canonical signaling likely account for its comparatively reduced representation in the literature [124,125]. There are a few reasons why we TIMP-4 remains understudied. It has restricted tissue expression as compared to TIMP-1, −2 and −3. It contains unique structural interactions that have not been fully characterized yet. Our current understanding is that it has ambiguous disease phenotypes. Unlike TIMP-3, for example, that is directly linked to human genetic disorders like Sorby's fundus dystrophy, there are no identified human diseases identified caused by TIMP-4 mutations [126]. Additionally, *Timp4*$^{-/-}$ mice did not yield dramatic, obvious or specific phenotype. Therefore, it is likely that TIMP-4 regulates specific biological processes in selective contexts that have yet to be further clarified [127,128].

6. TIMP-centered therapeutic strategies

The therapeutic exploitation of TIMPs has progressively shifted from indirect modulation of metalloproteinase activity toward direct targeting, engineering, and pharmacological modulation of TIMPs themselves. This paradigm reflects the recognition that TIMPs are not merely endogenous inhibitors, but multifunctional proteins whose structural plasticity and context-dependent signaling properties can be harnessed for therapeutic intervention.

6.1 Engineered TIMPs as therapeutic biologics

Rational engineering of TIMPs has enabled the generation of variants with tailored selectivity profiles and controlled functional outputs, with the goal of overcoming the limitations of endogenous TIMPs, such as broad activity spectra or undesired signaling functions [1,30,129]. A key strategy involves reprogramming protease selectivity through targeted mutations within the N-terminal inhibitory domain and surface-exposed loops, often guided by synergistic integration of computational modeling and yeast surface display (YSD) [130,131]. For example, substitution at position 98 in TIMP-1

(Thr98Leu) significantly enhances its ability to inhibit membrane-associated MMPs and ADAM proteases, thereby extending its therapeutic potential toward inflammatory and oncogenic contexts where these enzymes are dysregulated [24,37,40]. Beyond the N-terminus, recent evidence from directed evolution of full-length TIMP-1 reveals that the C-terminal domain acts cooperatively with the N-terminal domain to stabilize binding. Mutations such as Gly154Ala work together with N-terminal substitutions to form a “tyrosine clasp” at the enzyme-inhibitor interface [131,132]. This motif, defined in the TIMP-1–MMP-3 complex, arises from reciprocal interactions between Tyr35 of TIMP-1 (AB loop) and Tyr153 of the enzyme. In the wild-type protein, steric hindrance from Leu34 limits optimal hydrogen bonding, whereas the Leu34Gly mutation enables backbone rearrangement that shortens the Tyr-Tyr distance and tightens the interface. The resulting complex is stabilized by hydrogen bonds and edge-to-face π–π interactions and further reinforced by synergistic effects originating in the C-terminal domain, where the Gly154Ala substitution promotes the formation of a short α-helical segment spanning residues 154–157. This structural transition facilitates a more compact interdomain packing arrangement, introducing additional hydrogen bonds and hydrophobic contacts that collectively enhance binding affinity toward the target protease [131,132]. Similarly, identifying specificity signatures’ at positions 4, 38, and 71 has allowed N-TIMP-2 to achieve a 1000-fold preference for MMP-9 over MMP-14, providing precision tools for oncogenic contexts. Another design principle is the functional decoupling of inhibitory and signaling activities. N-terminally modified constructs, such as Ala+TIMP-2, exhibit a drastic reduction in MMP inhibition by disrupting zinc coordination, while retaining receptor-mediated signaling and pro-MMP complex formation [63,133,134]. This opens the possibility of developing TIMP-based biologics that selectively exploit either protease inhibition or signaling modulation, depending on the therapeutic need.

6.2 Pharmacological modulation of endogenous TIMPs

An alternative strategy involves modulating the expression or activity of endogenous TIMPs to restore proteolytic balance in pathological conditions.

6.2.1 Upregulation of TIMPs

Pharmacological induction of TIMP expression has been explored in inflammatory diseases. For instance, treatment with Infliximab and other anti-TNF agents has been shown to increase TIMP-1 levels in intestinal fibroblasts from patients with Crohn's disease, contributing to ECM remodeling, mucosal repair processes, and myofibroblast migration. Notably, recent structural and biochemical evidence reveals that Infliximab triggers a dose-dependent induction of TIMP-1 via a p38 MAPK-dependent signaling pathway, suggesting a specific transcriptional control activated by binding to transmembrane TNF. At the tissue level, effective anti-inflammatory therapy results in the restoration of the MMP/TIMP ratio, with TIMP-1 and TIMP-2 levels normalizing toward control values in clinical responders [29,135,136]. Furthermore, natural compounds like resveratrol act as allosteric activators of SIRT1, which in turn upregulate TIMP-3 expression. This SIRT1/TIMP-3 axis is critical for inhibiting TACE activity, thereby preventing the release of bioactive soluble TNF-α and effectively ameliorating experimental colitis [29,130,137,138].

6.2.2 Epigenetic reactivation

In cancer, TIMP-3 is frequently silenced through promoter hypermethylation, contributing to enhanced proteolytic and inflammatory signaling. To counteract the gene silencing observed in cancer, epigenetic and RNA-based strategies are currently being studied, aimed at "reactivating" the expression of endogenous TIMPs, particularly TIMP-3, by resetting the promoter methylation status or using shRNA (short hairpin RNA) to inhibit microRNAs (such as miR-221 or miR-222) that normally suppress their translation [30,139].

6.2.3 TIMP inhibition

In contrast to inflammatory settings, fibrotic diseases are often characterized by excessive TIMP activity, leading to impaired extracellular matrix turnover. In these contexts, the strategy is reversed through the use of neutralizing antibodies against TIMP-1 to reactivate matrix degradation, or the use of antifibrotic agents such as pirfenidone, which inhibits TIMP-1 and TGF-β signaling to reduce fibroblast proliferation [39,130,140,141].

6.3 TIMPs as biomarkers

In addition to their therapeutic potential, TIMPs have emerged as clinically relevant biomarkers across multiple disease areas. Elevated circulating levels of TIMP-1 are consistently associated with poor prognosis in several malignancies, reflecting its role in promoting cell survival and therapy resistance [142,143]. Moreover, ratios such as MMP-9/TIMP-1 provide sensitive indicators of disease activity in inflammatory conditions, sepsis, and cardiovascular disorders [144]. TIMP-3 expression levels are also of prognostic value, particularly in cancer, where their loss correlates with increased tumor aggressiveness and poor clinical outcome [30].

7. Conclusion

7.1 The imperative for TIMP research in understanding tissue homeostasis and pathology

TIMPs represent a fundamental biological system governing the fidelity of ECM remodeling and tissue preservation. Beyond their classical role as endogenous protease inhibitors, TIMPs function as critical regulators of cellular signaling and tissue integrity, acting as molecular sentinels that determine whether tissues progress toward regeneration and health or degenerate into chronic inflammation, fibrosis, cancer malignancy, and organ failure. The evolutionary conservation of TIMP genes across the animal kingdom—from bacteria to humans—coupled with the emergence of four structurally distinct isoforms in humans, reflects the importance of this regulatory system to multicellular life. Each isoform exhibits unique biochemical properties and tissue-specific localization patterns, suggesting specialized roles that have been refined through millions of years of evolution.

7.2 The knowledge gap and clinical significance

A paradox underscores the urgency of TIMP research: despite half a century of investigation, the precise molecular mechanisms by which TIMP dysregulation precipitates disease progression remain incompletely understood. This knowledge gap has profound implications for human health. The disruption of the MMP-TIMP equilibrium, arising through excessive protease activity, inadequate inhibitor expression, or compromised inhibitor function, can lead to uncontrolled ECM degradation, aberrant inflammatory responses, and pathological cell migration. These dysregulatory events converge in some of the most prevalent and therapeutically refractory diseases affecting modern populations: fibrosis, cancer, osteoarthritis, cardiovascular disease, neurodegenerative conditions and others. Emerging evidence further expands the biological significance of TIMPs. Beyond their well-characterized metalloproteinase inhibitory functions, TIMPs modulate cellular signaling pathways, regulate immune cell infiltration, and influence extracellular vesicle biogenesis. This expanded mechanistic understanding reveals that TIMP biology is substantially more complex than previously appreciated, and that manipulating this system may yield multifaceted therapeutic benefits.

7.3 Future research directions and therapeutic opportunities

Future TIMP research must pursue several interconnected objectives. First, structural and biochemical investigations employing state-of-the-art techniques, including cryo-electron microscopy, hydrogen-deuterium exchange mass spectrometry, and single-molecule imaging, should elucidate the molecular dynamics of TIMP-protease interactions and their dysregulation in specific disease contexts. Second, mechanistic studies must clarify how individual TIMP isoforms contribute to fibrotic pathways in tissue-specific contexts: TIMP-1 and TIMP-2 in hepatic fibrosis, TIMP-3 in pulmonary fibrosis, and TIMP-4 in cardiac remodeling. Third, therapeutic development should leverage the isoform-specific properties and tissue distributions of individual TIMPs rather than pursuing broad-spectrum approaches. The C-terminal ECM-binding domain of TIMP-3, for instance, positions this isoform as uniquely suited for localized interventions, while TIMP-4's preferential expression in cardiovascular tissues offers opportunities for cardioprotective strategies. Fourth, integration of TIMP research with multi-omics platforms, including transcriptomics, proteomics, and N-terminomics, may reveal dysregulatory signatures predictive of disease progression and amenable to personalized intervention [22,145,146]. Finally, translational research must transition from fundamental mechanistic understanding to clinical application through biomarker discovery, animal model validation, and ultimately human clinical trials testing TIMP-based interventions. As the molecular basis of TIMP dysfunction becomes increasingly well-defined, TIMPs could emerge as a central therapeutic target in many diseases.

References

[1] K. Brew, H. Nagase, The tissue inhibitors of metalloproteinases (TIMPs): an ancient family with structural and functional diversity, Biochim. Biophys. Acta 1803 (1) (2010) 55–71, doi:10.1016/j.bbamcr.2010.01.003.
[2] G. Murphy, H. Nagase, Progress in matrix metalloproteinase research, Mol. Aspects Med. 29 (5) (2008) 290–308, doi:10.1016/j.mam.2008.05.002.
[3] S. Porter, I.M. Clark, L. Kevorkian, D.R. Edwards, The ADAMTS metalloproteinases, Biochem. J. 386 (Pt 1) (2005) 15–27, doi:10.1042/BJ20040424.
[4] E.A. Bauer, G.P. Stricklin, J.J. Jeffrey, A.Z. Eisen, Collagenase production by human skin fibroblasts, Biochem. Biophys. Res. Commun. 64 (1) (1975) 232–240, doi:10.1016/0006-291x(75)90243-0.

[5] D.E. Woolley, D.R. Roberts, J.M. Evanson, Inhibition of human collagenase activity by a small molecular weight serum protein, Biochem. Biophys. Res. Commun. 66 (2) (1975) 747–754.
[6] K.E. Kuettner, J. Hiti, R. Eisenstein, E. Harper, Collagenase inhibition by cationic proteins derived from cartilage and aorta, Biochem. Biophys. Res. Commun. 72 (1) (1976) 40–46, doi:10.1016/0006-291x(76)90957-8.
[7] C.A. Vater, C.L. Mainardi, E.D. Harris Jr., Inhibitor of human collagenase from cultures of human tendon, J. Biol. Chem. 254 (8) (1979) 3045–3053.
[8] H.G. Welgus, G.P. Stricklin, A.Z. Eisen, E.A. Bauer, R.V. Cooney, J.J. Jeffrey, A specific inhibitor of vertebrate collagenase produced by human skin fibroblasts, J. Biol. Chem. 254 (6) (1979) 1938–1943.
[9] A.J. Docherty, A. Lyons, B.J. Smith, E.M. Wright, P.E. Stephens, T.J. Harris, G. Murphy, J.J. Reynolds, Sequence of human tissue inhibitor of metalloproteinases and its identity to erythroid-potentiating activity, Nature 318 (6041) (1985) 66–69, doi:10.1038/318066a0.
[10] E. Niskanen, J.C. Gasson, C.D. Teates, D.W. Golde, In vivo effect of human erythroid-potentiating activity on hematopoiesis in mice, Blood 72 (2) (1988) 806–810.
[11] T.C. Boone, M.J. Johnson, Y.A. De Clerck, K.E. Langley, cDNA cloning and expression of a metalloproteinase inhibitor related to tissue inhibitor of metalloproteinases, Proc. Natl. Acad. Sci. U.S.A. 87 (7) (1990) 2800–2804, doi:10.1073/pnas.87.7.2800.
[12] W.G. Stetler-Stevenson, P.D. Brown, M. Onisto, A.T. Levy, L.A. Liotta, Tissue inhibitor of metalloproteinases-2 (TIMP-2) mRNA expression in tumor cell lines and human tumor tissues, J. Biol. Chem. 265 (23) (1990) 13933–13938.
[13] S.S. Apte, B.R. Olsen, G. Murphy, The gene structure of tissue inhibitor of metalloproteinases (TIMP)-3 and its inhibitory activities define the distinct TIMP gene family, J. Biol. Chem. 270 (24) (1995) 14313–14318 Erratum in: J Biol Chem. 1996 Feb 2;271(5):2874, doi:10.1074/jbc.270.24.14313.
[14] J. Greene, M. Wang, Y.E. Liu, L.A. Raymond, C. Rosen, Y.E. Shi, Molecular cloning and characterization of human tissue inhibitor of metalloproteinase 4, J. Biol. Chem. 271 (48) (1996) 30375–30380, doi:10.1074/jbc.271.48.30375.
[15] J. Stetefeld, M. Jenny, T. Schulthess, R. Landwehr, B. Schumacher, S. Frank, M.A. Rüegg, J. Engel, R.A. Kammerer, The laminin-binding domain of agrin is structurally related to N-TIMP-1, Nat. Struct. Biol. 8 (8) (2001) 705–709, doi:10.1038/90422.
[16] N. Pohar, T.A. Godenschwege, E. Buchner, Invertebrate tissue inhibitor of metalloproteinase: structure and nested gene organization within the synapsin locus is conserved from Drosophila to human, Genomics 57 (2) (1999) 293–296, doi:10.1006/geno.
[17] L.G.N. de Almeida, H. Thode, Y. Eslambolchi, S. Chopra, D. Young, S. Gill, L. Devel, A. Dufour, Matrix metalloproteinases: from molecular mechanisms to physiology, pathophysiology, and pharmacology, Pharmacol. Rev. 74 (3) (2022) 712–768, doi:10.1124/pharmrev.121.000349.
[18] R. Khokha, A. Murthy, A. Weiss, Metalloproteinases and their natural inhibitors in inflammation and immunity, Nat. Rev. Immunol. 13 (9) (2013) 649–665, doi:10.1038/nri3499.
[19] M.G. Masciantonio, C.K.S. Lee, V. Arpino, S. Mehta, S.E. Gill, The balance between metalloproteinases and TIMPs: critical regulator of microvascular endothelial cell function in health and disease, Prog. Mol. Biol. Transl. Sci. 147 (2017) 101–131.
[20] K. Brew, D. Dinakarpandian, H. Nagase, Tissue inhibitors of metalloproteinases: evolution, structure and function, Biochim. Biophys. Acta 1477 (1-2) (2000) 267–283, doi:10.1016/s0167-4838(99)00279-4.
[21] F.X. Gomis-Rüth, K. Maskos, M. Betz, A. Bergner, R. Huber, K. Suzuki, N. Yoshida, H. Nagase, K. Brew, G.P. Bourenkov, H. Bartunik, W. Bode, Mechanism of inhibition

of the human matrix metalloproteinase stromelysin-1 by TIMP-1, Nature 389 (6646) (1997) 77–81, doi:10.1038/37995.

[22] R.A. Williamson, G. Martorell, M.D. Carr, G. Murphy, A.J. Docherty, R.B. Freedman, J. Feeney, Solution structure of the active domain of tissue inhibitor of metalloproteinases-2. A new member of the OB fold protein family, Biochemistry 33 (39) (1994) 11745–11759, doi:10.1021/bi00205a010.

[23] C. Fernandez-Catalan, W. Bode, R. Huber, D. Turk, J.J. Calvete, A. Lichte, H. Tschesche, K. Maskos, Crystal structure of the complex formed by the membrane type 1-matrix metalloproteinase with the tissue inhibitor of metalloproteinases-2, the soluble progelatinase A receptor, EMBO J. 17 (17) (1998) 5238–5248, doi:10.1093/emboj/17.17.5238.

[24] H. Nagase, R. Visse, G. Murphy, Structure and function of matrix metalloproteinases and TIMPs, Cardiovasc. Res. 69 (3) (2006) 562–573, doi:10.1016/j.cardiores.2005.12.002.

[25] W. Bode, C. Fernandez-Catalan, H. Tschesche, F. Grams, H. Nagase, K. Maskos, Structural properties of matrix metalloproteinases, Cell. Mol. Life Sci. 55 (4) (1999) 639–652, doi:10.1007/s000180050320.

[26] M. Kashiwagi, M. Tortorella, H. Nagase, K. Brew, TIMP-3 is a potent inhibitor of aggrecanase 1 (ADAM-TS4) and aggrecanase 2 (ADAM-TS5), J. Biol. Chem. 276 (16) (2001) 12501–12504, doi:10.1074/jbc.C000848200.

[27] N.H. Lim, M. Kashiwagi, R. Visse, J. Jones, J.J. Enghild, K. Brew, H. Nagase, Reactive-site mutants of N-TIMP-3 that selectively inhibit ADAMTS-4 and ADAMTS-5: biological and structural implications, Biochem. J. 431 (1) (2010) 113–122, doi:10.1042/BJ20100725.

[28] H. Nagase, K. Brew, Engineering of tissue inhibitor of metalloproteinases mutants as potential therapeutics, Arthritis Res. 4 (Suppl 3) (2002) S51–S61, doi:10.1186/ar573.

[29] M. de Bruyn, J. Vandooren, E. Ugarte-Berzal, I. Arijs, S. Vermeire, G. Opdenakker, The molecular biology of matrix metalloproteinases and tissue inhibitors of metalloproteinases in inflammatory bowel diseases, Crit. Rev. Biochem. Mol. Biol. 51 (5) (2016) 295–358, doi:10.1080/10409238.2016.1199535.

[30] H.W. Jackson, V. Defamie, P. Waterhouse, R. Khokha, TIMPs: versatile extracellular regulators in cancer, Nat. Rev. Cancer 17 (1) (2017) 38–53, doi:10.1038/nrc.2016.115.

[31] V. Arpino, M. Brock, S.E. Gill, The role of TIMPs in regulation of extracellular matrix proteolysis, Matrix Biol. 44–46 (2015) 247–254, doi:10.1016/j.matbio.2015.03.005.

[32] E. Morgunova, A. Tuuttila, U. Bergmann, K. Tryggvason, Structural insight into the complex formation of latent matrix metalloproteinase 2 with tissue inhibitor of metalloproteinase 2, Proc. Natl. Acad. Sci. U.S.A. 99 (11) (2002) 7414–7419, doi:10.1073/pnas.102185399.

[33] A. Amour, P.M. Slocombe, A. Webster, M. Butler, C.G. Knight, B.J. Smith, P.E. Stephens, C. Shelley, M. Hutton, V. Knäuper, A.J. Docherty, G. Murphy, TNF-alpha converting enzyme (TACE) is inhibited by TIMP-3, FEBS Lett. 435 (1) (1998) 39–44, doi:10.1016/s0014-5793(98)01031-x.

[34] A.B. Hamze, S. Wei, H. Bahudhanapati, S. Kota, K.R. Acharya, K. Brew, Constraining specificity in the N-domain of tissue inhibitor of metalloproteinases-1; gelatinase-selective inhibitors, Protein Sci. 16 (9) (2007) 1905–1913, doi:10.1110/ps.072978507.

[35] A. Amour, C.G. Knight, A. Webster, P.M. Slocombe, P.E. Stephens, V. Knäuper, A.J. Docherty, G. Murphy, The in vitro activity of ADAM-10 is inhibited by TIMP-1 and TIMP-3, FEBS Lett. 473 (3) (2000) 275–279, doi:10.1016/s0014-5793(00)01528-3.

[36] G. Hashimoto, T. Aoki, H. Nakamura, K. Tanzawa, Y. Okada, Inhibition of ADAMTS4 (aggrecanase-1) by tissue inhibitors of metalloproteinases (TIMP-1, 2, 3 and 4), FEBS Lett. 494 (3) (2001) 192–195, doi:10.1016/s0014-5793(01)02323-7.

[37] Q. Meng, V. Malinovskii, W. Huang, Y. Hu, L. Chung, H. Nagase, W. Bode, K. Maskos, K. Brew, Residue 2 of TIMP-1 is a major determinant of affinity and specificity for matrix metalloproteinases but effects of substitutions do not correlate with those of the corresponding P1′ residue of substrate, J. Biol. Chem. 274 (15) (1999) 10184–10189, doi:10.1074/jbc.274.15.10184.
[38] S. Wei, Y. Chen, L. Chung, H. Nagase, K. Brew, Protein engineering of the tissue inhibitor of metalloproteinase 1 (TIMP-1) inhibitory domain. In search of selective matrix metalloproteinase inhibitors, J. Biol. Chem. 278 (11) (2003) 9831–9834, doi:10.1074/jbc.M211793200.
[39] J. Batra, A.S. Soares, C. Mehner, E.S. Radisky, Matrix metalloproteinase-10/TIMP-2 structure and analyses define conserved core interactions and diverse exosite interactions in MMP/TIMP complexes, PLoS One 8 (9) (2013) e75836, doi:10.1371/journal.pone.0075836.
[40] M.H. Lee, M. Rapti, V. Knaüper, G. Murphy, Threonine 98, the pivotal residue of tissue inhibitor of metalloproteinases (TIMP)-1 in metalloproteinase recognition, J. Biol. Chem. 279 (17) (2004) 17562–17569, doi:10.1074/jbc.M312589200.
[41] M.H. Lee, M. Rapti, G. Murphy, Unveiling the surface epitopes that render tissue inhibitor of metalloproteinase-1 inactive against membrane type 1-matrix metalloproteinase, J. Biol. Chem. 278 (41) (2003) 40224–40230, doi:10.1074/jbc.M305678200.
[42] R.A. Williamson, M. Hutton, G. Vogt, M. Rapti, V. Knäuper, M.D. Carr, G. Murphy, Tyrosine 36 plays a critical role in the interaction of the AB loop of tissue inhibitor of metalloproteinases-2 with matrix metalloproteinase-14, J. Biol. Chem. 276 (35) (2001) 32966–32970, doi:10.1074/jbc.M101843200.
[43] M. Hutton, F. Willenbrock, K. Brocklehurst, G. Murphy, Kinetic analysis of the mechanism of interaction of full-length TIMP-2 and gelatinase A: evidence for the existence of a low-affinity intermediate, Biochemistry 37 (28) (1998) 10094–10098, doi:10.1021/bi980616p.
[44] M.W. Olson, D.C. Gervasi, S. Mobashery, R. Fridman, Kinetic analysis of the binding of human matrix metalloproteinase-2 and -9 to tissue inhibitor of metalloproteinase (TIMP)-1 and TIMP-2, J. Biol. Chem. 272 (47) (1997) 29975–29983, doi:10.1074/jbc.272.47.29975.
[45] M. Rapti, S.J. Atkinson, M.H. Lee, A. Trim, M. Moss, G. Murphy, The isolated N-terminal domains of TIMP-1 and TIMP-3 are insufficient for ADAM10 inhibition, Biochem. J. 411 (2) (2008) 433–439, doi:10.1042/BJ20071430.
[46] M. Wisniewska, P. Goettig, K. Maskos, E. Belouski, D. Winters, R. Hecht, R. Black, W. Bode, Structural determinants of the ADAM inhibition by TIMP-3: crystal structure of the TACE-N-TIMP-3 complex, J. Mol. Biol. 381 (5) (2008) 1307–1319, doi:10.1016/j.jmb.2008.06.088.
[47] L. Troeberg, K. Fushimi, S.D. Scilabra, H. Nakamura, V. Dive, I.B. Thøgersen, J.J. Enghild, H. Nagase, The C-terminal domains of ADAMTS-4 and ADAMTS-5 promote association with N-TIMP-3, Matrix Biol. 28 (8) (2009) 463–469, doi:10.1016/j.matbio.2009.07.005.
[48] S. Wei, M. Kashiwagi, S. Kota, Z. Xie, H. Nagase, K. Brew, Reactive site mutations in tissue inhibitor of domains of ADAMTS-4 and ADAMTS-5 promote association tumor necrosis factor-alpha-converting enzyme, J. Biol. Chem. 280 (38) (2005) 32877–32882, doi:10.1074/jbc.C500220200.
[49] C.S. Moore, S.J. Crocker, An alternate perspective on the roles of TIMPs and MMPs in pathology, Am. J. Pathol. 180 (1) (2012) 12–16, doi:10.1016/j.ajpath.2011.09.008.
[50] E. Lambert, E. Dassé, B. Haye, E. Petitfrère, TIMPs as multifacial proteins, Crit. Rev. Oncol. Hematol. 49 (3) (2004) 187–198, doi:10.1016/j.critrevonc.2003.09.008.

[51] S.J. Crocker, A. Pagenstecher, I.L. Campbell, The TIMPs tango with MMPs and more in the central nervous system, J. Neurosci. Res. 75 (1) (2004) 1–11, doi:10.1002/jnr.10836.IL.
[52] W.G. Stetler-Stevenson, Tissue inhibitors of metalloproteinases in cell signaling: metalloproteinase-independent biological activities, Sci. Signal. 1 (27) (2008) re6, doi:10.1126/scisignal.127re6.
[53] X.W. Liu, M.M. Bernardo, R. Fridman, H.R. Kim, Tissue inhibitor of metalloproteinase-1 protects human breast epithelial cells against intrinsic apoptotic cell death via the focal adhesion kinase/phosphatidylinositol 3-kinase and MAPK signaling pathway, J. Biol. Chem. 278 (41) (2003) 40364–40372, doi:10.1074/jbc.M302999200.
[54] K.K. Jung, X.W. Liu, R. Chirco, R. Fridman, H.R. Kim, Identification of CD63 as a tissue inhibitor of metalloproteinase-1 interacting cell surface protein, EMBO J. 25 (17) (2006) 3934–3942, doi:10.1038/sj.emboj.7601281.
[55] M. Toricelli, F.H. Melo, G.B. Peres, D.C. Silva, M.G. Jasiulionis, Timp1 interacts with beta-1 integrin and CD63 along melanoma genesis and confers anoikis resistance by activating PI3-K signaling pathway independently of Akt phosphorylation, Mol. Cancer 12 (2013) 22 Erratum in: Mol Cancer. 2015 Aug 22;14:161., doi:10.1186/1476-4598-12-22.
[56] B. Grünwald, V. Harant, S. Schaten, M. Frühschütz, R. Spallek, B. Höchst, K. Stutzer, S. Berchtold, M. Erkan, O. Prokopchuk, M. Martignoni, I. Esposito, M. Heikenwalder, A. Gupta, J. Siveke, P. Saftig, P. Knolle, D. Wohlleber, A. Krüger, Pancreatic premalignant lesions secrete tissue inhibitor of metalloproteinases-1, which activates hepatic stellate cells via CD63 signaling to create a premetastatic niche in the liver, Gastroenterology 151 (5) (2016) e7, doi:10.1053/j.gastro.2016.07.043.
[57] T. Wang, J.H. Lv, X.F. Zhang, C.J. Li, X. Han, Y.J. Sun, Tissue inhibitor of metalloproteinase-1 protects MCF-7 breast cancer cells from paclitaxel-induced apoptosis by decreasing the stability of cyclin B1, Int. J. Cancer 126 (2) (2010) 362–370, doi:10.1002/ijc.24753.
[58] L. Guedez, W.G. Stetler-Stevenson, L. Wolff, J. Wang, P. Fukushima, A. Mansoor, M. Stetler-Stevenson, In vitro suppression of programmed cell death of B cells by tissue inhibitor of metalloproteinases-1, J. Clin. Invest. 102 (11) (1998) 2002–2010, doi:10.1172/JCI2881.
[59] I. Tsagaraki, E.C. Tsilibary, A.K. Tzinia, TIMP-1 interaction with $\alpha v\beta 3$ integrin confers resistance to human osteosarcoma cell line MG-63 against TNF-α-induced apoptosis, Cell Tissue Res. 342 (1) (2010) 87–96, doi:10.1007/s00441-010-1025-1.
[60] L. Guedez, A. Martinez, S. Zhao, A. Vivero, S. Pittaluga, M. Stetler-Stevenson, M. Raffeld, W.G. Stetler-Stevenson, Tissue inhibitor of metalloproteinase 1 (TIMP-1) promotes plasmablastic differentiation of a Burkitt lymphoma cell line: implications in the pathogenesis of plasmacytic/plasmablastic tumors, Blood 105 (4) (2005) 1660–1668, doi:10.1182/blood-2004-04-1385.
[61] T. Hayakawa, K. Yamashita, E. Ohuchi, A. Shinagawa, Cell growth-promoting activity of tissue inhibitor of metalloproteinases-2 (TIMP-2), J. Cell Sci. 107 (Pt 9) (1994) 2373–2379, doi:10.1242/jcs.107.9.2373.
[62] L. Chesler, D.W. Golde, N. Bersch, M.D. Johnson, Metalloproteinase inhibition and erythroid potentiation are independent activities of tissue inhibitor of metalloproteinases-1, Blood 86 (12) (1995) 4506–4515.
[63] D.W. Seo, H. Li, L. Guedez, P.T. Wingfield, T. Diaz, R. Salloum, B.Y. Wei, W.G. Stetler-Stevenson, TIMP-2 mediated inhibition of angiogenesis: an MMP-independent mechanism, Cell 114 (2) (2003) 171–180.
[64] S.J. Lee, P.S. Tsang, T.M. Diaz, B.Y. Wei, W.G. Stetler-Stevenson, TIMP-2 modulates VEGFR-2 phosphorylation and enhances phosphodiesterase activity in endothelial cells, Lab. Invest. 90 (3) (2010) 374–382, doi:10.1038/labinvest.2009.136.

[65] S. Hernandez-Barrantes, M. Toth, M.M. Bernardo, M. Yurkova, D.C. Gervasi, Y. Raz, Q.A. Sang, R. Fridman, Binding of active (57 kDa) membrane type 1-matrix metalloproteinase (MT1-MMP) to tissue inhibitor of metalloproteinase (TIMP)-2 regulates MT1-MMP processing and pro-MMP-2 activation, J. Biol. Chem. 275 (16) (2000) 12080–12089, doi:10.1074/jbc.275.16.12080.
[66] Y. Itoh, M. Seiki, MT1-MMP: a potent modifier of pericellular microenvironment, J. Cell. Physiol. 206 (1) (2006) 1–8, doi:10.1002/jcp.20431.
[67] A.Y. Strongin, I. Collier, G. Bannikov, B.L. Marmer, G.A. Grant, G.I. Goldberg, Mechanism of cell surface activation of 72-kDa type IV collagenase. Isolation of the activated form of the membrane metalloprotease, J. Biol. Chem. 270 (10) (1995) 5331–5338, doi:10.1074/jbc.270.10.5331.
[68] Y. Itoh, A. Takamura, N. Ito, Y. Maru, H. Sato, N. Suenaga, T. Aoki, M. Seiki, Homophilic complex formation of MT1-MMP facilitates proMMP-2 activation on the cell surface and promotes tumor cell invasion, EMBO J. 20 (17) (2001) 4782–4793, doi:10.1093/emboj/20.17.4782.
[69] S.J. Atkinson, T. Crabbe, S. Cowell, R.V. Ward, M.J. Butler, H. Sato, M. Seiki, J.J. Reynolds, G. Murphy, Intermolecular autolytic cleavage can contribute to the activation of progelatinase A by cell membranes, J. Biol. Chem. 270 (51) (1995) 30479–30485, doi:10.1074/jbc.270.51.30479.
[70] J.R. Worley, P.B. Thompkins, M.H. Lee, M. Hutton, P. Soloway, D.R. Edwards, G. Murphy, V. Knäuper, Sequence motifs of tissue inhibitor of metalloproteinases 2 (TIMP-2) determining progelatinase A (proMMP-2) binding and activation by membrane-type metalloproteinase 1 (MT1-MMP), Biochem. J. 372 (Pt 3) (2003) 799–809, doi:10.1042/BJ20021573.
[71] H.S. Kai, G.S. Butler, C.J. Morrison, A.E. King, G.R. Pelman, C.M. Overall, Utilization of a novel recombinant myoglobin fusion protein expression system to characterize the tissue inhibitor of metalloproteinase (TIMP)-4 and TIMP-2 C-terminal domain and tails by mutagenesis. The importance of acidic residues in binding the MMP-2 hemopexin C-domain, J. Biol. Chem. 277 (50) (2002) 48696–48707, doi:10.1074/jbc.M209177200.
[72] G.S. Butler, M.J. Butler, S.J. Atkinson, H. Will, T. Tamura, S. Schade van Westrum, T. Crabbe, J. Clements, M.P. d'Ortho, G. Murphy, The TIMP2 membrane type 1 metalloproteinase "receptor" regulates the concentration and efficient activation of progelatinase A. A kinetic study, J. Biol. Chem. 273 (2) (1998) 871–880, doi:10.1074/jbc.273.2.871.
[73] R.V. Ward, S.J. Atkinson, J.J. Reynolds, G. Murphy, Cell surface-mediated activation of progelatinase A: demonstration of the involvement of the C-terminal domain of progelatinase A in cell surface binding and activation of progelatinase A by primary fibroblasts, Biochem. J. 304 (Pt 1) (1994) 263–269, doi:10.1042/bj3040263.
[74] W.H. Yu, S. Yu, Q. Meng, K. Brew, J.F. Woessner Jr., TIMP-3 binds to sulfated glycosaminoglycans of the extracellular matrix, J. Biol. Chem. 275 (40) (2000) 31226–31232, doi:10.1074/jbc.M000907200.
[75] G.S. Butler, S.S. Apte, F. Willenbrock, G. Murphy, Human tissue inhibitor of metalloproteinases 3 interacts with both the N- and C-terminal domains of gelatinases A and B. Regulation by polyanions, J. Biol. Chem. 274 (16) (1999) 10846–10851, doi:10.1074/jbc.274.16.10846.
[76] K.J. Leco, R. Khokha, N. Pavloff, S.P. Hawkes, D.R. Edwards, Tissue inhibitor of metalloproteinases-3 (TIMP-3) is an extracellular matrix-associated protein with a distinctive pattern of expression in mouse cells and tissues, J. Biol. Chem. 269 (12) (1994) 9352–9360.
[77] J.H. Qi, Q. Ebrahem, N. Moore, G. Murphy, L. Claesson-Welsh, M. Bond, A. Baker, B. Anand-Apte, A novel function for tissue inhibitor of metalloproteinases-3 (TIMP3):

inhibition of angiogenesis by blockage of VEGF binding to VEGF receptor-2, Nat. Med. 9 (4) (2003) 407–415, doi:10.1038/nm846.
[78] B.H. Weber, G. Vogt, R.C. Pruett, H. Stöhr, U. Felbor, Mutations in the tissue inhibitor of metalloproteinases-3 (TIMP3) in patients with Sorsby's fundus dystrophy, Nat. Genet. 8 (4) (1994) 352–356, doi:10.1038/ng1294-352.
[79] Z. Li, M.P. Clarke, M.D. Barker, N. McKie, TIMP3 mutation in Sorsby's fundus dystrophy: molecular insights, Expert Rev. Mol. Med. 7 (24) (2005) 1–15, doi:10.1017/S1462399405010045.
[80] Y. Tabata, Y. Isashiki, K. Kamimura, K. Nakao, N. Ohba, A novel splice site mutation in the tissue inhibitor of the metalloproteinases-3 gene in Sorsby's fundus dystrophy with unusual clinical features, Hum. Genet. 103 (2) (1998) 179–182, doi:10.1007/pl00008707.
[81] A.H. Baker, A.B. Zaltsman, S.J. George, A.C. Newby, Divergent effects of tissue inhibitor of metalloproteinase-1, -2, or -3 overexpression on rat vascular smooth muscle cell invasion, proliferation, and death in vitro. TIMP-3 promotes apoptosis, J. Clin. Invest. 101 (6) (1998) 1478–1487, doi:10.1172/JCI1584.
[82] M. Ahonen, M. Poukkula, A.H. Baker, M. Kashiwagi, H. Nagase, J.E. Eriksson, V.M. Kähäri, Tissue inhibitor of metalloproteinases-3 induces apoptosis in melanoma cells by stabilization of death receptors, Oncogene 22 (14) (2003) 2121–2134, doi:10.1038/sj.onc.1206292.
[83] J.E. Fata, K.J. Leco, E.B. Voura, H.Y. Yu, P. Waterhouse, G. Murphy, R.A. Moorehead, R. Khokha, Accelerated apoptosis in the Timp-3-deficient mammary gland, J. Clin. Invest. 108 (6) (2001) 831–841, doi:10.1172/JCI13171.
[84] R.O. Hynes, The extracellular matrix: not just pretty fibrils, Science 326 (5957) (2009) 1216–1219, doi:10.1126/science.1176009.
[85] C.T.E. Jr, O. KG, The matrix metalloproteinase system: changes, regulation, and impact throughout the ovarian and uterine reproductive cycle, Endocr. Rev. 24 (4) (2003) 428–465, doi:10.1210/er.2002-0005.
[86] A. Page-McCaw, A.J. Ewald, Z. Werb, Matrix metalloproteinases and the regulation of tissue remodelling, Nat. Rev. Mol. Cell Biol. 8 (3) (2007) 221–233, doi:10.1038/nrm2125.
[87] T.A. Godenschwege, N. Pohar, S. Buchner, E. Buchner, Inflated wings, tissue autolysis and early death in tissue inhibitor of metalloproteinases mutants of Drosophila, Eur. J. Cell Biol. 79 (7) (2000) 495–501, doi:10.1078/0171-9335-00072.
[88] M.A. Nieuwesteeg, L.A. Walsh, M.A. Fox, S. Damjanovski, Domain specific over-expression of TIMP-2 and TIMP-3 reveals MMP-independent functions of TIMPs during Xenopus laevis development, Biochem. Cell Biol. 90 (4) (2012) 585–595, doi:10.1139/O2012-014.
[89] T. Sobue, Y. Hakeda, Y. Kobayashi, H. Hayakawa, K. Yamashita, T. Aoki, M. Kumegawa, T. Noguchi, T. Hayakawa, Tissue inhibitor of metalloproteinases 1 and 2 directly stimulate the bone-resorbing activity of isolated mature osteoclasts, J. Bone Miner. Res. 16 (12) (2001) 2205–2214, doi:10.1359/jbmr.2001.16.12.2205.
[90] T. Hayakawa, Multiple functions of tissue inhibitors of metalloproteinases (TIMPs): a new aspect involving osteoclastic bone resorption, J. Bone Miner. Metab. 20 (1) (2002) 1–13, doi:10.1007/s774-002-8440-0.
[91] T.H. Vu, J.M. Shipley, G. Bergers, J.E. Berger, J.A. Helms, D. Hanahan, S.D. Shapiro, R.M. Senior, Z. Werb, MMP-9/gelatinase B is a key regulator of growth plate angiogenesis and apoptosis of hypertrophic chondrocytes, Cell 93 (3) (1998) 411–422, doi:10.1016/s0092-8674(00)81169-1.
[92] L. Roten, S. Nemoto, J. Simsic, M.L. Coker, V. Rao, S. Baicu, G. Defreyte, P.J. Soloway, M.R. Zile, F.G. Spinale, Effects of gene deletion of the tissue inhibitor of the matrix

metalloproteinase-type 1 (TIMP-1) on left ventricular geometry and function in mice, J. Mol. Cell. Cardiol. 32 (1) (2000) 109–120, doi:10.1006/jmcc.1999.1052.
[93] E.E. Creemers, J.N. Davis, A.M. Parkhurst, P. Leenders, K.B. Dowdy, E. Hapke, A.M. Hauet, P.G. Escobar, J.P. Cleutjens, J.F. Smits, M.J. Daemen, M.R. Zile, F.G. Spinale, Deficiency of TIMP-1 exacerbates LV remodeling after myocardial infarction in mice, Am. J. Physiol. Heart Circ. Physiol. 284 (1) (2003) H364–H371, doi:10.1152/ajpheart.00511.2002.
[94] J.S. Ikonomidis, J.W. Hendrick, A.M. Parkhurst, A.R. Herron, P.G. Escobar, K.B. Dowdy, R.E. Stroud, E. Hapke, M.R. Zile, F.G. Spinale, Accelerated LV remodeling after myocardial infarction in TIMP-1-deficient mice: effects of exogenous MMP inhibition, Am. J. Physiol. Heart Circ. Physiol. 288 (1) (2005) H149–H158, doi:10.1152/ajpheart.00370.2004.
[95] P.W. Fedak, D.S. Smookler, Z. Kassiri, N. Ohno, K.J. Leco, S. Verma, D.A. Mickle, K.L. Watson, C.V. Hojilla, W. Cruz, R.D. Weisel, R.K. Li, R. Khokha, TIMP-3 deficiency leads to dilated cardiomyopathy, Circulation 110 (16) (2004) 2401–2409, doi:10.1161/01.CIR.0000134959.83967.2D.
[96] Z. Kassiri, V. Defamie, M. Hariri, G.Y. Oudit, S. Anthwal, F. Dawood, P. Liu, R. Khokha, Simultaneous transforming growth factor beta-tumor necrosis factor activation and cross-talk cause aberrant remodeling response and myocardial fibrosis in Timp3-deficient heart, J. Biol. Chem. 284 (43) (2009) 29893–29904, doi:10.1074/jbc.M109.028449.
[97] D.W. Seo, S.H. Kim, S.H. Eom, H.J. Yoon, Y.R. Cho, P.H. Kim, Y.K. Kim, J.W. Han, T. Diaz, B.Y. Wei, W.G. Stetler-Stevenson, TIMP-2 disrupts FGF-2-induced downstream signaling pathways, Microvasc. Res. 76 (3) (2008) 145–151, doi:10.1016/j.mvr.2008.07.003.
[98] T. Akahane, M. Akahane, A. Shah, C.M. Connor, U.P. Thorgeirsson, TIMP-1 inhibits microvascular endothelial cell migration by MMP-dependent and MMP-independent mechanisms, Exp. Cell Res. 301 (2) (2004) 158–167, doi:10.1016/j.yexcr.2004.08.002.
[99] A.C. Newby, Metalloproteinase production from macrophages—a perfect storm leading to atherosclerotic plaque rupture and myocardial infarction, Exp. Physiol. 101 (11) (2016) 1327–1337, doi:10.1113/EP085567.
[100] M. Gooz, ADAM-17: the enzyme that does it all, Crit. Rev. Biochem. Mol. Biol. 45 (2) (2010) 146–169, doi:10.3109/10409231003628015.
[101] J.L. Johnson, S.J. George, A.C. Newby, C.L. Jackson, Divergent effects of matrix metalloproteinases 3, 7, 9, and 12 on atherosclerotic plaque stability in mouse brachiocephalic arteries, Proc. Natl. Acad. Sci. U.S.A. 102 (43) (2005) 15575–15580, doi:10.1073/pnas.0506201102.
[102] R.A. Black, C.T. Rauch, C.J. Kozlosky, J.J. Peschon, J.L. Slack, M.F. Wolfson, B.J. Castner, K.L. Stocking, P. Reddy, S. Srinivasan, N. Nelson, N. Boiani, K.A. Schooley, M. Gerhart, R. Davis, J.N. Fitzner, R.S. Johnson, R.J. Paxton, C.J. March, D.P. Cerretti, A metalloproteinase disintegrin that releases tumour-necrosis factor-alpha from cells, Nature 385 (6618) (1997) 729–733, doi:10.1038/385729a0.
[103] V. Nagy, O. Bozdagi, A. Matynia, M. Balcerzyk, P. Okulski, J. Dzwonek, R.M. Costa, A.J. Silva, L. Kaczmarek, G.W. Huntley, Matrix metalloproteinase-9 is required for hippocampal late-phase long-term potentiation and memory, J. Neurosci. 26 (7) (2006) 1923–1934, doi:10.1523/JNEUROSCI.4359-05.2006.
[104] F.A. Chaillan, S. Rivera, E. Marchetti, J. Jourquin, Z. Werb, P.D. Soloway, M. Khrestchatisky, F.S. Roman, Involvement of tissue inhibition of metalloproteinases-1 in learning and memory in mice, Behav. Brain Res. 173 (2) (2006) 191–198, doi:10.1016/j.bbr.2006.06.020.
[105] J.M. Castellano, K.I. Mosher, R.J. Abbey, A.A. McBride, M.L. James, D. Berdnik, J.C. Shen, B. Zou, X.S. Xie, M. Tingle, I.V. Hinkson, M.S. Angst, T. Wyss-Coray, Human

umbilical cord plasma proteins revitalize hippocampal function in aged mice, Nature 544 (7651) (2017) 488–492, doi:10.1038/nature22067.
[106] L. Pérez-Martínez, D.M. Jaworski, Tissue inhibitor of metalloproteinase-2 promotes neuronal differentiation by acting as an anti-mitogenic signal, J. Neurosci. 25 (20) (2005) 4917–4929, doi:10.1523/JNEUROSCI.5066-04.2005.
[107] A. Dityatev, M. Schachner, Extracellular matrix molecules and synaptic plasticity, Nat. Rev. Neurosci. 4 (6) (2003) 456–468, doi:10.1038/nrn1115.
[108] V.W. Yong, C.A. Krekoski, P.A. Forsyth, R. Bell, D.R. Edwards, Matrix metalloproteinases and diseases of the CNS, Trends Neurosci. 21 (2) (1998) 75–80, doi:10.1016/s0166-2236(97)01169-7.
[109] M. Wetzel, L. Li, K.M. Harms, T. Roitbak, P.B. Ventura, G.A. Rosenberg, R. Khokha, L.A. Cunningham, Tissue inhibitor of metalloproteinases-3 facilitates Fas-mediated neuronal cell death following mild ischemia, Cell Death Differ. 15 (1) (2008) 143–151, doi:10.1038/sj.cdd.4402246.
[110] J. Bian, Y. Wang, M.R. Smith, H. Kim, C. Jacobs, J. Jackman, H.F. Kung, N.H. Colburn, Y. Sun, Suppression of in vivo tumor growth and induction of suspension cell death by tissue inhibitor of metalloproteinases (TIMP)-3, Carcinogenesis 17 (9) (1996) 1805–1811, doi:10.1093/carcin/17.9.1805.
[111] M.R. Smith, H. Kung, S.K. Durum, N.H. Colburn, Y. Sun, TIMP-3 induces cell death by stabilizing TNF-alpha receptors on the surface of human colon carcinoma cells, Cytokine 9 (10) (1997) 770–780, doi:10.1006/cyto.1997.0233.
[112] A.H. Baker, S.J. George, A.B. Zaltsman, G. Murphy, A.C. Newby, Inhibition of invasion and induction of apoptotic cell death of cancer cell lines by overexpression of TIMP-3, Br. J. Cancer 79 (9-10) (1999) 1347–1355, doi:10.1038/sj.bjc.6690217.
[113] D.W. Wu, L.H. Tsai, P.M. Chen, M.C. Lee, L. Wang, C.Y. Chen, Y.W. Cheng, H. Lee, Loss of TIMP-3 promotes tumor invasion via elevated IL-6 production and predicts poor survival and relapse in HPV-infected non-small cell lung cancer, Am. J. Pathol. 181 (5) (2012) 1796–1806, doi:10.1016/j.ajpath.2012.07.032.
[114] V. Egea, S. Zahler, N. Rieth, P. Neth, T. Popp, K. Kehe, M. Jochum, C. Ries, Tissue inhibitor of metalloproteinase-1 (TIMP-1) regulates mesenchymal stem cells through let-7f microRNA and Wnt/β-catenin signaling, Proc. Natl. Acad. Sci. U.S.A. 109 (6) (2012) E309–E316, doi:10.1073/pnas.1115083109.
[115] N. Johansson, M. Ahonen, V.M. Kähäri, Matrix metalloproteinases in tumor invasion, Cell. Mol. Life Sci. 57 (1) (2000) 5–15, doi:10.1007/s000180050495.
[116] Y. Itoh, H. Nagase, Matrix metalloproteinases in cancer, Essays Biochem. 38 (2002) 21–36, doi:10.1042/bse0380021.
[117] G. Murphy, The ADAMs: signalling scissors in the tumour microenvironment, Nat. Rev. Cancer 8 (12) (2008) 929–941, doi:10.1038/nrc2459.
[118] D.S. Smookler, F.F. Mohammed, Z. Kassiri, G.S. Duncan, T.W. Mak, R. Khokha, Tissue inhibitor of metalloproteinase 3 regulates TNF-dependent systemic inflammation, J. Immunol. 176 (2) (2006) 721–725, doi:10.4049/jimmunol.176.2.721.
[119] R. Chirco, X.W. Liu, K.K. Jung, H.R. Kim, Novel functions of TIMPs in cell signaling, Cancer Metastasis Rev. 25 (1) (2006) 99–113, doi:10.1007/s10555-006-7893-x.
[120] B. Meissburger, L. Stachorski, E. Röder, G. Rudofsky, C. Wolfrum, Tissue inhibitor of matrix metalloproteinase 1 (TIMP1) controls adipogenesis in obesity in mice and in humans, Diabetologia 54 (6) (2011) 1468–1479, doi:10.1007/s00125-011-2093-9.
[121] R. Menghini, S. Menini, R. Amoruso, L. Fiorentino, V. Casagrande, V. Marzano, F. Tornei, P. Bertucci, C. Iacobini, M. Serino, O. Porzio, M.L. Hribal, F. Folli, R. Khokha, A. Urbani, R. Lauro, G. Pugliese, M. Federici, Tissue inhibitor of metalloproteinase 3 deficiency causes hepatic steatosis and adipose tissue inflammation in mice, Gastroenterology 136 (2) (2009) 663–672.e4, doi:10.1053/j.gastro.2008.10.079.

[122] S.S. Glasson, R. Askew, B. Sheppard, B. Carito, T. Blanchet, H.L. Ma, C.R. Flannery, D. Peluso, K. Kanki, Z. Yang, M.K. Majumdar, E.A. Morris, Deletion of active ADAMTS5 prevents cartilage degradation in a murine model of osteoarthritis, Nature 434 (7033) (2005) 644–648 Erratum in: Nature. 2007 Mar 1;446(7131):102, doi:10.1038/nature03369.

[123] M.J. Arthur, I.I. Fibrogenesis, Metalloproteinases and their inhibitors in liver fibrosis, Am. J. Physiol. Gastrointest. Liver Physiol. 279 (2) (2000) G245–G249, doi:10.1152/ajpgi.2000.279.2.G245.

[124] I. Koskivirta, O. Rahkonen, M. Mäyränpää, S. Pakkanen, M. Husheem, A. Sainio, H. Hakovirta, J. Laine, E. Jokinen, E. Vuorio, P. Kovanen, H. Järveläinen, Tissue inhibitor of metalloproteinases 4 (TIMP4) is involved in inflammatory processes of human cardiovascular pathology, Histochem. Cell Biol. 126 (3) (2006) 335–342, doi:10.1007/s00418-006-0163-8.

[125] A. Radomski, P. Jurasz, E.J. Sanders, C.M. Overall, H.F. Bigg, D.R. Edwards, M.W. Radomski, Identification, regulation and role of tissue inhibitor of metalloproteinases-4 (TIMP-4) in human platelets, Br. J. Pharmacol. 137 (8) (2002) 1330–1338, doi:10.1038/sj.bjp.0704936.

[126] J.M. Dewing, R.O. Carare, A.J. Lotery, J.A. Ratnayaka, The diverse roles of timp-3: insights into degenerative diseases of the senescent retina and brain, Cells 9 (1) (2019) 39, doi:10.3390/cells9010039.

[127] I. Koskivirta, Z. Kassiri, O. Rahkonen, R. Kiviranta, G.Y. Oudit, T.D. McKee, V. Kytö, A. Saraste, E. Jokinen, P.P. Liu, E. Vuorio, R. Khokha, Mice with tissue inhibitor of metalloproteinases 4 (Timp4) deletion succumb to induced myocardial infarction but not to cardiac pressure overload, J. Biol. Chem. 285 (32) (2010) 24487–24493, doi:10.1074/jbc.M110.136820.

[128] M. Shimoda, S. Principe, H.W. Jackson, V. Luga, H. Fang, S.D. Molyneux, Y.W. Shao, A. Aiken, P.D. Waterhouse, C. Karamboulas, F.M. Hess, T. Ohtsuka, Y. Okada, L. Ailles, A. Ludwig, J.L. Wrana, T. Kislinger, R. Khokha, Loss of the Timp gene family is sufficient for the acquisition of the CAF-like cell state, Nat. Cell Biol. 16 (9) (2014) 889–901, doi:10.1038/ncb3021.

[129] S. Costa, M.A. Ragusa, G. Lo Buglio, S.D. Scilabra, A. Nicosia, The repertoire of tissue inhibitors of metalloproteases: evolution, regulation of extracellular matrix proteolysis, engineering and therapeutic challenges, Life (Basel) 12 (8) (2022) 1145, doi:10.3390/life12081145.

[130] J.Y. Park, Y. Choi, H.D. Kim, H.H. Kuo, Y.C. Chang, C.H. Kim, Matrix metalloproteinases and their inhibitors in the pathogenesis of epithelial differentiation, vascular disease, endometriosis, and ocular fibrotic pterygium, Int. J. Mol. Sci. 26 (12) (2025) 5553, doi:10.3390/ijms26125553.

[131] A. Shoari, G. Khalili-Tanha, M.A. Coban, E.S. Radisky, Structure and computation-guided yeast surface display for the evolution of TIMP-based matrix metalloproteinase inhibitors, Front. Mol. Biosci. 10 (2023) 1321956, doi:10.3389/fmolb.2023.1321956.

[132] M. Raeeszadeh-Sarmazdeh, K.A. Greene, B. Sankaran, G.P. Downey, D.C. Radisky, E.S. Radisky, Directed evolution of the metalloproteinase inhibitor TIMP-1 reveals that its N- and C-terminal domains cooperate in matrix metalloproteinase recognition, J. Biol. Chem. 294 (24) (2019) 9476–9488, doi:10.1074/jbc.RA119.008321.

[133] J. Shirian, V. Arkadash, I. Cohen, T. Sapir, E.S. Radisky, N. Papo, J.M. Shifman, Converting a broad matrix metalloproteinase family inhibitor into a specific inhibitor of MMP-9 and MMP-14, FEBS Lett. 592 (7) (2018) 1122–1134, doi:10.1002/1873-3468.13016.

[134] P.T. Wingfield, J.K. Sax, S.J. Stahl, J. Kaufman, I. Palmer, V. Chung, M.L. Corcoran, D.E. Kleiner, W.G. Stetler-Stevenson, Biophysical and functional characterization of full-length, recombinant human tissue inhibitor of metalloproteinases-2 (TIMP-2) produced in Escherichia coli. Comparison of wild type and amino-terminal alanine

appended variant with implications for the mechanism of TIMP functions, J. Biol. Chem. 274 (30) (1999) 21362–21368, doi:10.1074/jbc.274.30.21362.
[135] M. de Bruyn, K. Machiels, J. Vandooren, B. Lemmens, L. Van Lommel, C. Breynaert, J. Van der Goten, D. Staelens, T. Billiet, G. De Hertogh, M. Ferrante, G. Van Assche, S. Vermeire, G. Opdenakker, F. Schuit, P. Rutgeerts, I. Arijs, Infliximab restores the dysfunctional matrix remodeling protein and growth factor gene expression in patients with inflammatory bowel disease, Inflamm. Bowel Dis. 20 (2) (2014) 339–352, doi:10.1097/01.MIB.0000438430.15553.90.
[136] A. Di Sabatino, S.L. Pender, C.L. Jackson, J.D. Prothero, J.N. Gordon, L. Picariello, L. Rovedatti, G. Docena, G. Monteleone, D.S. Rampton, F. Tonelli, G.R. Corazza, T.T. MacDonald, Functional modulation of Crohn's disease myofibroblasts by anti-tumor necrosis factor antibodies, Gastroenterology 133 (1) (2007) 137–149, doi:10.1053/j.gastro.2007.04.069.
[137] M. Sharma, J. Mohapatra, A. Wagh, H.M. Patel, D. Pandey, S. Kadam, A. Argade, S.S. Deshpande, G.B. Shah, A. Chatterjee, M.R. Jain, Involvement of TACE in colon inflammation: a novel mechanism of regulation via SIRT-1 activation, Cytokine 66 (1) (2014) 30–39, doi:10.1016/j.cyto.2013.12.010.
[138] M. Lagouge, C. Argmann, Z. Gerhart-Hines, H. Meziane, C. Lerin, F. Daussin, N. Messadeq, J. Milne, P. Lambert, P. Elliott, B. Geny, M. Laakso, P. Puigserver, J. Auwerx, Resveratrol improves mitochondrial function and protects against metabolic disease by activating SIRT1 and PGC-1alpha, Cell 127 (6) (2006) 1109–1122, doi:10.1016/j.cell.2006.11.013.
[139] R. Gan, Y. Yang, X. Yang, L. Zhao, J. Lu, Q.H. Meng, Downregulation of miR-221/222 enhances sensitivity of breast cancer cells to tamoxifen through upregulation of TIMP3, Cancer Gene Ther. 21 (7) (2014) 290–296 Epub 2014 Jun 13. Erratum in: Cancer Gene Ther. 2023 Nov;30(11):1582., doi:10.1038/cgt.2014.29.
[140] S.V. McLennan, X.Y. Wang, V. Moreno, D.K. Yue, S.M. Twigg, Connective tissue growth factor mediates high glucose effects on matrix degradation through tissue inhibitor of matrix metalloproteinase type 1: implications for diabetic nephropathy, Endocrinology 145 (12) (2004) 5646–5655, doi:10.1210/en.2004-0436.
[141] Y. Tao, Q. Chen, C. Zhao, X. Yang, Q. Cun, W. Yang, Y. Zhang, Y. Zhu, H. Zhong, The in vitro anti-fibrotic effect of pirfenidone on human pterygium fibroblasts is associated with down-regulation of autocrine TGF-β and MMP-1, Int. J. Med. Sci. 17 (6) (2020) 734–744, doi:10.7150/ijms.43238.
[142] P. Kuvaja, A. Talvensaari-Mattila, T. Turpeenniemi-Hujanen, High preoperative plasma TIMP-1 is prognostic for early relapse in primary breast carcinoma, Int. J. Cancer 123 (4) (2008) 846–851, doi:10.1002/ijc.23531.
[143] O. Hekmat, S. Munk, L. Fogh, R. Yadav, C. Francavilla, H. Horn, S.Ø Würtz, A.S. Schrohl, B. Damsgaard, M.U. Rømer, K.C. Belling, N.F. Jensen, I. Gromova, D.B. Bekker-Jensen, J.M. Moreira, L.J. Jensen, R. Gupta, U. Lademann, N. Brünner, J.V. Olsen, J. Stenvang, TIMP-1 increases expression and phosphorylation of proteins associated with drug resistance in breast cancer cells, J. Proteome Res. 12 (9) (2013) 4136–4151, doi:10.1021/pr400457u.
[144] L. Lorente, M.M. Martín, J. Solé-Violán, J. Blanquer, L. Labarta, C. Díaz, J.M. Borreguero-León, J. Orbe, J.A. Rodríguez, A. Jiménez, J.A. Páramo, Association of sepsis-related mortality with early increase of TIMP-1/MMP-9 ratio, PLoS One 9 (4) (2014) e94318, doi:10.1371/journal.pone.0094318.
[145] L. Wang, K. Main, H. Wang, O. Julien, A. Dufour, Biochemical tools for tracking proteolysis, J. Proteome Res. 20 (12) (2021) 5264–5279, doi:10.1021/acs.jproteome.1c00289.
[146] A. Derakhshani, M. Bulluss, R. Penner, A. Dufour, N-terminomics/TAILS of human tumor biopsies and cancer cell lines, Methods Mol. Biol. 2747 (2024) 19–28, doi:10.1007/978-1-0716-3589-6_2.

CHAPTER SIX

MMPs in oral diseases

Krista Juurikka[a,b], Tuula Salo[c,d,e], Pirjo Åström[a,f,*,#]

[a]Research Unit of Biomedicine and Internal Medicine, Faculty of Medicine, University of Oulu, Oulu, Finland
[b]Faculty of Biochemistry and Molecular Medicine, University of Oulu, Oulu, Finland
[c]Oral Pathology, Research Unit of Population Health, Faculty of Medicine, Medical Research Center, University of Oulu and Oulu University Hospital, Oulu, Finland
[d]Department of Oral and Maxillofacial Diseases, Translational Immunology Program (TRIMM), iCAN Digital Precision Cancer Medicine Flagship, University of Helsinki, Helsinki, Finland
[e]Department of Pathology, Helsinki University Hospital, Helsinki, Finland
[f]Biocenter Oulu, University of Oulu, Oulu, Finland
*Corresponding author. e-mail address: pirjo.astrom@oulu.fi.

Contents

Abstract

MMPs are essential for the physiological tissue turnover within the oral cavity and maxillofacial structures but their dysregulation may lead to pathogenic, tissue-destructive oral diseases. Within oral tissues, matrix metalloproteinases (MMPs) are produced by a variety of cells, including fibroblasts, keratinocytes, epithelial cells, immune cells, osteoclasts, osteoblasts, and odontoblasts. Activation of MMPs in periodontal tissues degrades extracellular matrix components which may lead to periodontal attachment loss and alveolar bone destruction. Various MMPs serve as putative biomarkers in dental diseases,

[#]Senior author.

Fungal and Protozoan Carbonic Anhydrases, Volume 60
ISSN 1874-6047
https://doi.org/10.1016/bs.enz.2026.07.001

and their levels can be easily measured from oral fluids. In addition to their role in periodontal manifestations, MMPs are associated with caries, pulpal inflammation, oral potentially malignant disorders and cancers. MMP inhibitors and chairside test measuring active MMP8 have demonstrated potential clinical usability in adjunctive periodontal therapy and diagnostics. In the next chapters, we discuss the importance and function of MMPs and their use as biomarkers in oral diseases.

Keywords: Caries, Dental bonding, Matrix metalloproteinase, Oral neoplasms, Oral potentially malignant disorders, Oral cancer, Periapical lesion, Peri-implantitis, Periodontitis, Pulpitis

1. Preface

MMPs are responsible for the remodeling and degradation of most extracellular matrix (ECM) proteins during organogenesis, development and physiological tissue turnover, and thus are also essential for the health of the oral cavity and maxillofacial structures. They play roles throughout jaws and tooth development, and participate in cell differentiation, morphogenesis, and the remodeling of periodontal and dental tissues, including even enamel maturation [1]. For example, MMP20 (enamelysin) is secreted by ameloblasts and processes enamel matrix proteins during early enamel formation, enabling proper mineralization and normal enamel structure [2,3]. However, MMP20 is also expressed in odontogenic tumors and oral carcinomas [4–6]. Mutations in *MMP20* lead to enamel defects such as amelogenesis imperfecta [7]. Dysregulation of various MMPs contributes to the pathogenesis of tissue-destructive oral diseases, including gingivitis, periodontitis, peri–implantitis and dental caries. Excessive MMP activation promotes breakdown of ECM components and contributes to periodontal attachment loss, alveolar bone destruction, and caries progression [1]. They also serve as putative biomarkers associated with dental tissue degradation, reversible and irreversible pulpitis, (peri)apical periodontitis, and periodontal diseases. Increased expression of various MMPs is also associated with oral potentially malignant disorders (OPMD) as well as in aggressive tumor behavior and poor prognosis. MMPs contribute to the progression of OPMDs and oral squamous cell carcinoma (OSCC) by promoting basement membrane degradation, tumor invasion, angiogenesis, and metastasis [8,9].

In oral cavity and maxillofacial tissues, MMPs are produced by multiple cell types, including neutrophils, macrophages, fibroblasts, keratinocytes, osteoclasts, osteoblasts, odontoblasts, mast cells and epithelial cells [10–12]. Collagenases (mainly MMP8 and −13) and gelatinases (mainly MMP9)

are the most studied MMPs, and they are significantly associated with periodontal inflammation and connective tissue destruction. MMP8, predominantly released by neutrophils, cleaves fibrillar collagen within gingival connective tissue and the periodontal ligament. MMP9 is mainly produced by inflammatory and epithelial cells, and it degrades basement membrane components and facilitates inflammatory cell migration. Proteolytic cascades between different MMPs further amplify tissue destruction and periodontal attachment loss [11,13]. In the oral environment, inflammatory cytokines, reactive oxygen species, and bacterial virulence factors derived from the dental biofilm enhance the expression, granular release, and activation of MMPs. Once activated, MMPs remodel and degrade the crucial ECM components that maintain the integrity of oral tissues, including type I and III collagen, gelatin, laminin, elastin, and fibronectin but also various cytokines, chemokines, and growth factors. The activity of MMPs is regulated by their endogenous regulators, tissue inhibitors of metalloproteinases (TIMP), to maintain homeostasis and prevent excess degradation. Chronic inflammation disrupts the MMP/TIMP balance, resulting in matrix degradation, impaired tissue repair, and alveolar bone resorption [11,14].

Saliva and oral fluids (mouth rinse, gingival crevicular fluid [GCF], and peri–implant crevicular fluid [PICF]) are highly useful in measuring disease-specific oral and systemic biomarkers [9]. Saliva is composed mainly of water, ions, bioactive molecules as well as micro-organisms and micro-organism-derived products. It contributes to antimicrobial defense, buffering, lubrication, digestion, and maintenance of oral tissues. Unstimulated whole saliva is commonly preferred for diagnostic applications because stimulated saliva may dilute biomarker concentrations. Various host and microbial factors may affect the levels of salivary biomarkers. Yet, saliva has several benefits as a source of disease biomarkers: it contains several measurable compounds; it is easy to collect noninvasively, transport and store. Saliva also contains various MMPs which may originate from GCF or from the salivary glands' secretions [15]. In homeostasis, GCF α2-macroglobulin keeps the MMPs in an inactive form [16].

Elevated levels of MMPs have been detected in saliva, GCF, PICF, oral rinse and tissue samples, making them promising biomarkers for diagnostics and disease monitoring. Especially active MMP8 (aMMP8) chairside tests have demonstrated potential clinical utility in several studies. It has been used to identify active periodontal inflammation and assess treatment responses. A meta-analysis comprising of six studies found that salivary aMMP8 point-of-care test has fair accuracy and can be useful in combination with other diagnostics methods. However, the study concluded that additional research would be needed to confirm its clinical utility for detecting periodontitis [17]. Therapeutically, modulation of MMP activity represents an important target in oral medicine [8,9,18]. Subantimicrobial dose doxycycline, a broad-spectrum antibiotic and MMP inhibitor, has been used as adjunctive periodontal therapy to reduce collagen degradation and inflammation [19]. Current research focuses on selective MMP inhibitors, host-modulation therapies, and biomarker-guided precision dentistry in aiming to improve prevention, diagnosis, and treatment of oral diseases.

2. Oral potentially malignant disorders and neoplasms

In 2022, almost 390,000 new cases of lip and oral cavity cancers and almost 190,000 cancer deaths occurred worldwide, making it the fifth most common and sixth most deadly cancer in the world [20]. 90 percent of oral cancers originate from mucosal epithelial cells and are defined as oral OSCCs. They belong to a larger family of head and neck squamous cell carcinomas (HNSCC), which also covers tumors originating from the mucosa of oropharynx, larynx and hypopharynx. OSCCs most often occur in the lateral mobile tongue (~50 percent of cases) with regional metastases forming in cervical lymph nodes and distant metastases in the lung. Risk factors for OSCC include smoking, betel quid chewing and alcohol consumption and these risk factors are also thought to lead to the generation of oral potentially malignant disorders (OPMDs, discussed in detail later), which have the potential to transform into OSCC later on [21,22].

The oncogenesis of OSCC is complex and multifactorial, but one of the key factors is a dysregulated tumor microenvironment. Alteration of the stroma, hypoxia, suppressed local immunity and unbalanced oral microbiome are all thought to contribute to the disease's emergence and progression but at the same time, they provide targets for novel treatment modalities [22]. Tumor stroma consists of both stromal cells, such as fibroblasts, vascular and immune cells and the ECM, which is actively modified during carcinogenesis. ECM components in the OSCC include molecules such as laminins, proteoglycans, hyaluronic acid, and collagens. Degradation of ECM is one of the key steps in invasion and metastasis, and as MMPs participate in ECM turnover via their proteinase activity, increased MMP activity is thought to help tumor cell motility by degrading ECM. Yet ECM degradation also releases ECM-bound growth factors and angiogenic modulators, which leads to changes in cell signaling, which in turn leads to changes in cancer cell behavior. One example of such a change in cell signaling in OSCC is the activation of non-receptor tyrosine kinase Src, which activates PI3K/Akt pathways and leads to changes in cell motility that enable invasion and metastasis [22,23]. Another example is the release of angiogenesis inhibitor endostatin from type XVIII collagen in vitro. Endostatin blocks the activities of certain tumor-associated MMPs, such as pro-MMP2, −9, and −13, which may partially explain its antitumor effects [24], although such detailed interactions are challenging to show in vivo [23,25]. Even though MMPs and their proteolytic activity have been canonically considered as tumor promoting factors, the reality is much more complex as the above example of endostatin shows.

The following paragraphs describe how MMPs have been studied as biomarkers for oral cancer and oral potentially malignant disorders, what cellular and molecular mechanisms of MMPs in OSCC have been identified and how the field nowadays treats changes in certain MMP levels (mostly MMP2 and −9) as indicators of invasion and metastasis (see for example recent studies on β2-adrenergic receptor and cyclooxygenase-2 inhibitors in OSCC [26], or the role of transcriptional coactivator MED1 in OSCC [27]).

2.1 Genetic predisposition—*MMP* genes and gene expression in oral cancer predisposition

Genetic variants, especially those that affect the activity of *MMP* gene promoters, have been studied in OSCC. This research has focused on *MMPs 1, −2, −3, −7* and *−9*, and has been mostly performed before 2020 and the results vary or are negative for most polymorphisms. A systematic review evaluated the association between OSCC and genetic polymorphisms in *MMP1, −2*, and *−3* and found that only a promoter polymorphism in *MMP1* (−1607 1G/2G) significantly associated with increased risk of OSCC (OR = 1.47) [28]. *MMP7* promoter polymorphism −181A/G, shown to increase gene transcription, was associated with only early stages of OSCC in European populations and not at all in an Asian population [29,30]. *MMP9* promoter polymorphism (−1562 C>T) did not associate with overall risk of OSCC, but associated with increased OSCC risk in specific patient groups (e.g., patients younger than 51 years old) [31]. In a European population, this polymorphism, but not polymorphisms in *MMP1* or *−3*, could be associated with OSCC risk [32]. Some studies have also investigated *MMP8, −11, −13* and *−14* polymorphisms, but the results have been equally unfavorable. *MMP8* promoter polymorphisms did not associate with increased risk of OSCC [33] and *MMP13* promoter polymorphism was only associated with advanced stages of OSCC [34]. Genetic variants that likely affect the expression of *MMP11*, did not associate with the risk of OSCC, although some genotypes were more frequently present with certain etiological factors such as smoking or betel quid chewing [35]. In contrary, a combination of four *MMP14* exon 5 and exon 6 polymorphisms, which might alter the function of the protein or the mRNA splicing/translation efficiency of the mRNA, was associated with the 2.3-fold increased risk of OSCC [36].

Performing analyses on oral cancer subtypes and anatomic locations separately could better identify subtype-specific risk factors. When *MMP7* gene promoter (181 A/G) and *MMP9* (−1562 C/T) polymorphisms were studied in oral tongue squamous cell carcinoma (OTSCC), those variants linked with higher enzyme activity were more often identified in OTSCC patients compared to healthy controls and when combined together, associated with increased risk of OTSCC and metastases [37]. Similarly, in a study of salivary gland carcinomas, *MMP2* (−1306 C>T) and *−9* (−1562 C>T) promoter polymorphisms associated with approximately 2-fold increased risk of salivary gland cancer [38]. Yet these studies need to be repeated to be able to conclude whether a MMP genetic variant can predict cancer risk for a specific subtype of OSCC.

OSCC tissue mRNA expression of *MMPs* has been also investigated. Interestingly, an elevation of *MMP7, −10* and *−13* and reduction of *MMP9* and *−12* mRNA expression was detected in small cohort of OSCC (mostly tongue) tumors from patients with metastases compared to those without metastases [39]. Analysis of over 500 HNSCC patients (TCGA), demonstrated that expression of *MMP1, −2, −3, −8, −9, −10* or *−13* could not be used to estimate overall survival [40], even if gene expression might be significantly higher in tumor compared to healthy tissue in various HNSCCs, which is the case for *MMP1* [41]. Interestingly, even if an *MMP (−1, −2, −7 or −9)* was clearly expressed (30–60 percent cancer cells positive) by cancer cells in the primary tumor, no expression was detected in circulating tumor cells [42], which could suggest phenotypic plasticity and/or temporal changes in *MMP* expression. Overall, *MMP* gene expression, as also demonstrated by the varying or negative findings in genetic variants studies described above, does not seem to convey risk for OSCC.

2.2 MMPs as biomarkers in oral cancers

2.2.1 MMP levels as tissue biomarkers in oral cancer

In general, high tumoral MMP levels are seen as a detrimental prognostic indication for OSCC patients. A meta-analysis that pooled together 15 studies and 1266 patients, showed that combined overexpression of MMPs negatively impacted overall survival (HR $= 2.18$, $P = 0.080$) and disease-free time (HR $= 2.45$, $P = 0.061$). This study also identified high MMP2 and −9 levels as markers for poor prognosis for OSCC, and MMP11 and −13 as factors associated with overall survival [43]. Other analyses have also investigated multiple MMPs at the same time. The MMPs −3, −7, −9, −12 and −13 were generally absent in less aggressive oral malignancy, verrucous carcinoma tumors, compared to OSCC, whereas MMP10 was localized in the epithelia in all samples, MMP19 was absent in cancer cells but present in hyperproliferative cells, and MMP26 was present in both hyperproliferative and cancer cells [44]. In one of the early OSCC MMP studies, a notable expression of MMP1, −2, −3, −9 and MMP14 was frequently observed in invasive and metastatic OSCC cases and high MMP1, −2 and −9 expression was associated with decreased ECM staining, suggesting that these MMPs participate in ECM degradation [45].

Tissue biomarker research has especially focused on levels of MMP2 and −9. Their levels were higher in OSCC tumor cells compared to surrounding healthy tissue [46]. Additionally, MMP2 levels increased with tumor grade in OSCC cases [47] and MMP9 levels increased with malignancy when oral dysplasia and OSCC cases were compared [48]. A meta-analysis with 419 OSCC patients summarized that high MMP9 levels are associated with worse overall survival (effect size $= 2.10$, $P < 0.001$) as well as higher T stage and presence of lymph node metastasis (LNM) [49]. MMP2 and −9 activities were elevated in OSCC tumor tissues, and MMP2 activity could be used to predict LNM, and either of them could be used to predict overall survival [50]. MMP2 level and activity were higher in OSCC tissues (with and without LNM) compared to adjacent healthy tissue and the expression correlated with MMP14 expression [51], which is also elevated in OSCC tumors. Furthermore, reduction of MMP expression or protein levels in vitro leads to decreased motility: A knockdown of MMP9 in OSCC cells led to reduced migration and proliferation as well as expression of RhoC,

Src, and F-actin in vitro. Furthermore, MMP9 knockdown reduced tumor growth, angiogenesis and LNM in an orthotopic tongue cancer mouse model [52]. With the promising effect of MMP9 in OSCC aggressiveness, compounds targeting MMP9 have also been identified. In silico screening for MMP9 inhibitors identified Kojic acid, which also showed cytotoxicity against OSCC cells in vitro [53]. Apart from intrinsic tumoral expression, treatment may affect tissue MMP levels as MMP2 and −9 levels in oral mucosa elevated after chemotherapy [54]. Whether the effect is direct (i.e., treatment itself affects MMP levels or activity) or transmitted by changes in tumor or the microenvironment (i.e., the death of the cells producing or activating MMPs) remains to be determined.

Other MMPs may also affect OSCC progression or be usable biomarkers. Cytoplasmic MMP1 staining localizes in epithelial cells and connective tissues both in normal oral mucosa and OSCCs, but the staining intensity is higher in OSCCs [55]. Low MMP1 staining in HNSCC tumors predicted better overall and disease-free survival, and high tumoral MMP1 level correlated with tumor size, grade and LNM [41]. Similar to MMP1, high tumoral MMP11 levels were detected in 42.3 percent of OSCC cases, correlated with advanced grade, LNM and predicted poor overall survival [56]. MMP11 knockdown invitro led to decreased migration in OSCC cells [57] and overexpression increased the migration and invasion of OSCC cells, yet this increased motility could be inhibited by using focal adhesion kinase (FAK) and Src inhibitors, suggesting the changes caused by MMP11 overexpression are mediated via common pathways disrupted in OSCC [56]. MMP14 staining was stronger and localized to the cytoplasm and membrane of cancer cells more than healthy mucosa, especially in tongue cancers and cases with neural invasion [58]. Another study reported high-risk for patients with MMP14 positive tumor staining, especially in LNM with extranodal extension, which is an indicator for worse prognosis. The risk assessment was based on the co-scoring system in which the staining in both tumor cells and cancer-associated fibroblasts is assessed [59]. MMP21 levels were higher in OSCC tissues compared to normal gingiva samples and the expression was localized to cytoplasm and cell membrane [60].

Several subtypes of OSCC exist based on anatomical location. Yet the differences are not only in anatomical location, but also in etiological and molecular factors (e.g., MMPs) affecting these sites and tumors. MMP1 was prominently expressed in gingival squamous cell carcinomas [61]. In tongue cancer, high MMP2 and −14 correlated with LNM and poor prognosis as well as activation of MMP2 [62,63]. On the other hand, high MMP8 levels in tongue tumors predicted better overall survival [64]. Overexpression of MMP8 in vitro led to decreased cell motility [65], which could be attenuated by the cleavage of anti-adhesive FXYD5 molecule by MMP8 [66]. On the other hand, MMP10 overexpression increased the proliferation, migration and invasion of tongue cancer cells and increased the expression of metastasis-related genes in vitro. In vivo patient data also showed higher levels of MMP10 in the primary tumor of metastatic tongue cancer patients and a higher number of LNM metastases for MMP10 overexpressing cancer cells in mice [67]. MMP2/9 staining was present in most cancer cells in early-stage tongue cancer [68] and MMP9 was expressed by stromal and tumor cells at the invasive front of tongue cancers and in the metastatic lymph nodes, increasing with tumor stage [63,69]. In the cancers of the floor of the mouth, MMP2 staining localized into fibroblasts in the tumor stroma and correlated with N status, whereas MMP9 staining localized into macrophages in the inflammatory infiltrate and did not correlate with clinicopathological factors but associated with worse overall survival [70]. Mucoepidermoid carcinomas (a rare type of salivary gland carcinoma) were found to have higher levels of MMP14 compared to normal salivary glands and this increased level could be descriptive of higher invasive capacity [71].

MMPs are not only produced by cancer cells but other cell types in the primary tumor and metastatic site as well, which might also affect their function and targets (Fig. 1). These multifactorial microenvironment and crosstalk between cell types have also been modelled in vitro. Interestingly, treatment of natural killer cells (innate immune cells with known anti-tumorigenic function) with recombinant human MMP2 or MMP9 could decrease their lytic activity against OSCC cells [72]. On the other hand, CAFs in bone invasive OSCCs associated with worse survival and bone invasion, and could secrete MMP9 in vitro to enhance CAF motility [73]. MMPs can also be secreted by bone marrow cells as a reaction to growth factors secreted by the oral cancer cells, as was shown for MMP1 in vitro co-culture model [61]. To add, microbes can also change MMP levels in epithelial cells and thus contribute to disease progression as shown for *Porphyromonas gingivalis* which in vitro caused the increased expression of MMPs in primary oral epithelial cells [74].

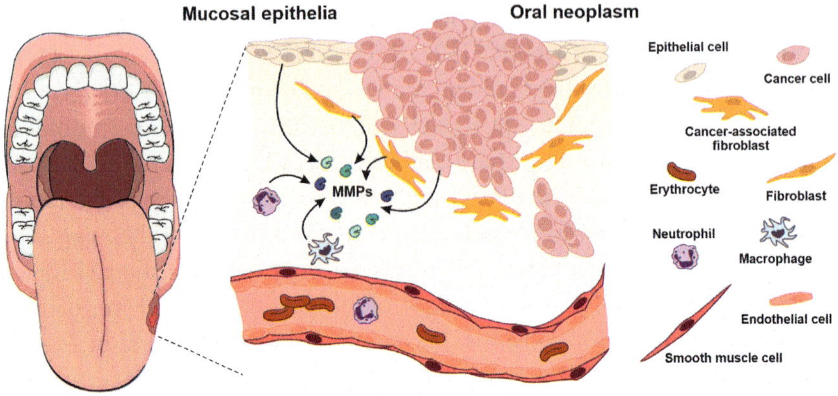

Fig. 1 MMPs participate in various cellular and molecular events in oral malignancies, which most often appear in the mobile tongue. Many cell types produce MMPs in the oral cancer tumor microenvironment, as illustrated in the right side of the figure.

2.2.2 *MMPs as salivary oral cancer biomarkers*

Salivary MMP levels mostly increase with pathological oral conditions. Overall, more proteases could be detected in the saliva of OSCC patients compared to healthy control or patients with benign lesions [75]. The levels of MMP1, −3, −10, −12 and −13 were higher in the saliva of OSCC patients than healthy controls in studies with limited sample cohorts, yet the levels of MMP8 and −9 variated between the studies and increased levels of MMP2 and −12 were only seen in one of the studies [40,75,76]. A larger OSCC (mostly tongue cancer) patient cohort revealed increased MMP2 and −9 levels and MMP activity in saliva compared to controls [46]. MMP9 is one of the most frequently studied salivary MMP biomarkers for OSCC. The sensitivity of MMP9 is great (up to 100 percent) [77–79], but specificity is lacking (as low as 26.7 percent) [80], meaning that the test can correctly identify cases but also finds false positives. A systematic review and meta-analysis that combined seven studies and 795 patients of early stage OSCC found that MMP9 has good specificity, sensitivity and area under curve, yet also raised concerns on study heterogeneity and high risk of bias, which validates further testing with improved setting [81]. Apart from tumorigenesis itself, cancer therapy is also known to affect MMP levels. Acute damage caused by radiation therapy decreased salivary flow rate, pH and buffer capacity in early phases of radiation therapy. Furthermore, low salivary pH correlated with endogenous activity of MMP9 even if presence of mucosal lesions did not correlate with salivary MMP8 or −9 levels [82].

MMP1 is another promising biomarker for OSCC in saliva. MMP1 levels are generally higher in the saliva of OSCC patients compared to healthy controls [83] and MMP1 in saliva was found to be useful in separating between OSCC and non-cancerous groups (healthy controls and OPMD) in a study of 460 participants and in a separate clinical trial of 1160 participants (NCT05049408). Interestingly, salivary MMP1 levels increased with malignancy (T and N status) [84,85] and in vitro data supports the role of MMP1 in OSCC aggressiveness: Silencing MMP1 expression in HNSCC cells, mostly tongue cancer, led to decreased invasion and migration as well as increased apoptosis [41]. Furthermore, in silico screened potential MMP1 inhibitors Kaempferol and Quercetin were found to decrease tongue cancer cell migration and proliferation as well as MMP expression [86]. When proteomic analysis was applied to saliva samples from OSCC patients and healthy controls, elevated levels of MMP1 were detected in the case of OSCC [87]. Promisingly, low variation, high accuracy and successful detection of MMP1 in OSCC patients (but not in healthy controls) was seen in a trial of detecting MMP1 from dry saliva spots with immunoenrichment-coupled Matrix-Assisted Laser Desorption Ionization- Time Of Flight (MALDI-TOF) mass spectrometry [88]. The promising findings in using salivary MMP1 levels to detect OSCC led to a rapid strip test being developed and tested in an OSCC patient cohort of over 600 participants. Initial testing of this rapid strip test found that the strip test is sensitive, can identify OSCC patients with different sites and stages, and the results are comparable to ELISA [89].

Salivary biomarkers are being actively studied in OSCC, because of the previously mentioned benefits, but especially to enable early detection and screening for recurrence [90], which are known clinical challenges of OSCC. For example, a clinical trial (NCT06701058) investigated using active MMP8 levels from a mouth rinse as a detection method for OSCC, yet no results have been posted. Interestingly salivary MMP8 levels might even differ from other MMPs: Salivary MMP8 levels were lower in OSCC patients compared to healthy controls [91]. Yet, the current studies still lack sufficient cohort numbers to reliably assess the validity of MMPs as salivary biomarkers for OSCC. Larger patient cohorts are needed. Furthermore, a single biomarker is likely not enough to differentiate between OSCC cases and healthy persons or assess the risk of recurrence, so biomarker panels, possibly consisting of metabolic, microbiome, proteomic and/or genetic factors, might be needed [90].

2.2.3 MMPs as systemic oral cancer biomarkers

Some analyses have investigated concomitant changes of local (saliva, tissue) and systemic (blood) MMP levels. For example, a systematic and combined analysis of MMP2, −7, and −9 levels in OSCC patients revealed a systemic (saliva and serum) and local (tumor) increase in mean difference (1.62 and 0.085, respectively) compared to controls [92]. MMP9 in saliva and serum were 15- and 6-fold higher (respectively) in OSCC patients compared to healthy controls, suggesting usability as a biomarker [93]. Similarly, elevated MMP3 levels in saliva and serum of OSCC patients were found and these levels correlated well with each other [94].

MMP2 and −9 are found among the most studied MMPs in serum samples from OSCC patients and are generally elevated in OSCC and especially in advanced cases. MMP2 and −9 levels were elevated in the serum of OSCC patients compared to healthy controls and associated with LNM [95], yet another similar study could only identify elevated levels for MMP9 [72]. Treatment might affect systemic MMP levels: MMP9 levels were elevated in pre-operative serum samples in OSCC patients, and high levels were associated with worse OS, LNM and serum VEGF, and MMP9 levels decreased after surgery [96]. Serum levels and their diagnostic potential have also been studied for other MMPs. MMP3 was elevated in the serum of OSCC patients compared to healthy controls, but no association to clinic-pathological features such as tumor stage or LNM could be identified [97]. Antibodies against MMP7 were detected in serum samples and the levels of these antibodies increased in patients compared to healthy controls and especially with malignancy (higher stage, LNM). Furthermore, high serum anti-MMP7 antibodies predicted worse overall survival [98] In vitro studies have shed light into a molecular mechanism that facilitates elevated MMP1 expression in OSCC. Transcription factor Jun (c-Jun) and related pathways were identified as one of the upregulated pathways in OSCC patients, and in silico analysis predicted that c-Jun could regulate MMP7 expression. Silencing c-Jun in OSCC cells led to decreased expression of MMP7 and lowered cell migration, which highlights MMP7 and modulation of its expression by transcription factors as one of the potential mechanisms of OSCC development [99].

MMPs have also been analyzed from plasma samples with similar results to serum samples. MMP1 levels in plasma and saliva samples did not correlate although an increase in MMP1 levels was detected in OSCC patients compared to healthy controls in both sample types. Larger differences between MMP1 in OSCC and in healthy persons were seen in saliva samples, indicating that saliva samples might give more accurate picture of the MMP levels than systemic measurement via blood [87]. MMP9 plasma levels were elevated in OSCC patients and proved to be usable in OSCC detection, especially combined with other markers such as MMP2 and TIMP2 [100]. Proform, active and total MMP2 and −9 levels were also significantly higher in the plasma of oral cancer patients compared to the controls and correlated with differentiation and LNM (for MMP2) [101]. MMP9 levels, but not MMP2, varied during treatment and between responders and non-responders [50]. Similarly, within male OSCC patients, MMP11 levels in plasma increased in conjunction with LNM and advanced stage [57].

2.3 Oral potentially malignant disorders

Oral potentially malignant disorders (OPMDs) include leukoplakia, erythroplakia, proliferative verrucous leukoplakia, oral lichen planus (OLP), oral submucous fibrosis, palatal lesions in reverse smokers, lupus erythematosus, epidermolysis bullosa, and dyskeratosis congenita, as well as oral lichenoid lesions and oral manifestations of chronic graft-versus-host disease. Other diseases, such as oral exophytic verrucous hyperplasia, have also been suggested as risk factors for oral cancer, but reliable evidence on the increased risk of oral cancer exists only for the above-listed diseases [102]. Gene expression analysis in OPMDs has shown low expression levels for many MMPs or varying results between studies (especially for MMP2) [103–105], yet MMP protein levels seem to be more consistently increased in OLP tissues [106]. In general, there is a simultaneous increase in MMPs and progression from healthy tissue to OSCC via OPMDs. This increase could be usable as a prognostic biomarker to predict which OPMDs will evolve into OSCC. This is hinted at by a study that used a combination of 4 protein markers (MMP1 included) from saliva to assess the risk of malignancy and showed that 77.8 percent of OPMDs that progressed to OSCC within 5 years had a risk score above 0.4 initially [85].

MMP9 has been the most studied MMP in OLP. MMP9 levels and expression are increased in samples from OLP patients compared to samples from healthy persons or gingival overgrowth, both in tissue gene expression [107] and protein level [108] as well as in crevicular fluid samples [109]. As sais, many cell types produce MMPs: T cells have been suggested as a potential source for MMP9 in OLP tissues [110]. Increased MMP9 levels were also detected in the saliva of vesiculoerosive (including OLP) lesions and OPMD patients compared to healthy controls [78,111], yet no differences in MMP9 activity was detected between these groups [111]. Similarly, saliva and serum MMP9 levels increased, via OPMD to OSCC patients, and salivary MMP9 levels could be used to differentiate between patients with premalignant lesions and OSCC [93].

Other proteases have also been studied, but mostly in singular studies. MMP14 levels [112] and MMP7 gene expression [113] increased from different forms of OLP (e.g., erosive) to OSCC tissue samples and similar gradual increase was detected for MMP3 in serum and saliva samples [94] and MMP12 levels in tissue, saliva and serum samples [114]. Mechanistic studies of MMPs and OPMDs are very limited. In vitro assays showed that macrophage-secreted MMP12 can cleave fibronectin, leading to disruptions in keratinocyte barrier integrity, which could facilitate the development of OLP [114].

2.4 Oral submucous fibrosis

Oral submucous fibrosis (OSF) is characterized by the stiffness of the mucosa due to excessive collagen cross-linking and deposition related most to the use of areca nuts. Fibrosis itself is part of a normal healing process, where various cell types infiltrate damaged epithelia and stroma to repair wound bed and epithelium via the formation of active granulation tissue. Fibroblasts produce ECM, which, in normal healing results in the formation of a mature ECM, but in fibrosis, is excessive and differs from the ECM in healthy tissue. Matrix stiffness has been suggested as one of the driving factors of OSF and the subsequent malignant transformation, which has been investigated since the 1930's [115]. As a bidirectional relationship between MMPs and matrix stiffness exists, it would be tempting to assume MMPs have a role in OSF, yet the evidence is limited, as can be seen below.

Some studies have estimated the effect of genetic variants of MMPs on OSF, mostly promoter polymorphisms that could affect the expression levels of these proteases. *MMP1* gene promoter (−1607 1G/2G) polymorphisms were highly present in OSF cases and associated with higher enzymatic activity [116]. *MMP3* promoter polymorphism 5A allele was found to be more prevalent in OSF fibrosis patients compared to OSCC patients and healthy controls (with or without use of areca nut) in a study with a small cohort [117]. Other studies looking into *MMP3* promoter polymorphism and with bigger cohorts, found this polymorphism more frequently in OSF and OSCC patients compared to controls and could estimate that this 5A allele associates with a 2.3-fold increased risk for OSF [118,119]. In contrary, MMP9 promoter polymorphism (−1562 C>T) did not associate with risk of OSF in multiple studies [31,120] nor did MMP2 promoter polymorphisms (−1306 C/T and −168 G/T) [120].

Salivary MMP levels in OSF patients have only been estimated in a few studies. A study with 30 OSF patients found increased levels of MMP12 in the saliva compared to healthy controls [76]. A study with a similar experimental setup discovered that salivary MMP8 levels are notably lower in OSF patients compared to healthy controls [91]. Furthermore, MMP8 levels decreased in saliva with disease progression and analysis of salivary DNA detected various MMP8 SNPs, that affect the expression of MMP8 [121]. Similar to OSCC, MMP8 seems to be one of the MMPs that are lowered during disease progression and are thus not contributing to the disease emergence.

In vitro, OSF has been induced in oral fibroblasts by the addition of areca nut active compounds such as arecoline. Anti-fibrotic miRNAs, a potential future therapy, were shown to decrease wound closure as well as collagen type I and MMP1 expression in arecoline-induced oral fibroblast culture [122]. The potential link between OSF and OSCC has been studied in vitro by stimulating oral cancer cells with the secretome of fibrotic buccal mucosal fibroblasts. This study found that this conditioned media could induce epithelial mesenchymal transition (EMT) in OSCC cells, increase invasion and elevate MMP2 levels, suggesting MMP2 as a mediator of malignant transformation caused by OSF [123]. In OSF tissue samples, MMP2 expression increased from early stages to advanced stages [124]. Fibrosis is associated with senescence in fibroblasts and in vitro fibroblast senescence is associated with elevated levels of MMP1 and −2 [125]. Tissue MMP1 and −2 protein expression has been shown to increase when OSF tissues are compared to healthy buccal mucosa [126,127], yet in contradiction, the gelatinolytic activity in OSF tissues was lower compared to healthy mucosa [127]. Interestingly, arecoline could also induce TIMP1 expression and reduce MMP2 levels and gelatinolytic activity in buccal mucosal fibroblasts in vitro [128]. These studies play on the idea of MMP2 as one of the drivers of OSF progression, yet the evidence is partly conflicting.

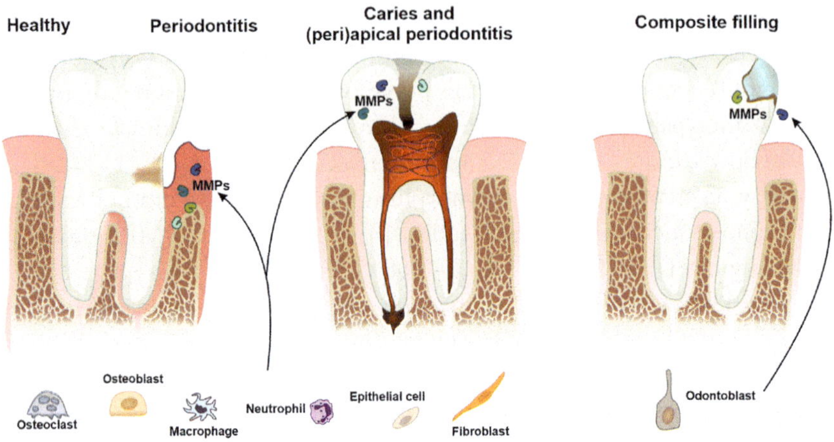

Fig. 2 MMPs play a role in periodontitis, caries and peri-apical lesions as well as dental bonding. The cell types producing MMPs in these conditions are illustrated in the bottom of the figure.

3. Dental diseases

3.1 Periodontal diseases and peri-implantitis

Periodontal diseases are inflammatory conditions involving the supporting structures of teeth. Although dental biofilm is the primary cause, genetic and environmental factors, especially smoking, affect disease risk and progression. Periodontal manifestations may also occur in systemic diseases such as hematological, immunological, dermatological, granulomatous, and neoplastic disorders. Gingivitis is a reversible inflammation of the gingiva caused by dental plaque without loss of supporting tissues. Periodontitis is a more severe, chronic inflammatory disease leading to irreversible destruction of connective tissue and alveolar bone, often resulting in tooth loss (Fig. 2). It arises from a disruption in the equilibrium between the host and the resident oral microbiome, resulting in microbial dysbiosis that promotes an exaggerated host inflammatory response and subsequent destruction of the periodontal ECM. Periodontitis has been associated with adverse pregnancy outcomes, cardiovascular disease, stroke, pulmonary disease, and diabetes [11,129–132]. In addition, severe periodontal disease has been connected to OSCC as a risk factor [133].

Peri-implantitis is a plaque-associated inflammatory condition around dental implants, characterized by peri–implant mucosal inflammation and progressive loss of supporting bone. The disease state can be further categorized as a continuum that exists between health and peri–implant diseases. Peri-implant mucositis represents a precursor stage for peri–implantitis which is more severe form of disease and is distinguished by progressive loss of supporting bone [134]. In general, oral mucositis refers to inflammation and ulceration of the mucosa, which may be caused by damage to the oral mucosa such as that occurring due to oral implant placement or cancer therapy [135]. Although the disease mechanisms in both periodontitis and peri–implantitis are similar, they occur in different structures and are considered different diseases.

Collagenases have been extensively studied in periodontal manifestations after their activity was first detected in human gingiva in 1966 [136]. Although MMPs undoubtedly play a role in gingivitis [109,137], most human studies exploring MMPs in periodontal diseases focus on periodontitis and peri–implantitis. MMP8, −9, and −13 are the most widely reported MMPs in GCF and are considered the most abundant MMPs in periodontal tissues [13,130]. The expression levels of both MMP8 and −9 are well documented to reflect periodontal disease severity and progression, treatment responses as well as severity of peri–implantitis [13]. Especially MMP8's role and value as a biomarker have been extensively studied in periodontal diseases. Chair side tests for active MMP8 has proven to be useful as an adjunctive diagnostic quantitative tool to monitor periodontitis and peri–implantitis [18,138–142]. In an experimental periodontitis model in mice, MMP8–deficiency leads to more severe disease after *P. gingivalis* infection compared to wild-type animals and is accompanied by reduced LIX/CXCL5 expression and impaired neutrophil infiltration [143,144]. This suggests that MMP8 has an important role in regulating the first-line defense and lack of its function leads to a defect in chemokine-mediated neutrophil recruitment to the biofilm interface. MMP-mediated activation also plays a role in periodontal ligament fibroblast migration and cytokine secretion [145]. The ability of several MMPs to cleave and inactivate CC motif chemokines to produce truncated proteins acting as potential receptor antagonists represents a potential regulatory feedback mechanism to restrict monocyte infiltration and inflammation [146].

The activation cascades of MMPs within the oral cavity are tissue and disease-specific and involve both host and microbial proteinases. The host immune response and inflammation in periodontal sites is initiated by a polymicrobial challenge of biofilm that includes, among other factors, exposure to bacterial lipopolysaccharide (LPS), such as that derived from

P. gingivalis [9,147]. In experimental models, *P. gingivalis* has been shown to directly affect the cytokine and MMP expressions in cells, with varying effects on periodontitis and peri–implantitis fibroblasts compared to cells from periodontally healthy individuals [148]. Pro-inflammatory cytokines such as IL-1β and TNF-α upregulate MMP expression in periodontal tissues, thereby promoting connective tissue destruction [149,150]. Components, including proteolytic enzymes, produced by bacteria such as *P. gingivalis* [151,152] and *Treponema denticola* [153–156] further release and activate MMPs [153] which degrade the structural components within periodontal tissues, contributing to gingival recession and tissue destruction [150,157]. Moreover, host proteases, such as serine proteases, contribute to complex proteinase signaling and activation cascades in periodontium to facilitate tissue destruction [158,].

3.2 Genetic predisposition—*MMP* genes in periodontitis and peri–implantitis

A growing body of literature has investigated the potential association between polymorphisms in various *MMP* genes in periodontitis patients. Although findings have been partly inconsistent, some of the recent systematic reviews and meta-analyses support the existence of a significant association with periodontitis. In a well-characterized cohort, a molecular biomarker profile comprising the *MMP8* −799 T allele, elevated *MMP8* mRNA expression, and increased aMMP8 protein levels served as independent predictors of peri–implantitis [159]. A meta-analysis found that *MMP8* −799 C/T and *MMP3* −1171 5A/6A polymorphisms increase the risk of periodontitis, while *MMP9* −753 C/T polymorphism reduces the risk. However, no associations were found for *MMP2* −753 C/T with risk of periodontitis [160]. *MMP1* g.−1607dupG polymorphism showed association with chronic periodontitis and disease severity in a meta-analysis of 11 studies [161]. Moreover, osseointegrated implant failure is associated with G-1607GG polymorphism in *MMP1* gene [162]. Differences in gene polymorphisms have been observed also in genotypes of *MMP3* (−1171 5A/6A) and *MMP7* (−181 A/G) in peri–implantitis patients compared to healthy controls [163], while a meta-analysis including 17 studies found no association of common variants of *MMP1* (−1607 1G/2G, −519 A/G, and −422 A/T), *MMP2* (−1575 G/A, −1306 C/T, −790 T/G, and −735 C/T), *MMP3* (−1171 5A/6A), *MMP8* (−381 A/G and +17 C/G), *MMP9* (−1562

C/T and +279 R/Q), *MMP12* (−357 Asn/Ser), and *MMP13* (−77 A/G, 11A/12A) to periodontitis [164].

3.3 MMPs as biomarkers in periodontitis and peri-implantitis

3.3.1 MMPs as salivary biomarkers in periodontitis and peri-implantitis

Periodontitis patients' saliva samples have been widely studied to provide biomarkers for the disease. MMPs such as MMP1, −2, −3, −8, −9 and −13 are widely detectable in human saliva [165,166]. MMP8 and MMP9 are by far the most extensively studied and their elevated levels, along with high IL-1β in saliva, were connected to periodontitis in meta-analysis of 27 studies [167].

Although individual MMPs show significantly increased levels in the saliva of periodontitis patients, the predictive accuracy between gingivitis or periodontitis patients and controls can be further improved if multiple parameters are tested in combination. A study analyzing 10 potential salivary biomarkers concluded that ratios of MMP8, −9 and TIMP-1 were more powerful in differentiating values compared to the single biomarkers [168]. A positive association with disease has also been reported for salivary MMP2/TIMP-2 complex in a cohort of chronic periodontitis patients [169]. It is noteworthy that saliva levels of MMPs might be affected by various factors, for example, smoking status. A systematic review including seven studies found that various molecules used as periodontal disease markers were changed in smokers, including an increase in TIMP-1, a decrease in MMP8/TIMP-1 ratio as well as combinations of MMP8 and carboxyterminal telopeptide of type I collagen (ICTP) [170]. The effect of smoking and other external factors may thus interfere with the interpretation of biomarker-based tests. In addition to cytokines and MMPs, the levels of oral microbes related to periodontal health have been used in biomarker panels. Although MMP8 and IL-1β as combination exhibited the best accuracy in discriminating between healthy subject from gingivitis patients, ICTP and *P. gingivalis*, in combination with IL-1β showed the best efficacy to discriminate periodontitis from both healthy and gingivitis groups [171].

Along with MMP8, increased levels of both MMP9 and −13 as well as high MMP9/TIMP-1 ratio are found in periodontitis patients' saliva [172–174]. However, MMP8 seems to serve as the best marker to differentiate subjects with severe alveolar bone loss from those with minor bone loss [174]. A cohort of Swedish periodontitis patients instead found high MMP13 in saliva to associate with higher clinical attachment loss and female gender. Unexpectedly, the study found no significant difference in MMP8, −9, or TIMP-1 levels in the saliva of periodontitis patients compared to controls [175]. However, high MMP9 has been connected to the depth of periodontal pocket, thus pointing out its potential as a biochemical indicator for periodontitis severity [173,176]. Also, salivary levels of MMP8 have been reported as a potential biomarker in monitoring the periodontitis disease progression, response to treatment and recovery following nonsurgical periodontal therapy [177–180]. Within the patient cohort of moderate to severe generalized chronic periodontitis, those individuals with type 2 diabetes showed the highest MMP8 levels, indicating the influence of diabetes on periodontal disease progression [181]. Interestingly, both saliva and serum levels of MMP8 are increased in periodontal patients [182]. As the link between periodontitis and cardiovascular diseases (CVD) is widely recognized, MMP8 may contribute to increased CVD susceptibility in periodontitis patients.

3.3.2 MMPs as gingival crevicular fluid and peri–implantitis fluid biomarker

MMP8 and −9 are the most studied MMPs from GCF [183–185]. MMP8 has demonstrated good sensitivity and specificity as a biomarker to diagnose periodontitis in systemically healthy subjects [186]. In addition, MMP13 is found with significantly higher levels in periodontitis compared to gingivitis sites and MMP14 higher in periodontitis than in healthy gingival sites [187], suggesting that these enzymes may act co-operatively within their activation cascades. Moreover, high accuracy for specific site-diagnosis of chronic periodontitis versus gingivitis and healthy sites has been demonstrated for both MPO and MMP8 [187]. A meta-analysis comparing the sensitivity and specificity of periodontitis biomarkers found that MMP8 was superior to other tested biomarkers, including elastase, cathepsin and trypsin [186,188]. Although MMP8 has been found to represent an accurate diagnostic biomarker for the identification of periodontal disease destruction, a meta-analysis of five articles found high variation in the measured levels and was not able to confirm the potential of MMP8 as biomarker to predict the outcome of nonsurgical periodontal therapy (NSPT) in patients with periodontitis after 3 months [189].

Both MMP8 and MMP9 levels in GCF during early pregnancy are increased in women with periodontitis and show an association with the risk of gestational diabetes mellitus development [190]. Based on the levels measured from both saliva and GCF, both MMP8 and MMP9 are considered as potential indicators of gingival inflammatory state during pregnancy [191]. The relevance of MMP9 in periodontitis is also supported by meta-analysis of six studies examining patients with periodontitis and periodontally healthy controls []. The study found statistically significant higher MMP9 levels in GCF (and in serum) of those patients with periodontal disease [185]. Interestingly, those with higher GCF MMP9 and periodontal disease were reported to also have lower serum vitamin D concentrations [192]. Patients with stage 2, grade B periodontitis with type 2 diabetes mellitus benefitted the adjunctive use of doxycycline subantimicrobial dose therapy together with scaling and root planing (SRP) and showed decreased levels of both MMP9 and −13 in GCF samples compared to those receiving SRP only [193].

In peri–implantitis, biomarkers may serve to assist in the early diagnosis of pathological bone loss that characterizes the disease, but do not at current stage replace other diagnostic methods. Increased MMP8, −9 and −13 levels in PICF are associated with peri–implantitis bone loss [184]. Of those, MMP8 is the most widely measured biomarker in peri–implantitis and meta-analysis has supported its value as a predictive disease biomarker [194]. A study measuring levels of MMP1, −2, −3, as well as the inflammatory mediators osteopontin (OPN), pentraxin-3 (PTX3), and thymic stromal lymphopoietin (TSLP) in GCF samples found both MMP2 and TSLP as potential biomarkers for not only periodontitis but also peri–implantitis [195]. Other MMPs reported as potential biomarkers in PICF include MMP9 and −13, among TIMP-2 and various inflammatory mediators, such as IL-1β and IL-8 [184,196].

Increased MMP8 levels are also reported in the peri–implant crevicular fluid from patients with peri–implant mucositis compared to healthy controls [197], yet interestingly no difference is observed when MMP8 levels are assessed in same sample type from peri–implantitis versus healthy implant [198]. This could indicate that MMP8 levels elevate in early, soft-tissue inflammation, but do not stay in high levels in later, hard-tissue destroying phases of the disease. Similarly, acute tissue damage may elevate MMP8, as increased MMP8 levels were detected in implant crevicular fluid two weeks after implantation compared to two months after the operation [199].

3.4 Caries, pulp and periapical pathogenesis

The integrity of dentin is essential for maintaining tooth structure and health. Unlike enamel, which contains less than one percent of organic matrix, dentin is composed of 30 percent of organic material, making it a more susceptible substrate for degradation by host-derived proteinases [200]. Collagen is an essential component of dentin as its crosslinking provides mechanical properties. MMPs have a significant role in dentinogenesis, dental bonding and the progression of caries [201] (Fig. 2). At least MMP2 [202,203], −3 [204], −8 [205], −9 [202,203], and −20 [206] have been identified in dentin [201]. MMP2 and −9 are known to cleave dentin sialophosphoprotein (DSPP), a non-collagenous protein secreted by odontoblasts, and play a central role in dentin mineralization [207,208]. Caries lesions are formed when demineralization is caused by microbial acids. The organic matrix of dentin is eventually degraded as the lesion progresses. The first evidence for the involvement of MMPs in this process is from end of 1990s when both pro- and active forms of MMP2, −8, and −9 were demonstrated in human caries lesions [209]. The MMPs involved in caries can be derived from odontoblasts, saliva or GCF. Acidic by-products from cariogenic bacteria demineralize dentin and pH changes occurring due to demineralization activate MMPs locally [209]. Inhibition of MMP activity has been shown to inhibit dentinal caries lesion progression [210]. Moreover, as MMPs degrade the collagenous matrix of the hybrid layers in dentin bonding, MMP inhibitors, including chlorhexidine, inhibit the weakening of the resin-dentin bond that occur over time [211].

In the dental pulp, invasion of bacteria and various inflammatory mediators stimulates the expression of MMPs, which are produced by pulpal odontoblasts, endothelial cells, inflammatory cells, and fibroblasts [1,12]. In pulpitis, MMP activity contributes to ECM degradation and inflammatory cell migration. Various MMPs and TIMPs are expressed by odontoblasts and in pulp cells [12]. At protein levels, MMP8, −9 and -13 have been demonstrated in inflamed dental pulp [212–214]. MMP2 and −9 show a strong correlation with the progression of apical periodontitis in an animal model [215]. MMP9 is produced by resorbing osteoclasts and thus may contribute to the degradation of the bone organic matrix in apical periodontitis. As MMPs play a central role in tissue destruction and repair in general, they are considered important biomarkers and potential therapeutic targets also in cariology and endodontics.

4. Systemic and developmental diseases affecting oral health

4.1 Oral ulcers

Oral mucosal wounds show faster re-epithelialization and heal more rapidly than cutaneous wounds due to unique physiological, anatomical and microbial features of oral cavity. Oral wounds also show attenuated inflammatory response with less fibrotic tissue and scar development [216]. In human oral mucosal wounds, MMP9, but not MMP2 expression increases, suggesting that MMP9 is involved in normal keratinocyte migration and tissue granulation [217]. However, in chronically inflamed human gingiva and subepithelial connective tissue, MMP2 expression is detected and assumed to play role in epithelial cell migration [218]. Evidence from mouse studies suggests that also MMP8 plays a role in oral wound healing by affecting tongue fibroblast migration and cleaving TGF-β1, possibly representing a mechanistic explanation for its effects in fibroblast function [219]. Likewise, MMP8 knockout mice show alterations in tooth socket healing suggesting that MMP8 participates in modulating the collagen metabolism and inflammatory response of alveolar bone and mucosa [220]. Based on the OSCC patient's mucositis study, salivary MMP8 or −9 did not correlate with the presence of radiation induced oral mucosal ulcerative lesions. However, the endogenously activated salivary MMP9 level correlated with low salivary pH typical in mucositis saliva samples [82].

One of the most common oral ulcers are aphtous ulcers (or canker sours). Although they affect the quality of life, they are usually harmless. However, they are painful and cause discomfort and impact eating and speaking. The etiology of aphtous ulcers is not completely understood but several factors, such as minor mechanical injuries as well as some systemic conditions such as nutritional deficiencies, celiac disease, inflammatory bowel diseases, and Behçet's disease have been associated with their development and exacerbation [221]. The expression of MMP8 has been localized intracellularly in the connective tissue and in association with the basement membrane laterally to the ulcer. MMP1 and −3 have been detected in epithelia close to the ulcers, capillary endothelium and postcapillary venules as well as in laterally to the ulcer in the lamina propria residing macrophage- and fibroblast-like mononuclear cells [222]. The use of low doses of doxycycline was shown to fasten the healing and reduce pain in patients with aphtous ulcers in a randomized, double-blind, placebo-controlled trial [223].

Beçhet's disease (BD) is a rare heterogeneous, multisystem, chronic systemic vasculitic disorder, likely caused by genetic predisposition and exposure to environmental triggers. Oral painful ulcers are the initial and persistent symptom of this disease and are present in 98 percent of cases [224]. MMPs, mostly MMP2 and −9 polymorphisms and levels, have been studied in BD, yet not specifically in oral ulcers or their formation nor recently (newest study from 2017). In general, an increase in MMP9 level is seen in the blood plasma [225], serum [226], skin lesions [227], cerebrospinal fluid [228] and aqueous humor of the eye [229] in BD patients compared to healthy controls and especially when the disease is active. As MMPs are known to induce leukocyte migration, it is not surprising that MMP9 levels positively correlate with the number of polymorphonuclear and mononuclear cells in CSF [228] or that platelet-derived inflammatory factors can increase MMP9 levels in megakaryoblasts (platelet precursors) [225]. For MMP2, the levels vary and no clear conclusion can be drawn [225–227]. Similarly, the analysis of genetic polymorphisms has focused on MMP9 and −2. For example, a promoter polymorphism for MMP9, −1562 C/T, is not associated with the risk of BD [230,231], but C/C polymorphism was higher in BD patients [227]. Similarly, a polymorphism in the MMP9 gene (rs17576), increased the risk of BD [232]. It should be noted though that gene expression or polymorphism does not necessarily associate with the protein level [230,233].

The role of MMPs in oral ulcers has also been studied in other diseases, such as systemic lupus erythematosus (SLE). SLE is an autoimmune disease affecting mostly females and presenting with multi-organ phenotype, including various oral manifestations [234]. MMP12 levels are elevated in blood polymorphonuclear cells in SLE patients [235] and similarly, MMP9, but not MMP2, levels are elevated in serum samples [236]. Of these, MMP9 level in serum is associated with the incidence of oral ulcers. MMP9 protein level in tongue ulcers of a rat model was found to be elevated compared to healthy tissue and could be lowered by treatment with active compounds derived from a medicinal plant *Thymus satureioides* [237].

4.2 Developmental defects in oral cavity

One of the most common birth defects worldwide affecting the oral cavity is the orofacial cleft (OFC). OFCs are the result of defective closure of facial tissues during embryonic development and are thought to be orchestrated by cues from ECM and defective signaling (e.g., TGF-beta signaling), especially related to neural crest cells. OFCs can manifest along with other syndromes, e.g., van der Woude syndrome, and are then often caused by a single disturbed gene, whereas nonsyndromic OFCs are likely caused by a combination of different genetic variants and risk factors [238]. A systematic review by Imani et al. pooled together all studies that evaluate *MMP* and *TIMP* variants in different orofacial cleft types. Although many *MMPs* and *TIMPs* have been studied, mostly in European or Brazilian populations, only few SNPs can be associated to OFCs: Namely those related to *MMP2* (rs243865, rs9923304), *MMP3* (rs522616, rs3025058), *MMP9* (rs17576, rs6094237) and *TIMP2* (multiple) [239]. Yet these associations are not supported by the meta-analysis they performed; thus, it is unlikely that genetic changes in *MMPs* alone have a causative role in development of OFC, but they could have a combinatory effect together with other factors. This sort of multifactorial role has been discussed in e.g., the case of MMP3 [240].

5. Final remarks

The connection between MMPs and oral health has been studied in past decades, focusing extensively on malignant oral neoplasms, teeth and periodontal diseases. The levels and activity of MMPs fluctuate in tissues and body fluids and are affected by many factors (such as systemic inflammation or different cell types), which create challenges in establishing them as disease-specific biomarkers. Currently, no MMP-based biomarker or test is in widespread active clinical use, although such inventions have been made. Still, it is likely that MMPs do contribute significantly to organogenesis, physiological and pathological events of many oral and maxillofacial conditions, together, but not solely, with other modulatory factors.

References

[1] M. Elgezawi, R. Haridy, K. Almas, et al., Matrix metalloproteinases in dental and periodontal tissues and their current inhibitors: developmental, degradational and pathological aspects, Int. J. Mol. Sci. 23 (2022) 8929.

[2] Y. Lu, P. Papagerakis, Y. Yamakoshi, et al., Functions of KLK4 and MMP-20 in dental enamel formation, Biol. Chem. 389 (2008) 695–700, doi:10.1515/BC.2008.080.

[3] J.D. Bartlett, J.P. Simmer, Proteinases in developing dental enamel, Crit. Rev. Oral Biol. Med. 10 (1999) 425–441, doi:10.1177/10454411990100040101.
[4] T. Takata, M. Zhao, T. Uchida, et al., Immunohistochemical detection and distribution of enamelysin (MMP-20) in human odontogenic tumors, J. Dent. Res. 79 (2000) 1608–1613, doi:10.1177/00220345000790081401.
[5] A. Väänänen, L. Tjäderhane, L. Eklund, et al., Expression of collagen XVIII and MMP-20 in developing teeth and odontogenic tumors, Matrix Biol. 23 (2004) 153–161, doi:10.1016/j.matbio.2004.04.003.
[6] A. Väänänen, R. Srinivas, M. Parikka, et al., Expression and regulation of MMP-20 in human tongue carcinoma cells, J. Dent. Res. 80 (2001) 1884–1889, doi:10.1177/00220345010800100501.
[7] J.-W. Kim, J.P Simmer, T.C Hart, et al., MMP-20 mutation in autosomal recessive pigmented hypomaturation amelogenesis imperfecta, J. Med. Genet. 42 (2005) 271–275, doi:10.1136/jmg.2004.024505.
[8] T. Atanasova, T. Stankova, A. Bivolarska, T. Vlaykova, Matrix metalloproteinases in oral health—special attention on MMP-8, Biomedicines 11 (2023) 1514.
[9] T. Sorsa, L. Tjäderhane, T. Salo, Matrix metalloproteinases (MMPs) in oral diseases, Oral Dis. 10 (2004) 311–318, doi:10.1111/j.1601-0825.2004.01038.x.
[10] J. Walsh Laurence, Mast cells and oral inflammation, Crit. Rev. Oral Biol. Med. 14 (2003) 188–198, doi:10.1177/154411130301400304.
[11] C. Franco, H.-R. Patricia, S. Timo, et al., Matrix metalloproteinases as regulators of periodontal inflammation, Int. J. Mol. Sci. 18 (2017) 440.
[12] H. Palosaari, C.J. Pennington, M. Larmas, et al., Expression profile of matrix metalloproteinases (MMPs) and tissue inhibitors of MMPs in mature human odontoblasts and pulp tissue, Eur. J. Oral Sci. 111 (2003) 117–127, doi:10.1034/j.1600-0722.2003.00026.x.
[13] I. Luchian, A. Goriuc, D. Sandu, M. Covasa, The role of matrix metalloproteinases (MMP-8, MMP-9, MMP-13) in periodontal and peri-implant pathological processes, Int. J. Mol. Sci. 23 (2022) 1806.
[14] A. Page-McCaw, A.J. Ewald, Z. Werb, Matrix metalloproteinases and the regulation of tissue remodelling, Nat. Rev. Mol. Cell Biol. 8 (2007) 221–233, doi:10.1038/nrm2125.
[15] D. Malamud, Saliva as a diagnostic fluid, Dent. Clin. North Am. 55 (2011) 159–178, doi:10.1016/j.cden.2010.08.004.
[16] H. Birkedal-Hansen, C.M. Cobb, R.E. Taylor, H.M. Fullmer, Serum inhibition of gingival collagenase, J. Oral Pathol. Med. 3 (1974) 284–290, doi:10.1111/j.1600-0714.1974.tb01723.x.
[17] S. Wei, T. Lin, G. Sáenz-Ravello, et al., Diagnostic accuracy of salivary active matrix metalloproteinase (aMMP)-8 point-of-care test for detecting periodontitis in adults: a systematic review and meta-analysis, J. Clin. Periodontol. 51 (2024) 1093–1108, doi:10.1111/jcpe.14000.
[18] L. Goyal, M. Gupta, S. Sareen, et al., Active matrixmetalloproteinase-8 in periodontal diagnosis: a scoping review, Diagnostics 15 (2025) 2932.
[19] P.M. Preshaw, A.F. Hefti, S. Jepsen, et al., Subantimicrobial dose doxycycline as adjunctive treatment for periodontitis, J. Clin. Periodontol. 31 (2004) 697–707, doi:10.1111/j.1600-051X.2004.00558.x.
[20] F. Bray, M. Laversanne, H. Sung, et al., Global cancer statistics 2022: GLOBOCAN estimates of incidence and mortality worldwide for 36 cancers in 185 countries, CA Cancer J. Clin. 74 (2024) 229–263, doi:10.3322/caac.21834.
[21] A. Bugshan, I. Farooq, Oral squamous cell carcinoma: metastasis, potentially associated malignant disorders, etiology and recent advancements in diagnosis, F1000Research 9 (2020) 229, doi:10.12688/f1000research.22941.1.

[22] Y. Tan, Z. Wang, M. Xu, et al., Oral squamous cell carcinomas: state of the field and emerging directions, Int. J. Oral Sci. 15 (2023) 44, doi:10.1038/s41368-023-00249-w.
[23] N.S. Mastronikolis, E. Kyrodimos, Z. Piperigkou, et al., Matrix-based molecular mechanisms, targeting and diagnostics in oral squamous cell carcinoma, IUBMB Life 76 (2024) 368–382, doi:10.1002/iub.2803.
[24] P. Nyberg, P. Heikkilä, T. Sorsa, et al., Endostatin inhibits human tongue carcinoma cell invasion and intravasation and blocks the activation of matrix metalloprotease-2, -9, and -13, J. Biol. Chem. 278 (2003) 22404–22411, doi:10.1074/jbc.M210325200.
[25] K.V. Kumar, K.N. Hema, Extracellular matrix in invasion and metastasis of oral squamous cell carcinoma, J. Oral Maxillofac. Pathol. 23 (2019) 10–16, doi:10.4103/jomfp.JOMFP_97_19.
[26] Z. Huang, L. Huang, C. Zhang, et al., Blocking β2-AR and inhibiting COX-2: a promising approach to suppress OSCC development, Int. Dent. J. 75 (2025) 807–816, doi:10.1016/j.identj.2024.06.014.
[27] Z. Li, M. Sun, R. Yang, et al., Mediator complex subunit 1 promotes oral squamous cell carcinoma progression by activating MMP9 transcription and suppressing CD8(+) T cell antitumor immunity, J. Exp. Clin. Cancer Res. 43 (2024) 270, doi:10.1186/s13046-024-03191-9.
[28] Y. Li, Y. Wang, H. Sun, et al., Association between matrix metalloproteinase-1, 2, 3 polymorphisms and oral cancer risk: a meta-analysis, Genet. Test Mol. Biomarkers 22 (2018) 456–464, doi:10.1089/gtmb.2018.0089.
[29] L.-C. Shih, C.-H. Li, K.-T. Sun, et al., Association of matrix metalloproteinase-7 genotypes to the risk of oral cancer in Taiwan, Anticancer Res. 38 (2018) 2087–2092, doi:10.21873/anticanres.12448.
[30] E. Vairaktaris, Z. Serefoglou, C. Yapijakis, et al., High gene expression of matrix metalloproteinase-7 is associated with early stages of oral cancer, Anticancer Res. 27 (2007) 2493–2498.
[31] H.-F. Tu, C.-H. Wu, S.-Y. Kao, et al., Functional −1562 C-to-T polymorphism in matrix metalloproteinase-9 (MMP-9) promoter is associated with the risk for oral squamous cell carcinoma in younger male areca users, J. Oral Pathol. Med. 36 (2007) 409–414, doi:10.1111/j.1600-0714.2007.00552.x.
[32] E. Vairaktaris, Z. Serefoglou, D. Avgoustidis, et al., Gene polymorphisms related to angiogenesis, inflammation and thrombosis that influence risk for oral cancer, Oral Oncol. 45 (2009) 247–253, doi:10.1016/j.oraloncology.2008.05.003.
[33] Y.-W. Hung, C.-W. Tsai, C.-N. Wu, et al., The contribution of matrix metalloproteinase-8 promoter polymorphism to oral cancer susceptibility, In Vivo (Brooklyn) 31 (2017) 585–590.
[34] E. Vairaktaris, C. Yapijakis, E. Nkenke, et al., A metalloproteinase-13 polymorphism affecting its gene expression is associated with advanced stages of oral cancer, Anticancer Res. 27 (2007) 4027–4030.
[35] C.-W. Lin, S.-F. Yang, C.-Y. Chuang, et al., Association of matrix metalloproteinase-11 polymorphisms with susceptibility and clinicopathologic characteristics for oral squamous cell carcinoma, Head Neck 37 (2015) 1425–1431, doi:10.1002/hed.23771.
[36] C.-J. Weng, M.-K. Chen, C.-W. Lin, et al., Single nucleotide polymorphisms and haplotypes of MMP-14 are associated with the risk and pathological development of oral cancer, Ann. Surg. Oncol. 19 (2012) 319–327, doi:10.1245/s10434-011-1736-x.
[37] G.J.F. do Nascimento, L.P. da Silva, F.R. de Matos, et al., Polymorphisms of matrix metalloproteinase-7 and -9 are associated with oral tongue squamous cell carcinoma, Braz. Oral. Res. 35 (2020) e019, doi:10.1590/1807-3107bor-2021.vol35.0019.

[38] M. Radunovic, N. Nikolic, S. Milenkovic, et al., The MMP-2 and MMP-9 promoter polymorphisms and susceptibility to salivary gland cancer, J. Buon. 21 (2016) 597–602.
[39] Z.-H. Ren, K. Wu, R. Yang, et al., Differential expression of matrix metalloproteinases and miRNAs in the metastasis of oral squamous cell carcinoma, BMC Oral Health 20 (2020) 24, doi:10.1186/s12903-020-1013-0.
[40] M. Cai, Z. Zheng, Z. Bai, et al., Overexpression of angiogenic factors and matrix metalloproteinases in the saliva of oral squamous cell carcinoma patients: potential non-invasive diagnostic and therapeutic biomarkers, BMC Cancer 22 (2022) 530, doi:10.1186/s12885-022-09630-0.
[41] W. Zhang, X. Huang, R. Huang, et al., MMP1 overexpression promotes cancer progression and associates with poor outcome in head and neck carcinoma, Comput. Math. Methods Med. 2022 (2022) 3058342, doi:10.1155/2022/3058342.
[42] H. Ito, M. Hatori, Y. Kinugasa, et al., Comparison of the expression profile of metastasis-associated genes between primary and circulating cancer cells in oral squamous cell carcinoma, Anticancer Res. 23 (2003) 1425–1431.
[43] W. Deng, W. Peng, T. Wang, et al., Overexpression of MMPs functions as a prognostic biomarker for oral cancer patients: a systematic review and meta-analysis, Oral Health Prev. Dent. 17 (2019) 505–514, doi:10.3290/j.ohpd.a43636.
[44] U. Impola, V.J. Uitto, J. Hietanen, et al., Differential expression of matrilysin-1 (MMP-7), 92 kD gelatinase (MMP-9), and metalloelastase (MMP-12) in oral verrucous and squamous cell cancer, J. Pathol. 202 (2004) 14–22, doi:10.1002/path.1479.
[45] S. Kurahara, M. Shinohara, T. Ikebe, et al., Expression of MMPS, MT-MMP, and TIMPs in squamous cell carcinoma of the oral cavity: correlations with tumor invasion and metastasis, Head Neck 21 (1999) 627–638, doi:10.1002/(sici)1097-0347(199910)21:7⟨627::aid-hed7⟩3.0.co;2-2.
[46] S.D. Cruz-Romero, A.M. Baldión, J.A.H. Tawil, et al., Saliva liquid biopsy: MMP2, MMP9, and TIMP2 as potential diagnostic biomarkers in oral squamous cell carcinoma. An observational case-control study, J. Appl. Oral Sci. 34 (2026) e20250555, doi:10.1590/1678-7765-2025-0555.
[47] A. Celentano, T. Yap, R. Paolini, et al., Inhibition of matrix metalloproteinase-2 modulates malignant behaviour of oral squamous cell carcinoma cells, J. Oral Pathol. Med. 50 (2021) 323–332, doi:10.1111/jop.12992.
[48] S.B. Maurya, D.C. Shetty, A.S. Rathore, et al., Immunolocalization of epithelial cell adhesion molecule and matrix metalloproteinase-9 in oral epithelial dysplasia and oral squamous cell carcinoma, J. Cancer Res. Ther. 19 (2023) 1775–1780, doi:10.4103/jcrt.jcrt_1012_21.
[49] W.-Y. Zheng, D.-T. Zhang, S.-Y. Yang, H. Li, Elevated matrix metalloproteinase-9 expression correlates with advanced stages of oral cancer and is linked to poor clinical outcomes, J. Oral Maxillofac. Surg. 73 (2015) 2334–2342, doi:10.1016/j.joms.2015.05.041.
[50] B.P. Patel, S.V. Shah, S.N. Shukla, et al., Clinical significance of MMP-2 and MMP-9 in patients with oral cancer, Head Neck 29 (2007) 564–572, doi:10.1002/hed.20561.
[51] K.K. Pramanik, R. Mishra, ERK-mediated upregulation of matrix metalloproteinase-2 promotes the invasiveness in human oral squamous cell carcinoma (OSCC), Exp. Cell Res. 411 (2022) 112984, doi:10.1016/j.yexcr.2021.112984.
[52] P. Yin, Y. Su, S. Chen, et al., MMP-9 knockdown inhibits oral squamous cell carcinoma lymph node metastasis in the nude mouse tongue-xenografted model through the RhoC/Src pathway, Anal. Cell Pathol. (Amst) 2021 (2021) 6683391, doi:10.1155/2021/6683391.
[53] M. Pourhajibagher, M. Alaeddini, S. Etemad-Moghadam, et al., Effects of kojic acid-mediated sonodynamic therapy as a matrix metalloprotease-9 inhibitor against oral squamous cell carcinoma: a bioinformatics screening and in vitro

analysis, Curr. Drug Discov. Technol. 21 (2024) e011223224137, doi:10.2174/0115701638266082231124055825.

[54] H.R. Wardill, R.M. Logan, J.M. Bowen, et al., Tight junction defects are seen in the buccal mucosa of patients receiving standard dose chemotherapy for cancer, Support Care Cancer 24 (2016) 1779–1788, doi:10.1007/s00520-015-2964-6.

[55] A. George, K. Ranganathan, U.K. Rao, Expression of MMP-1 in histopathological different grades of oral squamous cell carcinoma and in normal buccal mucosa—an immunohistochemical study, Cancer Biomark. 7 (2010) 275–283, doi:10.3233/CBM-2010-0191.

[56] C.-H. Hsin, Y.-E. Chou, S.-F. Yang, et al., MMP-11 promoted the oral cancer migration and Fak/Src activation, Oncotarget 8 (2017) 32783–32793, doi:10.18632/oncotarget.15824.

[57] C.-H. Hsin, M.-K. Chen, C.-H. Tang, et al., High level of plasma matrix metalloproteinase-11 is associated with clinicopathological characteristics in patients with oral squamous cell carcinoma, PLoS One 9 (2014) e113129, doi:10.1371/journal.pone.0113129.

[58] Y.V. Castañeda Rojas, S.D. Cruz-Romero, A.M. Baldión, et al., Prognostic significance of MMP-14 in oral squamous cell carcinoma: insights from a Colombian cohort, Oral Surg. Oral Med. Oral Pathol. Oral Radiol. 141 (2025) 654–666, doi:10.1016/j.oooo.2025.11.007.

[59] Y. Noda, M. Ishida, R. Yamaka, et al., MMP14 expression levels accurately predict the presence of extranodal extensions in oral squamous cell carcinoma: a retrospective cohort study, BMC Cancer 23 (2023) 142, doi:10.1186/s12885-023-10595-x.

[60] A. Baskaran, A. Santhadevy, N. Vezhavendhan, et al., Expression of matrix metalloproteinase-21 in oral squamous cell carcinoma, Cureus 15 (2023) e34256, doi:10.7759/cureus.34256.

[61] P.T. Nguyen, M. Miyauchi, A. Subarnbhesaj, et al., PlGF/Flt-1/MMP-1 axis in gingival carcinoma bone invasion, Histol. Histopathol. 40 (2025) 1757–1768, doi:10.14670/HH-18-911.

[62] T. Yoshizaki, Y. Maruyama, H. Sato, M. Furukawa, Expression of tissue inhibitor of matrix metalloproteinase-2 correlates with activation of matrix metalloproteinase-2 and predicts poor prognosis in tongue squamous cell carcinoma, Int. J. Cancer 95 (2001) 44–50, doi:10.1002/1097-0215(20010120)95:1⟨44::aid-ijc1008⟩3.0.co;2-m.

[63] F.I. Ciucă, P.C. Mărăşescu, M. Matei, et al., The prognostic value of CXCR4, MMP-2 and MMP-9 in tongue squamous carcinoma, Rom. J. Morphol. Embryol. 60 (2019) 59–68.

[64] J.T. Korpi, V. Kervinen, H. Mäklin, et al., Collagenase-2 (matrix metalloproteinase-8) plays a protective role in tongue cancer, Br. J. Cancer 98 (2008) 766–775, doi:10.1038/sj.bjc.6604239.

[65] P. Åström, K. Juurikka, E.S. Hadler-Olsen, et al., The interplay of matrix metalloproteinase-8, transforming growth factor-β1 and vascular endothelial growth factor-C cooperatively contributes to the aggressiveness of oral tongue squamous cell carcinoma, Br. J. Cancer 117 (2017) 1007–1016, doi:10.1038/bjc.2017.249.

[66] K. Juurikka, A. Dufour, K. Pehkonen, et al., MMP8 increases tongue carcinoma cell–cell adhesion and diminishes migration via cleavage of anti-adhesive FXYD5, Oncogenesis 10 (2021) 44, doi:10.1038/s41389-021-00334-x.

[67] B. Dharavath, A. Butle, A. Pal, et al., Role of miR-944/MMP10/AXL- axis in lymph node metastasis in tongue cancer, Commun. Biol. 6 (2023) 57, doi:10.1038/s42003-023-04437-6.

[68] S.-H. Kim, N.H. Cho, K. Kim, et al., Correlations of oral tongue cancer invasion with matrix metalloproteinases (MMPs) and vascular endothelial growth factor (VEGF) expression, J. Surg. Oncol. 93 (2006) 330–337, doi:10.1002/jso.20461.

[69] A.C.G. Henriques, F.R. de Matos, H.C. Galvão, R. de A Freitas, Immunohistochemical expression of MMP-9 and VEGF in squamous cell carcinoma of the tongue, J. Oral Sci. 54 (2012) 105–111, doi:10.2334/josnusd.54.105.
[70] F. Monteiro-Amado, I.I. Castro-Silva, C.J. de Lima, et al., Immunohistochemical evaluation of MMP-2, MMP-9 and CD31/microvascular density in squamous cell carcinomas of the floor of the mouth, Braz. Dent. J. 24 (2013) 3–9, doi:10.1590/0103-6440201301941.
[71] F.J.A. Loureiro, K.M. Balbinot, M.S. da Silva Kataoka, et al., Invadopodia related-proteins expression in mucoepidermoid carcinoma, Oral Dis. 31 (2025) 2427–2440, doi:10.1111/odi.15312.
[72] B.-K. Lee, M.-J. Kim, H.-S. Jang, et al., A high concentration of MMP-2/Gelatinase A and MMP-9/Gelatinase B reduce NK cell-mediated cytotoxicity against an oral squamous cell carcinoma cell line, In Vivo (Brooklyn) 22 (2008) 593–597.
[73] A. Gao, D.W. Kim, M. Pei, et al., Implications of TP-positive CAFs in the bone invasion of oral squamous cell carcinoma, Anticancer Res. 44 (2024) 3365–3374, doi:10.21873/anticanres.17157.
[74] J. Lee, J.S. Roberts, K.R. Atanasova, et al., Human primary epithelial cells acquire an epithelial-mesenchymal-transition phenotype during long-term infection by the oral opportunistic pathogen, Porphyromonas gingivalis, Front. Cell. Infect. Microbiol. 7 (2017) 493, doi:10.3389/fcimb.2017.00493.
[75] Y. Feng, Q. Li, J. Chen, et al., Salivary protease spectrum biomarkers of oral cancer, Int. J. Oral Sci. 11 (2019) 7, doi:10.1038/s41368-018-0032-z.
[76] Z. Saleem, A.H. Shaikh, U. Zaman, et al., Estimation of salivary matrix metalloproteinases- 12 (MMP- 12) levels among patients presenting with oral submucous fibrosis and oral squamous cell carcinoma, BMC Oral Health 21 (2021) 205, doi:10.1186/s12903-021-01571-7.
[77] T. Shpitzer, Y. Hamzany, G. Bahar, et al., Salivary analysis of oral cancer biomarkers, Br. J. Cancer 101 (2009) 1194–1198, doi:10.1038/sj.bjc.6605290.
[78] K. Smriti, M. Ray, T. Chatterjee, et al., Salivary MMP-9 as a biomarker for the diagnosis of oral potentially malignant disorders and oral squamous cell carcinoma, Asian Pac. J. Cancer Prev. 21 (2020) 233–238, doi:10.31557/APJCP.2020.21.1.233.
[79] Y.-J. Shin, H. Vu, J.-H. Lee, H.-D. Kim, Diagnostic and prognostic ability of salivary MMP-9 for oral squamous cell carcinoma: a pre-/post-surgery case and matched control study, PLoS One 16 (2021) e0248167, doi:10.1371/journal.pone.0248167.
[80] A. Peisker, G.-F. Raschke, M.-D. Fahmy, et al., Salivary MMP-9 in the detection of oral squamous cell carcinoma, Med. Oral Patol. Oral Cir. Bucal. 22 (2017) e270–e275, doi:10.4317/medoral.21626.
[81] A. Shaukat, S. Nisar, M. Asghar, et al., MMP-9 as a diagnostic salivary biomarker for early detection of oral cancers: systematic review and meta-analysis, BMC Oral Health 26 (2026) 219, doi:10.1186/s12903-025-07513-x.
[82] T. Vuotila, L. Ylikontiola, T. Sorsa, et al., The relationship between MMPs and pH in whole saliva of radiated head and neck cancer patients, J. Oral Pathol. Med. 31 (2002) 329–338, doi:10.1034/j.1600-0714.2002.310603.x.
[83] M. Stott-Miller, J.R. Houck, P. Lohavanichbutr, et al., Tumor and salivary matrix metalloproteinase levels are strong diagnostic markers of oral squamous cell carcinoma, Cancer Epidemiol. Biomarkers Prev. 20 (2011) 2628–2636, doi:10.1158/1055-9965.EPI-11-0503.
[84] Y.-T. Chang, L.J. Chu, Y.-C. Liu, et al., Verification of saliva matrix metalloproteinase-1 as a strong diagnostic marker of oral cavity cancer, Cancers (Basel) 12 (2020), doi:10.3390/cancers12082273.

[85] J.-S. Yu, Y.-T. Chen, W.-F. Chiang, et al., Saliva protein biomarkers to detect oral squamous cell carcinoma in a high-risk population in Taiwan, Proc. Natl. Acad. Sci. U.S.A. 113 (2016) 11549–11554, doi:10.1073/pnas.1612368113.
[86] V.P. Thakore, S.S. Uppalapati, S.P. Kumar, et al., Computational and experimental investigation of Kaempferol and Quercetin as potential inhibitors of MMP1 in oral cancer, Comput. Biol. Chem. 120 (2026) 108716, doi:10.1016/j.compbiolchem.2025.108716.
[87] L.-M. Chi, Y.-C. Hsiao, K.-Y. Chien, et al., Assessment of candidate biomarkers in paired saliva and plasma samples from oral cancer patients by targeted mass spectrometry, J. Proteomics 211 (2020) 103571, doi:10.1016/j.jprot.2019.103571.
[88] Y.-C. Hsiao, S.-Y. Lin, K.-Y. Chien, et al., An immuno-MALDI mass spectrometry assay for the oral cancer biomarker, matrix metalloproteinase-1, in dried saliva spot samples, Anal. Chim. Acta 1100 (2020) 118–130, doi:10.1016/j.aca.2019.12.006.
[89] L.J. Chu, Y.-T. Chang, C.-Y. Chien, et al., Clinical validation of a saliva-based matrix metalloproteinase-1 rapid strip test for detection of oral cavity cancer, Biomed J. 47 (2024) 100594, doi:10.1016/j.bj.2023.04.002.
[90] I. Chattopadhyay, M. Panda, Recent trends of saliva omics biomarkers for the diagnosis and treatment of oral cancer, J. Oral Biosci. 61 (2019) 84–94, doi:10.1016/j.job.2019.03.002.
[91] A. Kazmi, Z. Abbas, Z. Saleem, et al., Relation of salivary MMP-8 with oral submucous fibrosis and oral squamous cell carcinoma: a cross sectional analytical study, BMJ Open 12 (2022) e060738, doi:10.1136/bmjopen-2021-060738.
[92] A.A. Abbasi, S. Delfani, S. Pourranjbar, E.S. Afraz, Matrix metalloproteinases as prognostic markers in oral squamous cell carcinoma: a systematic review, meta-analysis and meta regression of MMP-2, MMP-7, and MMP-9 expression and their serum and saliva concentrations, Int. J. Biol. Macromol. 336 (2026) 149340, doi:10.1016/j.ijbiomac.2025.149340.
[93] N.A. Ghallab, O.G. Shaker, Serum and salivary levels of chemerin and MMP-9 in oral squamous cell carcinoma and oral premalignant lesions, Clin. Oral Investig. 21 (2017) 937–947, doi:10.1007/s00784-016-1846-8.
[94] F. Agha-Hosseini, I. Mirzaii-Dizgah, N. Mahboobi, et al., Serum and saliva MMP-3 in patients with OLP and oral SCC, J. Contemp. Dent. Pract. 16 (2015) 107–111, doi:10.5005/jp-journals-10024-1645.
[95] A. Lotfi, G. Mohammadi, A. Tavassoli, et al., Serum levels of MMP9 and MMP2 in patients with oral squamous cell carcinoma, Asian Pac. J. Cancer Prev. 16 (2015) 1327–1330, doi:10.7314/apjcp.2015.16.4.1327.
[96] C.-J. Liu, K.-W. Chang, S.-C. Lin, H.-W. Cheng, Presurgical serum levels of matrix metalloproteinase-9 and vascular endothelial growth factor in oral squamous cell carcinoma, Oral Oncol. 45 (2009) 920–925, doi:10.1016/j.oraloncology.2009.04.007.
[97] A.A. Tadbir, S. Purshahidi, H. Ebrahimi, et al., Serum level of MMP-3 in patients with oral squamous cell carcinoma—lack of association with clinico-pathological features, Asian Pac. J. Cancer Prev. 13 (2012) 4545–4548, doi:10.7314/apjcp.2012.13.9.4545.
[98] T. Jiang, P. Xie, H. Liu, Circulating anti–matrix metalloproteinase-7 antibodies may be a potential biomarker for oral squamous cell carcinoma, J. Oral Maxillofac. Surg. 74 (2016) 650–657, doi:10.1016/j.joms.2015.09.016.
[99] J. Wei, X. Jiang, Y. Xu, et al., MMP7, regulated by c-Jun, is involved in oral squamous cell carcinoma and associated with cancer-related fibroblasts infiltration, Curr. Cancer Drug Targets 25 (2025) 1260–1270, doi:10.2174/0115680096309161240821171847.
[100] R.D. Singh, H. Nilayangode, J.B. Patel, et al., Combined evaluation of matrix metalloproteinases and their inhibitors has better clinical utility in oral cancer, Int. J. Biol. Markers 26 (2011) 27–36, doi:10.5301/jbm.2011.6359.

[101] R.D. Singh, N. Haridas, J.B. Patel, et al., Matrix metalloproteinases and their inhibitors: correlation with invasion and metastasis in oral cancer, Indian J. Clin. Biochem. 25 (2010) 250–259, doi:10.1007/s12291-010-0060-8.
[102] S. Warnakulasuriya, O. Kujan, J.M. Aguirre-Urizar, et al., Oral potentially malignant disorders: a consensus report from an international seminar on nomenclature and classification, convened by the WHO Collaborating Centre for Oral Cancer, Oral Dis. 27 (2021) 1862–1880, doi:10.1111/odi.13704.
[103] M. Sutinen, T. Kainulainen, T. Hurskainen, et al., Expression of matrix metalloproteinases (MMP-1 and -2) and their inhibitors (TIMP-1, -2 and -3) in oral lichen planus, dysplasia, squamous cell carcinoma and lymph node metastasis, Br. J. Cancer 77 (1998) 2239–2245, doi:10.1038/bjc.1998.372.
[104] J. Wang, H. Luo, Y. Xiao, L. Wang, miR-125b inhibits keratinocyte proliferation and promotes keratinocyte apoptosis in oral lichen planus by targeting MMP-2 expression through PI3K/Akt/mTOR pathway, Biomed. Pharmacother. 80 (2016) 373–380, doi:10.1016/j.biopha.2016.02.043.
[105] N. Mazzarella, F. Femiano, F. Gombos, et al., Matrix metalloproteinase gene expression in oral lichen planus: erosive vs. reticular forms, J. Eur. Acad. Dermatol. Venereol. 20 (2006) 953–957, doi:10.1111/j.1468-3083.2006.01693.x.
[106] A.H. Rubaci, H.O. Kazancioglu, V. Olgac, G. Ak, The roles of matrix metalloproteinases-2, -7, -10 and tissue inhibitor of metalloproteinase-1 in the pathogenesis of oral lichen planus, J. Oral Pathol. Med. 41 (2012) 689–696, doi:10.1111/j.1600-0714.2012.01160.x.
[107] H. Wang, X. Guan, Z. Luo, et al., The association and potentially destructive role of Th9/IL-9 is synergistic with Th17 cells by elevating MMP9 production in local lesions of oral lichen planus, J. Oral Pathol. Med. 47 (2018) 425–433, doi:10.1111/jop.12690.
[108] H.H. Hazzaa, M.A.M. El Shiekh, O. Elkashty, et al., A critical influence of HIF-1 on MMP-9 and Galectin-3 in oral lichen planus, BMC Oral Health 24 (2024) 756, doi:10.1186/s12903-024-04457-6.
[109] F. Romano, P.G. Arduino, M. Maggiora, et al., Effect of a structured plaque control on MMP-1 and MMP-9 crevicular levels in patients with desquamative gingivitis associated with oral lichen planus, Clin. Oral Investig. 23 (2019) 2651–2658, doi:10.1007/s00784-018-2704-7.
[110] X.J. Zhou, P.B. Sugerman, N.W. Savage, L.J. Walsh, Matrix metalloproteinases and their inhibitors in oral lichen planus, J. Cutan. Pathol. 28 (2001) 72–82, doi:10.1034/j.1600-0560.2001.280203.x.
[111] Y.H. Mair, T. Jhamb, M.B. Visser, et al., Sera and salivary matrix metalloproteinases are elevated in patients with vesiculoerosive disease: a pilot study, Oral Surg. Oral Med. Oral Pathol. Oral Radiol. 121 (2016) 520–529, doi:10.1016/j.oooo.2016.01.002.
[112] O.M. Condurache Hrițcu, D.G. Ciobanu Apostol, V. Toader Ștefan, et al., Immunohistochemical assessment of Maspin, β-Catenin, and MMP-14 in oral potentially malignant lesions and oral squamous cell carcinoma: a retrospective observational study, Medicina (B. Aires). 61 (2025) 1037.
[113] T.-J. Li, J. Cui, COX-2, MMP-7 expression in oral lichen planus and oral squamous cell carcinoma, Asian Pac. J. Trop. Med. 6 (2013) 640–643, doi:10.1016/S1995-7645(13)60110-8.
[114] Z. Xie, X. Wang, X. Ren, X. Ge, MMP12 disrupts epithelial barrier integrity in oral lichen planus by degrading fibronectin, Sci. Rep. 14 (2024) 27511, doi:10.1038/s41598-024-78930-1.
[115] S.S. Choudhari, D.G. Kulkarni, S.C. Sarode, et al., Is matrix stiffness a cause for malignant transformation of oral submucous fibrosis? J. Oral Pathol. Med. 47 (2018) 627–628, doi:10.1111/jop.12714.

[116] A.K. Chaudhary, S. Pandya, R. Mehrotra, et al., Functional polymorphism of the MMP-1 promoter (-1607 1G/2G) in potentially malignant and malignant head and neck lesions in an Indian population, Biomarkers 15 (2010) 684–692, doi:10.3109/1354750X.2010.511267.
[117] P.R. Zade, S.R. Gosavi, V.K. Hazarey, S.M. Ganvir, Matrix metalloproteinases-3 gene-promoter polymorphism as a risk factor in oral submucous fibrosis in an Indian population: a pilot study, J. Investig. Clin. Dent. 8 (2017) e12228, doi:10.1111/jicd.12228.
[118] A.K. Chaudhary, M. Singh, A.C. Bharti, et al., Synergistic effect of stromelysin-1 (matrix metalloproteinase-3) promoter (-1171 5A->6A) polymorphism in oral submucous fibrosis and head and neck lesions, BMC Cancer 10 (2010) 369, doi:10.1186/1471-2407-10-369.
[119] H.-F. Tu, C.-J. Liu, C.-S. Chang, et al., The functional (-1171 5A–>6A) polymorphisms of matrix metalloproteinase 3 gene as a risk factor for oral submucous fibrosis among male areca users, J. Oral Pathol. Med. 35 (2006) 99–103, doi:10.1111/j.1600-0714.2006.00370.x.
[120] A.K. Chaudhary, S. Pandya, R. Mehrotra, et al., Role of functional polymorphism of matrix metalloproteinase-2 (-1306 C/T and -168 G/T) and MMP-9 (-1562 C/T) promoter in oral submucous fibrosis and head and neck squamous cell carcinoma in an Indian population, Biomark. Biochem. 16 (2011) 577–586, doi:10.3109/1354750X.2011.609602.
[121] R. Qamar, A. Khatoon, T. Mirza, et al., Association of salivary mmp-8 gene polymorphism with its protein expression in oral submucous fibrosis patients, Odontology 113 (2025) 1491–1501, doi:10.1007/s10266-025-01074-9.
[122] C.-Y. Fang, Y.-J. Chen, K.-C. Lin, et al., Co-transfection with anti-fibrotic microRNAs for treating oral submucous fibrosis, J. Oral Pathol. Med. 54 (2025) 574–583, doi:10.1111/jop.70009.
[123] P.-N. Chen, C.-W. Lin, S.-F. Yang, Y.-C. Chang, Oral submucous fibrosis stimulates invasion and epithelial-mesenchymal transition in oral squamous cell carcinoma by activating MMP-2 and IGF-IR, J. Cell. Mol. Med. 25 (2021) 9814–9825, doi:10.1111/jcmm.16929.
[124] A. Shrestha, S. Carnelio, Evaluation of matrix metalloproteinases-2 (MMP-2) and tissue inhibitors of metalloproteinases-2 (TIMP-2) in oral submucous fibrosis and their correlation with disease severity, Kathmandu Univ. Med. J. (KUMJ) 11 (2013) 274–281, doi:10.3126/kumj.v11i4.12521.
[125] G.N. Pitiyage, P. Slijepcevic, A. Gabrani, et al., Senescent mesenchymal cells accumulate in human fibrosis by a telomere-independent mechanism and ameliorate fibrosis through matrix metalloproteinases, J. Pathol. 223 (2011) 604–617, doi:10.1002/path.2839.
[126] G. Mishra, K. Ranganathan, Matrix metalloproteinase-1 expression in oral submucous fibrosis: an immunohistochemical study, Indian J. Dent. Res. 21 (2010) 320–325, doi:10.4103/0970-9290.70785.
[127] R. Rajendran, M.P.K. Rajeesh, S. Shaikh, et al., Expression of matrix metalloproteinases (MMP-1, MMP-2 and MMP-9) and their inhibitors (TIMP-1 and TIMP-2) in oral submucous fibrosis, Indian J. Dent. Res. 17 (2006) 161–166, doi:10.4103/0970-9290.29870.
[128] Y.-C. Chang, S.-F. Yang, K.-W. Tai, et al., Increased tissue inhibitor of metalloproteinase-1 expression and inhibition of gelatinase A activity in buccal mucosal fibroblasts by arecoline as possible mechanisms for oral submucous fibrosis, Oral Oncol. 38 (2002) 195–200, doi:10.1016/s1368-8375(01)00045-8.
[129] R. Tattar, B.D.C. da Costa, V.C.M. Neves, The interrelationship between periodontal disease and systemic health, Br. Dent. J. 239 (2025) 103–108, doi:10.1038/s41415-025-8642-2.

[130] D. Radzki, A. Negri, A. Kusiak, M. Obuchowski, Matrix metalloproteinases in the periodontium—vital in tissue turnover and unfortunate in periodontitis, Int. J. Mol. Sci. 25 (2024) 2763.
[131] C. Rodríguez-Medina, S. Amaya Sánchez, A. Contreras, J.E. Botero, Grading the strength and certainty of the scientific evidence of the bidirectional association between periodontitis and noncommunicable diseases: an umbrella review, Evid. Based Dent. 26 (2025) 147, doi:10.1038/s41432-025-01132-9.
[132] F. Hasan, A. Tandon, H. AlQallaf, et al., Inflammatory association between periodontal disease and systemic health, Inflammation 48 (2025) 3763–3775, doi:10.1007/s10753-025-02317-1.
[133] Y. Ma, N. Tuerxun, G. Maimaitili, Periodontitis and the risk of oral cancer: a meta-analysis of case-control studies, Acta Odontol. Scand. 83 (2024), doi:10.2340/aos.v83.40478.
[134] I. Ghassib, Z. Chen, J. Zhu, H.L. Wang, Use of IL-1 β, IL-6, TNF-α, and MMP-8 biomarkers to distinguish peri-implant diseases: a systematic review and meta-analysis, Clin. Implant Dent. Relat. Res. 21 (2019) 190–207, doi:10.1111/cid.12694.
[135] S.T. Sonis, The pathobiology of mucositis, Nat. Rev. Cancer 4 (2004) 277–284, doi:10.1038/nrc1318.
[136] H.M. Fullmer, W. Gibson, Collagenolytic activity in gingivae of man, Nature 209 (1966) 728–729, doi:10.1038/209728a0.
[137] M. Dahan, B. Nawrocki, R. Elkaïm, et al., Expression of matrix metalloproteinases in healthy and diseased human gingiva, J. Clin. Periodontol. 28 (2001) 128–136, doi:10.1034/j.1600-051x.2001.028002128.x.
[138] V. Xanthopoulou, I.T. Räisänen, T. Sorsa, et al., Diagnostic value of aMMP-8 and azurocidin in peri-implant sulcular fluid as biomarkers of peri-implant health or disease, Clin. Exp. Dent. Res. 10 (2024) e883, doi:10.1002/cre2.883.
[139] M.J. Kivelä-Rajamäki, O.P. Teronen, P. Maisi, et al., Laminin-5 gamma2-chain and collagenase-2 (MMP-8) in human peri-implant sulcular fluid, Clin. Oral Implants Res. 14 (2003) 158–165, doi:10.1034/j.1600-0501.2003.140204.x.
[140] P. Mäntylä, M. Stenman, D.F. Kinane, et al., Gingival crevicular fluid collagenase-2 (MMP-8) test stick for chair-side monitoring of periodontitis, J. Periodontal. Res. 38 (2003) 436–439, doi:10.1034/j.1600-0765.2003.00677.x.
[141] T. Sorsa, P. Mäntylä, H. Rönkä, et al., Scientific basis of a matrix metalloproteinase-8 specific chair-side test for monitoring periodontal and peri-implant health and disease, Ann. N. Y. Acad. Sci. 878 (1999) 130–140, doi:10.1111/j.1749-6632.1999.tb07679.x.
[142] M. Kivelä-Rajamäki, P. Maisi, R. Srinivas, et al., Levels and molecular forms of MMP-7 (matrilysin-1) and MMP-8 (collagenase-2) in diseased human peri-implant sulcular fluid, J. Periodontal. Res. 38 (2003) 583–590, doi:10.1034/j.1600-0765.2003.00688.x.
[143] M. Hernández, J. Gamonal, T. Salo, et al., Reduced expression of lipopolysaccharide-induced CXC chemokine in Porphyromonas gingivalis-induced experimental periodontitis in matrix metalloproteinase-8 null mice, J. Periodontal. Res. 46 (2011) 58–66, doi:10.1111/j.1600-0765.2010.01310.x.
[144] H. Kuula, T. Salo, E. Pirilä, et al., Local and systemic responses in matrix metalloproteinase 8-deficient mice during Porphyromonas gingivalis-induced periodontitis, Infect. Immun. 77 (2009) 850–859, doi:10.1128/IAI.00873-08.
[145] F. Cavalla, C. Osorio, R. Paredes, et al., Matrix metalloproteinases regulate extracellular levels of SDF-1/CXCL12, IL-6 and VEGF in hydrogen peroxide-stimulated human periodontal ligament fibroblasts, Cytokine 73 (2015) 114–121, doi:10.1016/j.cyto.2015.02.001.
[146] G.A. McQuibban, J.-H. Gong, J.P. Wong, et al., Matrix metalloproteinase processing of monocyte chemoattractant proteins generates CC chemokine receptor antagonists

with anti-inflammatory properties in vivo, Blood 100 (2002) 1160–1167, doi:10.1182/blood.V100.4.1160.h81602001160_1160_1167.
[147] R.P. Darveau, Periodontitis: a polymicrobial disruption of host homeostasis, Nat. Rev. Micro. 8 (2010) 481–490, doi:10.1038/nrmicro2337.
[148] M. Irshad, N. Scheres, D. Anssari Moin, et al., Cytokine and matrix metalloproteinase expression in fibroblasts from peri-implantitis lesions in response to viable Porphyromonas gingivalis, J. Periodontal. Res. 48 (2013) 647–656, doi:10.1111/jre.12051.
[149] A. Beklen, M. Ainola, M. Hukkanen, et al., MMPs, IL-1, and TNF are regulated by IL-17 in periodontitis, J. Dent. Res. 86 (2007) 347–351, doi:10.1177/154405910708600409.
[150] T. Sorsa, L. Tjäderhane, Y.T. Konttinen, et al., Matrix metalloproteinases: contribution to pathogenesis, diagnosis and treatment of periodontal inflammation, Ann. Med. 38 (2006) 306–321, doi:10.1080/07853890600800103.
[151] A.A.J. DeCarlo, L.J. Windsor, M.K. Bodden, et al., Activation and novel processing of matrix metalloproteinases by a thiol-proteinase from the oral anaerobe Porphyromonas gingivalis, J. Dent. Res. 76 (1997) 1260–1270, doi:10.1177/00220345970760060501.
[152] R. Grayson, C.W.I. Douglas, J. Heath, et al., Activation of human matrix metalloproteinase 2 by gingival crevicular fluid and Porphyromonas gingivalis, J. Clin. Periodontol. 30 (2003) 542–550, doi:10.1034/j.1600-051x.2003.00301.x.
[153] Y. Ding, V.-J. Uitto, M. Haapasalo, et al., Membrane components of Treponema denticola trigger proteinase release from human polymorphonuclear leukocytes, J. Dent. Res. 75 (1996) 1986–1993, doi:10.1177/00220345960750121101.
[154] Y. Ding, M. Haapasalo, E. Kerosuo, et al., Release and activation of human neutrophil matrix metallo- and serine proteinases during phagocytosis of Fusobacterium nucleatum, Porphyromonas gingivalis and Treponema denticola, J. Clin. Periodontol. 24 (1997) 237–248, doi:10.1111/j.1600-051X.1997.tb01837.x.
[155] D. Miao, V. Godovikova, X. Qian, et al., Treponema denticola upregulates MMP-2 activation in periodontal ligament cells: interplay between epigenetics and periodontal infection, Arch. Oral Biol. 59 (2014) 1056–1064, doi:10.1016/j.archoralbio.2014.06.003.
[156] M. Di, F.J. Christopher, J.C. Timm, et al., The Treponema denticola chymotrypsin-like protease dentilisin induces matrix metalloproteinase-2-dependent fibronectin fragmentation in periodontal ligament cells, Infect. Immun. 79 (2011) 806–811, doi:10.1128/iai.01001-10.
[157] V. Checchi, T. Maravic, P. Bellini, et al., The role of matrix metalloproteinases in periodontal disease, Int. J. Environ. Res. Public Health 17 (2020) 4923.
[158] M. Ben-Eltriki, A.R. Ahmadi, Y. Nakao, et al., Granzyme B promotes matrix metalloproteinase-1 (MMP-1) release from gingival fibroblasts in a PAR1- and Erk1/2-dependent manner: a novel role in periodontal inflammation, J. Periodontal. Res. 59 (2024) 94–103, doi:10.1111/jre.13190.
[159] I. Fragkioudakis, C. Kottaridi, M. Paraskeva, D. Sakellari, MMP-8 in peri-implantitis: a cross-sectional study on genetic polymorphisms and enzymatic activation, J. Periodontal. Res. 61 (2026) 95–97, doi:10.1111/jre.70026.
[160] H. Weng, Y. Yan, Y.-H. Jin, et al., Matrix metalloproteinase gene polymorphisms and periodontitis susceptibility: a meta-analysis involving 6,162 individuals, Sci. Rep. 6 (2016) 24812, doi:10.1038/srep24812.
[161] D. Li, Q. Cai, L. Ma, et al., Association between MMP-1 g.-1607dupG polymorphism and periodontitis susceptibility: a meta-analysis, PLoS One 8 (2013) e59513.
[162] M.F.F. Leite, M.C.L.G. Santos, A.P. de Souza, S.R.P. Line, Osseointegrated implant failure associated with MMP-1 promotor polymorphisms (-1607 and -519), Int. J. Oral Maxillofac. Implants 23 (2008) 653–658.

[163] L. Saremi, S. Shahbazi, M.E. Ghaffari, et al., The association of matrix metalloproteinase-1, -2, -3, -7, and -13 gene polymorphisms with peri-implantitis in an Iranian population: a case–control study, Clin. Exp. Dent. Res. 10 (2024) e70049, doi:10.1002/cre2.70049.
[164] W. Li, Y. Zhu, P. Singh, et al., Association of common variants in MMPs with periodontitis risk, Dis. Markers 2016 (2016) 1545974, doi:10.1155/2016/1545974.
[165] E. Matuszczak, I. Cwalina, M. Tylicka, et al., Levels of selected matrix metalloproteinases—MMP-1, MMP-2 and fibronectin in the saliva of patients planned for endodontic treatment or surgical extraction, J. Clin. Med. 9 (2020) 3971.
[166] M. Monea, A.M. Pop, The use of salivary levels of matrix metalloproteinases as an adjuvant method in the early diagnosis of oral squamous cell carcinoma: a narrative literature review, Curr. Issues Mol. Biol. 44 (2022) 6306–6322.
[167] M.A. Alarcón-Sánchez, R. Rodríguez-Montaño, S.A. Mosaddad, A. Heboyan, Levels of IL-1β, MMP-8, and MMP-9 in the saliva of subjects with periodontitis: a systematic review and meta-analysis, J. Clin. Lab. Anal. 39 (2025) e70040, doi:10.1002/jcla.70040.
[168] N. Bostanci, K. Mitsakakis, B. Afacan, et al., Validation and verification of predictive salivary biomarkers for oral health, Sci. Rep. 11 (2021) 6406, doi:10.1038/s41598-021-85120-w.
[169] M.F. Moura, T.P. Navarro, T.A. Silva, et al., Periodontitis and endothelial dysfunction: periodontal clinical parameters and levels of salivary markers interleukin-1β, tumor necrosis factor-α, matrix metalloproteinase-2, tissue inhibitor of metalloproteinases-2 complex, and nitric oxide, J. Periodontol. 88 (2017) 778–787, doi:10.1902/jop.2017.170023.
[170] J. Noh, J.-H. Jang, H.-S. Yoon, et al., Evaluation of salivary biomarkers of periodontal disease based on smoking status: a systematic review, Int. J. Environ. Res. Public Health 19 (2022) 14619.
[171] Y. Zhang, N. Kang, F. Xue, et al., Evaluation of salivary biomarkers for the diagnosis of periodontitis, BMC Oral Health 21 (2021) 266, doi:10.1186/s12903-021-01600-5.
[172] D.M. Isaza-Guzmán, C. Arias-Osorio, M.C. Martínez-Pabón, S.I. Tobón-Arroyave, Salivary levels of matrix metalloproteinase (MMP)-9 and tissue inhibitor of matrix metalloproteinase (TIMP)-1: a pilot study about the relationship with periodontal status and MMP-9−1562C/T gene promoter polymorphism, Arch. Oral Biol. 56 (2011) 401–411, doi:10.1016/j.archoralbio.2010.10.021.
[173] M. Nędzi-Góra, J. Kostrzewa-Janicka, R. Górska, Elastase and metalloproteinase-9 concentrations in saliva in patients with chronic periodontitis, Cent. J. Immunol. 39 (2014) 357–364, doi:10.5114/ceji.2014.45948.
[174] U.K. Gursoy, E. Könönen, S. Huumonen, et al., Salivary type I collagen degradation end-products and related matrix metalloproteinases in periodontitis, J. Clin. Periodontol. 40 (2013) 18–25, doi:10.1111/jcpe.12020.
[175] E. Virtanen, M. Yakob, T. Tervahartiala, et al., Salivary MMP-13 gender differences in periodontitis: a cross-sectional study from Sweden, Clin. Exp. Dent. Res. 3 (2017) 165–170, doi:10.1002/cre2.76.
[176] N.E. Kushlinskii, E.A. Solovykh, T.B. Karaoglanova, et al., Content of matrix metalloproteinase-8 and matrix metalloproteinase-9 in oral fluid of patients with chronic generalized periodontitis, Bull. Exp. Biol. Med. 152 (2011) 240–244, doi:10.1007/s10517-011-1498-2.
[177] E. Özcan, N. Işıl Saygun, M.A. Serdar, et al., Non-surgical periodontal therapy reduces saliva adipokine and matrix metalloproteinase levels in periodontitis, J. Periodontol. 87 (2016) 934–943, doi:10.1902/jop.2016.160046.

[178] W.M. Sexton, Y. Lin, R.J. Kryscio, et al., Salivary biomarkers of periodontal disease in response to treatment, J. Clin. Periodontol. 38 (2011) 434–441, doi:10.1111/j.1600-051X.2011.01706.x.

[179] T. Sorsa, U.K. Gursoy, S. Nwhator, et al., Analysis of matrix metalloproteinases, especially MMP-8, in gingival crevicular fluid, mouthrinse and saliva for monitoring periodontal diseases, Periodontol 2000 70 (2016) 142–163, doi:10.1111/prd.12101.

[180] C.-H. Lee, Y.-W. Chen, Y.-K. Tu, et al., The potential of salivary biomarkers for predicting the sensitivity and monitoring the response to nonsurgical periodontal therapy: a preliminary assessment, J. Periodontal. Res. 53 (2018) 545–554, doi:10.1111/jre.12544.

[181] F. Tavakoli, M. Faramarzi, S. Salimnezhad, et al., Comparing the activity level of salivary matrix metalloproteinase-8 in patients with diabetes and moderate to severe chronic generalized periodontitis, Clin. Exp. Dent. Res. 10 (2024) e865, doi:10.1002/cre2.865.

[182] R. Lira-Junior, V.Ö. Öztürk, G. Emingil, et al., Salivary and serum markers related to innate immunity in generalized aggressive periodontitis, J. Periodontol. 88 (2017) 1339–1347, doi:10.1902/jop.2017.170287.

[183] E. Dursun, T.F. Tözüm, Peri-implant crevicular fluid analysis, enzymes and biomarkers: a systemetic review, J. Oral Maxillofac. Res. 7 (2016) e9.

[184] F. Delucchi, C. Canepa, L. Canullo, et al., Biomarkers from peri-implant crevicular fluid (PICF) as predictors of peri-implant bone loss: a systematic review, Int. J. Mol. Sci. 24 (2023) 3202.

[185] S. Yang, B. Gu, L. Zhao, et al., Meta-analysis of the association between serum and gingival crevicular fluid matrix metalloproteinase-9 and periodontitis, J. Am. Dent. Assoc. 150 (2019) 34–41, doi:10.1016/j.adaj.2018.08.025.

[186] N. Arias-Bujanda, A. Regueira-Iglesias, C. Balsa-Castro, et al., Accuracy of single molecular biomarkers in gingival crevicular fluid for the diagnosis of periodontitis: a systematic review and meta-analysis, J. Clin. Periodontol. 46 (2019) 1166–1182, doi:10.1111/jcpe.13188.

[187] J.M. Leppilahti, P.A. Hernández-Ríos, J.A. Gamonal, et al., Matrix metalloproteinases and myeloperoxidase in gingival crevicular fluid provide site-specific diagnostic value for chronic periodontitis, J. Clin. Periodontol. 41 (2014) 348–356, doi:10.1111/jcpe.12223.

[188] L.M. Golub, I.T. Räisänen, T. Sorsa, P.M. Preshaw, An unexplored pharmacologic/diagnostic strategy for peri-implantitis: a protocol proposal, Diagnostics 10 (2020) 1050.

[189] S.S. Gul, F.M. Zardawi, A.A. Abdulkareem, et al., Efficacy of MMP-8 level in gingival crevicular fluid to predict the outcome of nonsurgical periodontal treatment: a systematic review, Int. J. Environ. Res. Public Health 19 (2022) 3131.

[190] A. Chaparro, O. Realini, M. Hernández, et al., Early pregnancy levels of gingival crevicular fluid matrix metalloproteinases-8 and -9 are associated with the severity of periodontitis and the development of gestational diabetes mellitus, J. Periodontol. 92 (2021) 205–215, doi:10.1002/JPER.19-0743.

[191] V. Özgen Öztürk, P. Meriç, T. Sorsa, et al., Regulation of matrix metalloproteinases-8, -9 and endogenous tissue inhibitor-1 in oral biofluids during pregnancy and postpartum, Arch. Oral Biol. 124 (2021) 105065, doi:10.1016/j.archoralbio.2021.105065.

[192] Y.A. Yıldırım, A. Ozturk, F. Doğruel, et al., Serum vitamin D concentration is inversely associated with matrix metalloproteinase-9 level in periodontal diseases, J. Periodontol. 96 (2025) 663–674, doi:10.1002/JPER.24-0106.

[193] M.S. Attia, J.A. Alblowi, Effect of subantimicrobial dose doxycycline treatment on gingival crevicular fluid levels of MMP-9 and MMP-13 in periodontitis stage 2, Grade B in subjects with type 2 diabetes mellitus, J. Immunol. Res. 2020 (2020) 2807259, doi:10.1155/2020/2807259.

[194] A. Al-Majid, S. Alassiri, N. Rathnayake, et al., Matrix metalloproteinase-8 as an inflammatory and prevention biomarker in periodontal and peri-implant diseases, Int. J. Dent. 2018 (2018) 7891323, doi:10.1155/2018/7891323.

[195] L. Jansson, A. Lundmark, C. Modin, et al., Levels of matrix metalloproteinase-1 (MMP-1), MMP-2, MMP-3, osteopontin, pentraxin-3, and thymic stromal lymphopoietin in crevicular fluid samples from peri-implantitis, periodontitis, and healthy sites, J. Periodontal. Res. 60 (2025) 473–483, doi:10.1111/jre.13338.

[196] I.-C. Wang, J.V. Sugai, J. Majzoub, et al., Pro-inflammatory profiles in cardiovascular disease patients with peri-implantitis, J. Periodontol. 93 (2022) 824–836, doi:10.1002/JPER.21-0419.

[197] D. Gleiznys, A. Kriauciunas, J. Maminskas, et al., Expression of interleukin-17, tumor necrosis factor-alpha, and matrix metalloproteinase-8 in patients with chronic peri-implant mucositis, Med. Sci. Monit. 27 (2021) e932243, doi:10.12659/MSM.932243.

[198] H.-L. Wang, C. Garaicoa-Pazmino, A. Collins, et al., Protein biomarkers and microbial profiles in peri-implantitis, Clin. Oral Implants Res. 27 (2016) 1129–1136, doi:10.1111/clr.12708.

[199] K.E. Schmidt, T.M. Auschill, C. Heumann, et al., Clinical and laboratory evaluation of the effects of different treatment modalities on titanium healing caps: a randomized, controlled clinical trial, Clin. Oral Investig. 22 (2018) 2149–2160, doi:10.1007/s00784-017-2287-8.

[200] C. Chaussain-Miller, F. Fioretti, M. Goldberg, S. Menashi, The role of matrix metalloproteinases (MMPs) in human caries, J. Dent. Res. 85 (2006) 22–32, doi:10.1177/154405910608500104.

[201] L. Anumula, S. Ramesh, V.S.K. Kolaparthi, Matrix metalloproteinases in dentin: assessing their presence, activity, and inhibitors—a review of current trends, Dent. Mater. 40 (2024) 2051–2073, doi:10.1016/j.dental.2024.09.011.

[202] A. Mazzoni, F. Mannello, F.R. Tay, et al., Zymographic analysis and characterization of MMP-2 and -9 forms in human sound dentin, J. Dent. Res. 86 (2007) 436–440, doi:10.1177/154405910708600509.

[203] A. Mazzoni, D.H. Pashley, F.R. Tay, et al., Immunohistochemical identification of MMP-2 and MMP-9 in human dentin: correlative FEI-SEM/TEM analysis, J. Biomed. Mater. Res. A 88A (2009) 697–703, doi:10.1002/jbm.a.31920.

[204] A. Mazzoni, V. Papa, F. Nato, et al., Immunohistochemical and biochemical assay of MMP-3 in human dentine, J. Dent. 39 (2011) 231–237, doi:10.1016/j.jdent.2011.01.001.

[205] M. Sulkala, T. Tervahartiala, T. Sorsa, et al., Matrix metalloproteinase-8 (MMP-8) is the major collagenase in human dentin, Arch. Oral Biol. 52 (2007) 121–127, doi:10.1016/j.archoralbio.2006.08.009.

[206] M. Sulkala, M. Larmas, T. Sorsa, et al., The Localization of matrix metalloproteinase-20 (MMP-20, Enamelysin) in mature human teeth, J. Dent. Res. 81 (2002) 603–607, doi:10.1177/154405910208100905.

[207] M. Prasad, W.T. Butler, C. Qin, Dentin sialophosphoprotein in biomineralization, Connect. Tissue Res. 51 (2010) 404–417, doi:10.3109/03008200903329789.

[208] J. Jia, Z. Bian, Y. Song, Dspp mutations disrupt mineralization homeostasis during odontoblast differentiation, Am. J. Transl. Res. 7 (2015) 2379–2396.

[209] L. Tjäderhane, H. Larjava, T. Sorsa, et al., The activation and function of host matrix metalloproteinases in dentin matrix breakdown in caries lesions, J. Dent. Res. 77 (1998) 1622–1629, doi:10.1177/00220345980770081001.

[210] M. Sulkala, J. Wahlgren, M. Larmas, et al., The effects of MMP inhibitors on human salivary MMP activity and caries progression in rats, J. Dent. Res. 80 (2001) 1545–1549, doi:10.1177/00220345010800061301.

[211] O. Kiuru, J. Sinervo, H. Vähänikkilä, et al., MMP inhibitors and dentin bonding: systematic review and meta-analysis, Int. J. Dent. 2021 (2021) 9949699, doi:10.1155/2021/9949699.
[212] J. Wahlgren, P. Maisi, T. Sorsa, et al., Expression and induction of collagenases (MMP-8 and -13) in plasma cells associated with bone-destructive lesions, J. Pathol. 194 (2001) 217–224, doi:10.1002/path.854.
[213] J. Wahlgren, T. Salo, O. Teronen, et al., Matrix metalloproteinase-8 (MMP-8) in pulpal and periapical inflammation and periapical root-canal exudates, Int. Endod. J. 35 (2002) 897–904, doi:10.1046/j.1365-2591.2002.00587.x.
[214] C.-H. Tsai, Y.-J. Chen, F.-M. Huang, et al., The upregulation of matrix metalloproteinase-9 in inflamed human dental pulps, J. Endod. 31 (2005) 860–862, doi:10.1097/01.don.0000164851.55389.4e.
[215] D. Barreiros, P.F. Nelson, F.W.G. Paula-Silva, et al., MMP2 and MMP9 are associated with apical periodontitis progression and might be modulated by TLR2 and MyD88, Braz. Dent. J. 29 (2018) 43–47, doi:10.1590/0103-6440201801731.
[216] P. Chuhuaicura, C. Rodríguez-Niklitschek, G.H. Oporto, L.A. Salazar, Distinct molecular mechanisms in oral mucosal wound healing: translational insights and future directions, Int. J. Mol. Sci. 26 (2025) 10660.
[217] T. Salo, M. Mäkelä, M. Kylmäniemi, et al., Expression of matrix metalloproteinase-2 and -9 during early human wound healing, Lab. Invest. 70 (1994) 176–182.
[218] M. Mäkelä, H. Larjava, E. Pirilä, et al., Matrix metalloproteinase 2 (Gelatinase A) is related to migration of keratinocytes, Exp. Cell Res. 251 (1999) 67–78, doi:10.1006/excr.1999.4564.
[219] P. Åström, E. Pirilä, R. Lithovius, et al., Matrix metalloproteinase-8 regulates transforming growth factor-β1 levels in mouse tongue wounds and fibroblasts in vitro, Exp. Cell Res. 328 (2014) 217–227, doi:10.1016/j.yexcr.2014.07.010.
[220] J.T. Korpi, P. Åström, N. Lehtonen, et al., Healing of extraction sockets in collagenase-2 (matrix metalloproteinase-8)-deficient mice, Eur. J. Oral Sci. 117 (2009) 248–254, doi:10.1111/j.1600-0722.2009.00620.x.
[221] B. Vashishat, S. Sinha, T. Srivastava, et al., Management of oral aphthous ulcer: a review, Bioinformation 20 (2024) 434–438, doi:10.6026/973206300200434.
[222] R. Häyrinen-Immonen, T. Sorsa, D. Nordström, et al., Collagenase and stromelysin in recurrent aphthous ulcers (RAU), Int. J. Oral Maxillofac. Surg. 22 (1993) 46–49, doi:10.1016/S0901-5027(05)80357-1.
[223] S. Skulason, W.P. Holbrook, T. Kristmundsdottir, Clinical assessment of the effect of a matrix metalloproteinase inhibitor on aphthous ulcers, Acta Odontol. Scand. 67 (2009) 25–29, doi:10.1080/00016350802526559.
[224] S. David, B. Bahram, C. Patrice, Behçet's syndrome, N. Engl. J. Med. 390 (2024) 640–651, doi:10.1056/NEJMra2305712.
[225] T. Bello María, López-Longo, J. Francisco, Arias-Salgado, G. Elena, M. Martín, V. Jiménez-Yuste, A.R. de la Rúa, N.V.I.F.Á. Butta, Platelet soluble CD40L and matrix metalloproteinase 9 activity are proinflammatory mediators in Behçet disease patients, Thromb. Haemost. 107 (2012) 88–98, doi:10.1160/TH11-08-0556.
[226] S. Pay, T. Abbasov, H. Erdem, et al., Serum MMP-2 and MMP-9 in patients with Behçet's disease: do their higher levels correlate to vasculo-Behçet's disease associated with aneurysm formation? Clin. Exp. Rheumatol. 25 (2007) S70–S75.
[227] K.S Park, Y. Min, S.R Park, et al., Matrix metalloproteinase-2, -9, -12, and tissue inhibitor of metalloproteinase 2 gene polymorphisms and cutaneous expressions in patients with Behçet's disease, Tissue Antigens 79 (2012) 333–339, doi:10.1111/j.1399-0039.2012.01863.x.

[228] K. Hamzaoui, B. Maître, A. Hamzaoui, Elevated levels of MMP-9 and TIMP-1 in the cerebrospinal fluid of neuro-Behçet's disease, Clin. Exp. Rheumatol. 27 (2009) S52–S57.
[229] F.J. Descamps, D. Kangave, B. Cauwe, et al., Interphotoreceptor retinoid-binding protein as biomarker in systemic autoimmunity with eye inflictions, J. Cell. Mol. Med. 12 (2008) 2449–2456, doi:10.1111/j.1582-4934.2008.00264.x.
[230] A. Naouali, W. Kaabachi, K. Tizaoui, et al., Association of MMP-9 gene polymorphisms with Behçet's disease risk, Immunol. Lett. 164 (2015) 18–24, doi:10.1016/j.imlet.2015.01.005.
[231] Y.J. Lee, S.W. Kang, H.J. Baek, et al., Association between matrix metalloproteinase 9 promoter polymorphisms and Behçet's disease, Hum. Immunol. 71 (2010) 717–722, doi:10.1016/j.humimm.2010.03.009.
[232] A. Kamal, F.T. Elgengehy, M.M Abd Elaziz, et al., Matrix metalloproteinase-9 rs17576 gene polymorphism and Behçet's disease: is there an association? Immunol. Invest. 46 (2017) 460–468, doi:10.1080/08820139.2017.1296857.
[233] Y. Aksoy, A. Ercan, O. Dalmizrak, et al., The determination of matrix metalloproteinase 9 activity and gene expression levels in Behcet's disease patients with aneurysmal complications, Clin. Rheumatol. 30 (2011) 515–519, doi:10.1007/s10067-010-1559-3.
[234] P.-Y. Gegout, R. Wabbi, S. Jung, O. Huck, Oral consequences of systemic lupus erythematosus: an update, Curr. Oral Heal. Rep. 10 (2023) 184–195, doi:10.1007/s40496-023-00356-x.
[235] A. Dufour, C.L. Bellac, U. Eckhard, et al., C-terminal truncation of IFN-γ inhibits proinflammatory macrophage responses and is deficient in autoimmune disease, Nat. Commun. 9 (2018) 2416, doi:10.1038/s41467-018-04717-4.
[236] A. Faber-Elmann, Z. Sthoeger, A. Tcherniack, et al., Activity of matrix metalloproteinase-9 is elevated in sera of patients with systemic lupus erythematosus, Clin. Exp. Immunol. 127 (2002) 393–398, doi:10.1046/j.1365-2249.2002.01758.x.
[237] M.H. Elbatreek, A.M. Fathi, I. Mahdi, et al., Thymus satureioides Coss. combats oral ulcer via inhibition of inflammation, proteolysis, and apoptosis, Inflammopharmacology 31 (2023) 2557–2570, doi:10.1007/s10787-023-01285-y.
[238] K. Reynolds, S. Zhang, B. Sun, et al., Genetics and signaling mechanisms of orofacial clefts, Birth Defects Res. 112 (2020) 1588–1634, doi:10.1002/bdr2.1754.
[239] M.M. Imani, M. Shalchi, G. Ahmadabadi, M. Sadeghi, Evaluation of matrix metalloproteinases (MMPs) and tissue inhibitors of metalloproteinases (TIMPs) in human cases with orofacial clefts: a systematic review, Int. Orthod. 21 (2023) 100781, doi:10.1016/j.ortho.2023.100781.
[240] J.R. Evangelista, C.F. Bernardi, A.V.S. Faria, D.F. Bueno, Matrix metalloproteinase-3 (MMP3) in non-syndromic cleft lip and palate: extracellular matrix remodeling, developmental signaling, and molecular mechanisms, Biochem. Biophys. Res. Commun. 812 (2026) 153599, doi:10.1016/j.bbrc.2026.153599.

CHAPTER SEVEN

Matrix metalloproteinases (MMPs) in neurodegenerative diseases, an updated overview of the role of MT1-MMP and MT5-MMP in Alzheimer's disease

Marielle Frank[a], Cyrille Barthélémy[a], Antoine Dufour[b], Santiago Rivera[c], Kévin Baranger[a,*]

[a] La Rochelle Université, UMR 7266 CNRS LIENSs, La Rochelle, France
[b] Department of Physiology and Pharmacology, Department of Biochemistry and Molecular Biology, McCaig Institute for Bone and Joint Health, Hotchkiss Brain Institute, Snyder Institute for Chronic Diseases, Cumming School of Medicine, University of Calgary, Calgary, Alberta, Canada
[c] Aix-Marseille Univ, CNRS, INP, Inst Neurophysiopathol, Marseille, France
*Corresponding author. e-mail address: kevin.baranger@univ-lr.fr.

Contents

Abstract

Neurodegenerative diseases, including Alzheimer's disease, Parkinson's disease, amyotrophic lateral sclerosis and Huntington's disease, pose an increasing threat to global health due to ageing populations and the lack of effective treatments. Matrix metalloproteinases (MMPs), which are proteases that belong to the metzincin superfamily,

Fungal and Protozoan Carbonic Anhydrases, Volume 60
ISSN 1874-6047
https://doi.org/10.1016/bs.enz.2026.07.004

are critical regulators of neuroinflammation, blood-brain barrier integrity, protein misfolding, synaptic dysfunction and neuronal death. This chapter provides a comprehensive overview of the roles of MMPs at the intersection of the molecular and cellular mechanisms that underpin neurodegenerative processes. The focus is on their pathophysiological mechanisms, therapeutic potential, and recent advances. We highlight novel mechanistic insights in Alzheimer's disease, such as the roles of MT1-MMP and MT5-MMP in amyloid-beta peptide production and the dual functions of MMPs in both neuroprotection and neurotoxicity. This review emphasizes the need for further mechanistic research into the different MMPs involved in neurodegenerative diseases.

Keywords: Alzheimer's disease, Amyotrophic lateral sclerosis, Huntington's disease, MMPs, Parkinson's disease

1. Introduction

1.1 The global burden of neurodegenerative diseases

Neurodegenerative diseases (NDs) are characterized by the progressive loss of neuronal structure and function, leading to cognitive decline, motor dysfunction and, ultimately, death. As populations worldwide are ageing, and these disorders cannot yet be cured, reversed or significantly slowed down, their prevalence is expected to rise dramatically, posing unprecedented health, social and economic challenges [1,2]. NDs typically begin with mild symptoms that affect motor, behavioral, and cognitive functions. The gradual loss of large numbers of neurons inevitably worsens these symptoms, resulting in a loss of global autonomy and often the patient's death. Alzheimer's disease (AD) alone affects over 50 million people globally, and the World Health Organization projects that this number will triple by 2050. Similarly, approximately 10 million individuals are affected by Parkinson's disease (PD), while amyotrophic lateral sclerosis (ALS) and Huntington's disease (HD) also contribute to the growing burden of neurological disorders (https://www.who.int/publications/i/item/9789240116139). The lack of disease-modifying therapies for most NDs highlights the urgent need to understand their underlying mechanisms and identify new therapeutic targets.

The chronic nature of NDs is associated with genetic and environmental stimuli that act synergistically to cause unique dysfunctions for each disease, as well as fundamental commonalities such as gene mutations, increased free radical generation, mitochondrial dysfunction, immunosenescence, abnormal protein folding and inflammation. The confluence of these pathogenic driving forces makes it difficult to identify a single etiological cause that could be more easily targeted. Current treatments primarily focus on managing symptoms, such as using cholinesterase inhibitors for AD or L-DOPA for PD, but they fail to halt or reverse disease progression [3]. Thus, there is a critical need for mechanism-based interventions that can modify the course of the disease.

Clearly identifying molecular targets and designing effective therapeutic strategies requires an understanding of the underlying mechanisms of the pathology. Among the numerous biological factors that may contribute to these processes, matrix metalloproteinases (MMPs) are of particular interest because:

i) MMPs are expressed by most cell types in the central nervous system (CNS);

ii) MMP expression is modulated by potentially neurotoxic proteins, which can undergo cleavage and functional regulation by the same or other MMPs;

iii) MMPs contribute to chronic neuroinflammation by controlling the cleavage of inflammatory mediators (*e.g.*, cytokines and chemokines) and their receptors, as well as disrupting the tight junctions between endothelial cells that ensure the impermeability of the blood-brain barrier (BBB);

iv) MMPs are involved in processing the extracellular matrix (ECM) and transmembrane proteins that mediate cell-to-cell interactions;

v) MMPs can initiate or participate in pathways that amplify a web of interactions between substrates and other proteases.

For these reasons, summarized in Fig. 1, MMPs are involved in a broad spectrum of pathophysiological mechanisms, each with its own multifunctional peculiarities (Fig. 1). In this review, we provide a non-exhaustive overview of the knowledge on MMPs in NDs, focusing particularly on AD and two MT-MMPs: MT1- and MT5-MMP. While MMPs also play essential physiological roles, these aspects are beyond the scope of the present review and have been extensively covered in dedicated reviews on the topic [4–6].

1.2 The metzincin protease superfamily: MMPs, ADAMs and ADAMTS

The metzincin superfamily of metalloproteinases includes matrix metalloproteinases (MMPs), a disintegrin and metalloproteinases (ADAMs), and ADAM with thrombospondin motifs (ADAMTS). These enzymes share a catalytic domain containing a conserved Met-turn and a Zn^{2+} ion, which

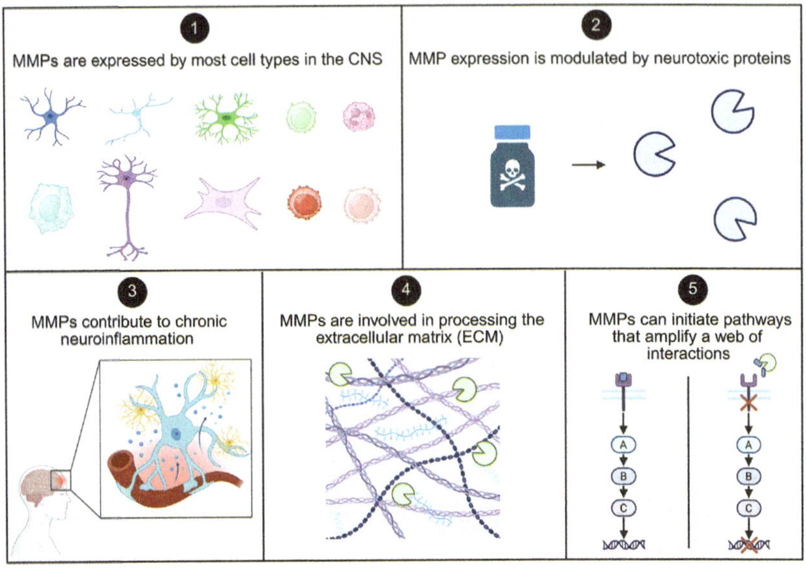

Fig. 1 A schematic overview of matrix metalloproteinases (MMPs) expression, regulation, and functions in the central nervous system.

is essential for their proteolytic activity (reviewed in [5]). Focusing on the first group, the MMPs are widely expressed in most human cells, and 23 members have been identified in humans (24 in mice and rats).

Based on their substrate specificity and their structure, they have been traditionally classified into five categories: collagenases (MMP-1, −8 and −13), gelatinases (−2 and −9), stromelysins (MMP-3 and −10), membrane-type MMPs (MT-MMPs, MMP-14, −15, −16, −17, −24 and −25) and the other MMPs (MMP-7, MMP-12 or MMP-28). MMPs not only degrade ECM components outside the cells, but they also process proteins at the plasma membrane and within cells (see [5] for review). ADAMs, such as ADAM10 and ADAM17, and ADAMTS, including ADAMTS-1 and ADAMTS-4, are structurally related but have distinct substrate specificities and biological functions [7]. ADAMs are involved in ectodomain shedding, such as the cleavage of amyloid precursor protein (APP) in the CNS, while ADAMTS play roles in ECM remodeling and proteoglycan processing [8].

MMPs share common structural features, which complicates the development of specific inhibitors difficult [5,6,9]. They all begin with a hydrophobic signal peptide comprising 17–29 amino acids, followed by a pro-domain comprising 77–87 amino acids. The latter interacts with a conserved cysteine residue and the Zn^{2+} cation of the catalytic site, thereby maintaining the enzyme in an inactive state, named zymogen. The conversion of inactive pro-MMPs into active MMPs is triggered by either the redox-mediated break-up of the interaction between "the cysteine" and the metal cation (known as the "cysteine switch"), or pro-domain removal by autocatalysis or other proteinases.

The catalytic domain, which contains the catalytic site, is composed of approximately 170 amino acids and harbors a Zn^{2+}-binding consensus sequence, HExxHxxGxxH. The three histidine residues are essential for full activity, as they ensure stability with respect to Zn^{2+}. A short hinge region followed by the hemopexin domain (approximately 200 amino acids) is located downstream of the catalytic site. This region exhibits a relatively variable amino acid sequence, which is crucial for the binding of substrates and thus determines their specificity. Notably, MMP-7, MMP-23 and MMP-26 lack a hemopexin domain. Additionally, the six members of the membrane-type MMP (MT-MMP) subfamily contain membrane-anchoring domains: a glycosylphosphatidylinositol domain in the case of MMP-17 (MT4-MMP) and MMP-25 (MT6-MMP), and an integral transmembrane domain with a short intracytoplasmic tail in the case of MMP-14 (MT1-MMP), MMP-15 (MT2-MMP), MMP-16 (MT3-MMP) and MMP-24 (MT5-MMP). These anchors enable localized proteolysis at the cell surface like ADAMs [10].

Of particular and functional interest are MMP-11, MMP-23 and MMP-28, as well as the MT-MMPs. These MMPs all contain a specific dibasic sequence at the C-terminus of their pro-domain that is recognized by intracellular proprotein convertases, such as the serine protease furin. This interaction leads to their activation in the trans-Golgi network before secretion [11]. However, this furin-mediated switch-on activation is not restricted to intracellular compartments and can also be orchestrated outside the cells [12]. Unlike MT-MMPs, MMPs are secreted as inactive zymogens (pro-MMPs) and require proteolytic removal of their pro-domain for activation. Key activators include other MMPs, *e.g.*, MMP-3 activating pro-MMP-9, serine proteases like plasmin and thrombin, and membrane-type MMPs, notably MT1-MMP-mediated activation of pro-MMP-2 (see below). Yet, the full repertoire of all proteases that can activate MMPs remains to be studied.

The wide-ranging activities of MMPs are tightly regulated at the transcriptional and post-translational levels, notably through pro-MMP activation (*i.e.*, removal of the pro-domain) and inhibition by endogenous tissue inhibitors of metalloproteinases (TIMPs) [5]. The four TIMPs

(TIMP-1 to TIMP-4) have been identified in humans and are widely distributed, except for TIMP-4, which is predominantly expressed in the heart, ovaries and brain [13]. They are pleiotropic secreted proteins sharing ~40 percent sequence homology and are best known for their reversible MMP inhibitory activities. *TIMP-1, TIMP-3* and *TIMP-4* genes are located within introns of genes that code for the synaptic proteins synapsins, *SYN1, SYN3* and *SYN2* respectively, while the *TIMP-2* encoding gene is located close to the *SYN4* gene [14]. This intriguing genetic link between TIMPs and synapsins —peripheral neuronal proteins coating synaptic vesicles— suggests a yet unknown functional relationship in the nervous system, warranting further exploration, given that many neuropathologies involve alterations in the expression of these proteins [15]. Structurally, TIMPs share a conserved backbone comprising twelve conserved cysteine residues and six disulfide bonds, which are essential for their biological activities. In addition to a signal peptide, the N-terminal part of TIMPs contains the MMP-inhibitory domain that directly interacts with the MMP catalytic site, while the C-terminal domain binds the hemopexin domain of MMPs.

TIMPs bind to MMPs in a 1:1 stoichiometric ratio, thereby inhibiting their proteolytic activity. However, the binding affinities and modes of interaction between TIMPs and MMPs differ significantly. The N-terminal domain of TIMPs is generally sufficient for MMP inhibition, as it binds the catalytic domain of MMPs, whereas the C-terminal domain can mediate additional interactions with substrates. TIMP-1, TIMP-2 and TIMP-4 are found as soluble or cell surface-associated proteins, whereas TIMP-3 is uniquely ECM-bound *via* heparan sulphate proteoglycans. Among the TIMPs, TIMP-3 displays the broadest inhibitory profile, as it is the only family member known to inhibit both MMPs and members of the ADAM and ADAMTS families. In contrast, most TIMPs inhibit several MMPs, while inhibition of ADAMTS proteases appears to be restricted mainly to TIMP-3 [5].

TIMP-1 preferentially inhibits secreted MMPs and does not efficiently inhibit MT-MMPs. In addition to its MMP inhibitory activity, TIMP-1 can inhibit ADAM10 through interactions involving its C-terminal domain. Similarly, TIMP-3 inhibits ADAM10, but unlike TIMP-1, it also targets ADAM17 and several ADAMTS proteases [5].

TIMP-2 and TIMP-4 share overlapping inhibitory profiles, targeting soluble MMPs and MT-MMPs. TIMP-4 is highly expressed in the CNS and displays high affinity for MMP-2 and MMP-9. TIMP-2 has a unique role in the activation of pro-MMP-2 by MT1-MMP. In this process, the

C-terminal domain of TIMP-2 binds the hemopexin domain of pro-MMP-2, anchoring the zymogen to the cell surface, while the N-terminal domain of TIMP-2 interacts with MT1-MMP. This fixes pro-MMP-2 to the membrane, enabling an adjacent active MT1-MMP to cleave the pro-domain of MMP-2, leading to MMP-2 activation [16].

1.3 Rationale for studying MMPs in NDs

MMPs are typically solely regarded as ECM-degrading enzymes, with collagens, laminins, fibronectin and aggrecan among their most-studied substrates. However, this view is now changing, as an increasing number of non-ECM substrates have been identified, including adhesion molecules such as ICAM-1, VCAM-1, N-cadherin, β-dystroglycan and many others (see [6] for review). MMPs also process cytokines and chemokines, including TNF-α, IL-1β and CXCL12, as well as neurotransmitter receptors such as NMDA and AMPA. An imbalance between MMPs and TIMPs is a hallmark of many pathological conditions, including neurodegeneration [17]. MMPs contribute to neuroinflammation by being upregulated in activated microglia, astrocytes, neurons and endothelial cells. They degrade tight junction proteins, such as occludin and claudin, as well as basement membrane components, such as laminin and collagen IV. This results in BBB leakage and neuronal damage [18]. Furthermore, MMPs contribute to neuronal death through mechanisms involving mitochondrial dysfunction and oxidative stress. In the context of neurodegeneration, MMP-3, for instance, cleaves α-synuclein, contributing to Lewy bodies (LBs) formation in PD [19]. Additionally, MMP-9 cleaves amyloid-beta (Aβ) peptide limiting amyloid plaques in AD but processes tau protein, thereby contributing to neurofibrillary tangle formation [20]. More recently, we demonstrated that MT1-MMP and MT5-MMP cleave the amyloid precursor protein (APP), resulting in the accumulation of Aβ and its immediate precursor, C99, that play a central role in Alzheimer's pathology [4,21,22].

In addition to their well-established catalytic activities, the non-proteolytic functions of MMPs are increasingly recognized in the literature [22–29]. Given their multifaceted roles, MMPs represent promising therapeutic targets and potential biomarkers for NDs. Although synthetic MMP inhibitors such as batimastat, marimastat and doxycycline have been developed, they lack selectivity for a single MMP, often leading to off-target effects such as musculoskeletal stiffness [9].

This chapter provides a non-exhaustive overview of MMPs in PD, ALS, HD and summarizes the current knowledge of their involvement in AD, with a particular focus on MT1- and MT5-MMP.

2. Disease-specific roles of MMPs

2.1 Parkinson's disease (PD)

PD is the second most common neurodegenerative disorder after AD. It is characterized by the accumulation of intracellular fibrillar α-synuclein (α-Syn) aggregates, known as Lewy Bodies (LBs), and the loss of dopaminergic neurons in both sporadic and genetic forms of the disease [30]. α-Syn is a 140 amino acid presynaptic protein for which three missense mutations (A30P, E46K and A53T), as well as wild-type gene amplification, have been associated with familial forms of PD [31,32]. In addition to α-syn, more than 290 other molecules have been identified in LBs, including parkin, tau, leucine-rich repeat kinase 2 (LRRK2), DJ-1 and APP [33,34].

Truncated forms of α-Syn contribute to PD pathology. The expression of MMPs is induced by α-Syn in cultured microglial cells [35], and in turn several MMPs, including MMP-1, MMP-2, MMP-3, MMP-9 and MT1-MMP can cleave α-Syn *in vitro* [19,36]. Notably, MMP-3 generates toxic α-Syn fragments that are prone to aggregation and exhibit higher cytotoxicity than full-length α-Syn, which aggregates into LBs. MMP-3 also triggers neuronal death [37] and the injection of N-methyl-4-phenyl-1,2,3,6-tetrahydropyridine (MPTP) into MMP-3 knockout (*Mmp3*$^{-/-}$) mice revealed reduced BBB disruption, reactive oxygen species (ROS) production, dopaminergic neuron death and improved locomotor activity [38,39]. MMP-3 levels are elevated in PD brains and correlate with disease severity [40]. Additionally, MMP-9 is upregulated in the substantia nigra of PD patients [41], where it activates microglia and promotes dopamine neuron death in the 1-methyl-4-phenyl-1,2,3,6-tetrahydropyridine mouse model of Parkinson's disease [42].

Mutations in the DJ-1 gene, which encodes a 189-amino acid protein deglycase, are associated with autosomal recessive early-onset PD [43]. MMP-1, −2, −8, −9 and MT1-MMP can efficiently cleave DJ-1 [44]. Interestingly, α-Syn promotes the expression of MT1-MMP and the migration of microglial cells *in vitro* and reduced expression of MT1-MMP impairs the migration of the BV-2 microglial cell line when transplanted into the injured substantia nigra of 6-OHDA-treated mice. This relates to both the interaction and the shedding of the cell adhesion molecule CD44 by MT1-MMP [27,45].

2.2 Amyotrophic lateral sclerosis (ALS)

The main target of ALS, an uncommon and rapidly progressing neurological illness, is motor neurons in the brain, brainstem, and spinal cord [46]. Around 40 percent of patients experience severe disability, muscle wasting, and cognitive impairment. This results in a bleak prognosis. Around 50 percent of patients have a life expectancy of less than 20 months after diagnosis, with most of them succumbing to ventilatory failure. The rapid development of the disease, combined with the lack of reliable biomarkers and the absence of preventative or curative treatments, makes therapeutic development extremely challenging. One major risk factor is ageing, particularly after the age of 55 [5,47]. Five to ten percent of patients have genetically inherited ALS, caused by damaging mutations in genes such as *C9orf72, SOD1, TDP-43* and *FUS*, which are involved in RNA processing, protein transport and clearance [48].

Most patients have sporadic ALS of unknown multifactorial origin. It is thought that ALS arises from a combination of genetic and environmental factors impacting vital biological functions, including neurotransmitter homeostasis, blood–brain barrier integrity, protein folding, oxidative stress and neuroinflammation. MMP-9 plays a crucial role in ALS pathology, particularly in determining subtype susceptibility to the selective loss of motor neurons characteristic of ALS. Its expression is associated with early endoplasmic reticulum stress and is elevated in fast-fatigable neurons [49]. While earlier research questioned the impact of MMP-9 deletion, more recent data demonstrate that lowering MMP-9 activity slows muscle denervation and preserves motor units in both SOD1 and TDP-43 models [49–51]. Another structural mechanism that makes motor neurons susceptible to oxidative injury is the phagocytosis of perineuronal nets by microglia and astrocytes that produce MMP-9 [52]. Further support for the contribution of MMP-9 to ALS comes from *in vitro* studies, which show that MMP-9 inhibition with glycoursodeoxycholic acid prevents cell degeneration in a $SOD1^{G93A}$ cell model [53,54]. However, the therapeutic potential of targeting MMP-9 remains complex: while MMP-9 knockdown in rNLS8 sporadic ALS mice preserves motor neurons and muscle function, long-term reduction in MMP-9 leads to the premature death of a subset of these mice

[50]. Thus, while MMP-9 represents a promising therapeutic target, further research is required to fully elucidate its role in ALS.

The detection of active MMP-9 in patient biofluids marks a significant translational advance. However, findings on its localization remain inconsistent: one study reports increased MMP-9 levels in the plasma of ALS patients, but not in the CSF [55], while another study shows increased MMP-9 levels in the CSF [56]. Additionally, enzyme-linked immunosorbent assay (ELISA) measurements have shown increased concentrations of MT1-MMP, MMP-2 and TIMP-1 in the CSF of ALS patients compared to healthy controls [57]. ALS serum also contains elevated levels of active MMP-9, which the dual MMP-2/MMP-9 inhibitor AQU-118–8 may suppress *in situ* [58]. This provides critical pharmacological evidence that active MMP-9 is accessible and can be targeted in patients. Other contributing factors include TIMP dysregulation. SOMAscan proteomics identified TIMP-3 as a uniquely elevated plasma protein in ALS patients. ADAM10 and ADAM17 also facilitate TNF-α shedding in SOD1 mutants [59]. These results are supported by previous observations of a positive correlation between TIMP-3 levels and neurodegeneration. $SOD1^{G93A}$ degenerating neurons exhibit increased TIMP-3 levels alongside increased interactions between Fas and the Fas-associated protein with a death domain (FADD). Mechanistically, TIMP-3 induces neuronal cell death by inhibiting the cleavage of FADD by MMP-3, thereby stabilizing Fas-FADD interactions and promoting Fas-mediated apoptosis [60].

2.3 Huntington's disease (HD)

HD is the most common inherited neurodegenerative polyglutamine disorder. Also known as Huntington's chorea, this incurable disease results in motor coordination deficiencies, mood and mental capacity disturbances, and eventually dementia. As symptoms typically manifest around the age of 40, HD is not strictly an age-related illness. It is caused by an expansion of the CAG triplet repeat in exon 1 of the huntingtin (Htt) gene, which produces a chain of polyglutamine (polyQ) residues at the N-terminus of Htt. This leads to the aggregation of Htt in the cytoplasm and nucleus of neurons, resulting in cell death that primarily affects the striatum and cortex [5,61–63].

N-terminal small fragments of Htt are particularly toxic [64–66]. MMPs, along with caspases and calpains, are some of the proteases responsible for its processing [67]. Among the MMPs investigated, MMP-10 and MT1-MMP appear to play particularly relevant roles. Both are upregulated in striatal

$Hdh^{111Q/111Q}$ cells compared to $Hdh^{7Q/7Q}$ cells and knockdown of either proteinase reduces caspase activation in $Hdh^{111Q/111Q}$ cells. Consistent with these findings, endogenous MMP inhibitors TIMP-1 and TIMP-3 inhibit Htt cytotoxicity in these cells. Mechanistically, MT1-MMP and MMP-2 do not directly cleave Htt, whereas MMP-10 cleaves at residue 402, generating a 48-kDa N-terminal fragment. Notably, MMP-10 knockdown prevents $Hdh^{111Q/111Q}$ cell death, highlighting MMP-10 as a key contributor to Htt-mediated toxicity [67].

Induced pluripotent stem cell (iPSC)-based neural stem cells from patients with HD also exhibit dysregulation of the MMP/TIMP balance. Specifically, MMP-3 and MMP-10 are upregulated, and their increased expression correlates with greater neurotoxicity. Beyond their direct effects on neuronal survival, MMPs also play a role in HD through vascular dysfunction and neuroinflammatory processes [68]. In HD post-mortem tissue and CSF, increased MMP-9 levels are correlated with increased concentrations of inflammatory mediators such as IL-6, as well as with disease severity [69]. Similarly, plasma levels of MMP-9 and IL-6 are higher in patients with HD than in healthy controls. This has also been observed in R6/2 HD mice compared with their wild-type littermates [70].

Recent therapeutic studies further support a pathogenic role for MMPs in HD. Combined inhibition of MMP-1 and PAR-1 using SCH79797 can restore BBB integrity in HD mouse models by decreasing MMP-9 expression and increasing claudin-5 levels [71]. Furthermore, a key neuroinflammatory axis in HD development involves the interaction between ADAMs/MMPs and the neuroprotective cytokine CX3CL1 [72]. Nevertheless, compared to ALS, the HD field has not reached the same stage of advancement despite the compelling nature of these pathways. Despite these promising mechanistic and therapeutic advances, the role of MMPs in HD remains considerably less well characterized compared to ALS, highlighting the need for further investigation of these pathways.

2.4 Alzheimer's disease (AD)

AD is the most common type of neurodegenerative disorder for which efficient cures, disease-modifying treatments, or preventive strategies are still lacking. It is pathologically characterized by the accumulation of amyloid plaques, mainly composed of Aβ, and neurofibrillary tangles (NFTs) — caused by tau protein hyperphosphorylation (p-Tau)— and neuroinflammation. Familial forms of AD, accounting for approximately 1 percent of cases, are caused by mutations in three genes: the amyloid precursor protein (APP) gene, and the presenilin 1 (*PSEN1*) and presenilin 2 (*PSEN2*) genes [73–75]. Aβ results from the sequential cleavage of amyloid precursor protein (APP) by two enzymes: first, β-secretase (BACE-1, or beta-site APP-cleaving enzyme-1), and then the γ-secretase complex. BACE-1 cleaves APP to generate a transmembrane fragment containing 99 amino acids (C99), which is subsequently cleaved by γ-secretase to release 4 kDa Aβ peptides, mainly containing 40 (Aβ40) and 42 (Aβ42) amino acids [4]. Small oligomers (*e.g.*, dimers and trimers) are increasingly recognized as the most toxic forms of Aβ. Regulated APP cleavage by β- and γ-secretase occurs following APP endocytosis into the endosomes or during its targeting to the cell surface through the trans-Golgi network, highlighting the importance of APP cellular trafficking in amyloidogenesis. Aβ accumulation results from a balance between its formation and degradation [4]. Among the various proteinases capable of cleaving Aβ (reviewed in [4,76]), some MMPs (*e.g.* MMP-2, MMP-3, MMP-9 and MT1-MMP) also exhibit Aβ-degrading activity [5].

APP can also undergo processing at the plasma membrane by α-secretases from the ADAM family. ADAM10 is the main constitutive α-secretase, while ADAM17 is responsible for regulated α-secretase activity. Cleavage by α-secretase in the middle of the Aβ sequence prevents its formation and generates a soluble N-terminal APP fragment (sAPPα) with neurotrophic and neuroprotective properties [4]. ADAM10 and ADAM17 are also implicated in AD through the shedding of other substrates. For example, ADAM17 sheds membrane-bound pro-TNF-α, resulting in the production of the pro-inflammatory soluble form [77]. There is growing evidence that TNF-α contributes to the pathogenesis of AD, and that TNF-α inhibitors, some of which are being tested in clinical trials, could improve pathological and cognitive outcomes [78]. More recently, triggering receptor expressed on myeloid cells (TREM2), a protein for which certain genetic variants are associated with an increased risk of AD [79–81] has been identified as a substrate for both ADAM-10 and −17 [82,83]. Shedding of TREM2 in microglia reduces their phagocytic activity, thereby contributing to Aβ accumulation. Similarly, the shedding of LRP-1 by different ADAMs (*e.g.*, ADAM12) may affect this receptor's ability to promote the clearance of Aβ via endocytosis [84]. The non-proteolytic activities of ADAMs, such as ADAM30, have also been identified. It interacts with and regulates the cellular trafficking of APP, thereby reducing Aβ production in a cathepsin D-dependent manner [29]. The impact of these non-proteolytic activities on APP metabolism highlights the need to unpick the molecular interactions underlying ADAM functions to improve our understanding of APP metabolism and the consequences for AD pathology. There is growing evidence that physiological Aβ is part of the innate immune system's arsenal for combating bacterial, fungal, and viral pathogens in the brain [85]. Similarly, tau hyperphosphorylation, another hallmark of AD alongside Aβ accumulation, is induced in neurons in response to herpes simplex virus 1 (HSV-1) infection and neutralizes the viral infectivity by directly binding to the viral capsid. Additionally, it was found that p-Tau helps microglia to phagocytose HSV-1-infected neurons [86]. Physiological Aβ also played a role in maintaining basal neurotransmitter release and synaptic activity [87], and blocking Aβ with antibodies in these conditions interferes with cGMP-induced enhancement of LTP and memory [88]. Taken together, these data suggest that the proteolytic regulation of Aβ goes beyond the concept of its neurotoxicity, raising the question of the potential clinical consequences of lowering Aβ levels below the physiological threshold.

MMPs have emerged as important modulators of AD pathogenesis through multiple mechanisms, including the regulation of APP processing, Aβ metabolism, tau pathology, neuroinflammation, BBB integrity, and synaptic function. Although several MMPs can promote the clearance of Aβ, others contribute to amyloidogenic processing of APP, highlighting the complex and context-dependent roles of these proteases in AD.

Several MMPs participate directly in APP processing. MT1-, MT3-, MT5-MMP, and MMP-2 have all been reported to cleave APP at the site later termed the η-secretase cleavage site. This cleavage, originally identified for MT1-MMP [89,90] and subsequently for MT3-MMP [91], was later characterized by Willem and colleagues, who demonstrated that MT5-MMP cleaves APP in neuronal cultures to generate soluble APPη (sAPPη) together with the membrane-bound η-C-terminal fragment (η-CTF) [92]. Additionally, it was reported that MT5-MMP knockout in 5xFAD mice, a transgenic mouse model that carries familial AD mutations in the human amyloid precursor protein *APP* and *PSEN1* genes, reduced the sAPPη fragment of 95 kDa, also termed sAPP95 [93], providing the first evidence for a physiological cleavage of APP by MT5-MMP in an *in vivo* AD context. This is consistent with early studies demonstrating that MT5-MMP is enriched around senile plaques in AD brains [94], prefiguring its potential relevance to AD pathology. Processing of η-CTF by α- and β-secretase generates the corresponding Aη-α and Aη-β peptides, which inhibit LTP [92,95]. In addition, η-CTF can also give rise to C99, thereby feeding into the canonical amyloidogenic pathway and ultimately increasing

$A\beta$ production [96]. Together, these findings establish η-processing as an important branch of APP metabolism that may contribute to AD through multiple mechanisms, including the production of synaptotoxic $A\eta$ peptides and the generation of amyloidogenic intermediates.

In contrast, several MMPs also exhibit $A\beta$-degrading activity. MMP-9 cleaves $A\beta40$ at three sites in the membrane-spanning domain and at the α-cleavage site (Lys16-Leu17) [97]. Consistent with this non-amyloidogenic activity, mice overexpressing MMP-9 exhibit increased levels of sAPPα together with enhanced LTP, learning, and memory capabilities [98]. Interestingly, like neprilysin and insulin-degrading enzyme, MMP-9 can also degrade preformed $A\beta$ fibrils and compact amyloid plaques in brain slices from APP/PS1 transgenic mice [99]. Similarly, MMP-2 exhibits $A\beta$-degrading properties, being more efficient at cleaving $A\beta42$ over $A\beta40$. The physiological relevance of these observations is supported by the finding that MMP-9 and MMP-2 knockout mice exhibit higher levels of $A\beta$ in the brain than wild-type mice *in vivo* [100,101]. Furthermore, incubation of recombinant $A\beta$ with conditioned media from cultured mouse astrocytes yields $A\beta$ C-terminal fragments, an effect that can be prevented by selective MMP-2 and MMP-9 chemical inhibitors or by TIMP-1 [100,101]. In addition to MMP-2 and MMP-9, MMP-3 and MMP-7 can also degrade $A\beta$ (reviewed in [5]).

Membrane-tethered MMPs also contribute to $A\beta$ degradation. MT1-MMP degrades exogenous soluble $A\beta40$ and $A\beta42$, generating six different $A\beta$-derived peptides, with cleavage sites mostly located around the α-site between residues 12 and 18 of the $A\beta$ sequence. Furthermore, a soluble variant of MT1-MMP lacking the transmembrane domain has been shown to degrade parenchymal amyloid plaques *in situ* in brain slices from Tg2576 mice [102], a feature otherwise shared only by MMP-9. Beyond APP and $A\beta$ metabolism, MMPs also regulate tau pathology. Elevated levels of active MMP-2 are found alongside p-Tau in NFTs and dystrophic neurites in the entorhinal cortex of AD patients [103]. Although MMP-2 can cleave recombinant Tau *in vitro*, it fails to cleave Tau contained in NFTs. This has led to speculation that the inactivation of MMP-2 in NFTs might prevent the formation of toxic truncated Tau assemblies.

MMP-9 and MMP-3 can also process tau, but only MMP-9 generates Tau fragments that promote the formation of oligomers, which are increasingly considered the most neurotoxic species of Tau [20]. Consistent with MMPs' implication in AD pathology, MMP-9 levels are increased in AD brains, where the proteinase contributes to microglial activation and the release of pro-inflammatory cytokines, including IL-1β and TNF-α. In addition, MMP-3 and MMP-10 levels in the CSF correlate with total Tau and p-Tau levels in AD patients [104,105] and, more recently, Vervuurt et al. found that decreased MMP/TIMP ratios in the CSF correlate with cerebral amyloid angiopathy and AD [106], suggesting their potential as AD progression or diagnostic biomarkers.

Collectively, these findings illustrate the multifaceted roles of MMPs in AD. Depending on the protease involved, its substrate specificity, and the stage of disease, MMPs can exert either protective effects by promoting Aβ clearance or detrimental effects by enhancing APP processing, Tau pathology, and neuroinflammation. Defining the temporal and spatial regulation of these proteolytic events in human AD will be essential for determining whether individual MMPs represent viable therapeutic targets.

3. Contribution of membrane-bound metalloproteases to Alzheimer's pathogenesis

In addition to the α-secretases of the ADAM family, several soluble and membrane-bound MMPs have been implicated in APP processing and metabolism. The earliest evidence indicating a functional link between APP and MMPs/MT-MMPs was provided by Miyazaki and colleagues in 1993 [107], who identified MMP-2 as an α-secretase that can hydrolyze the Lys16-Leu17 bond within the Aβ sequence. The same study identified above the α-site a decapeptide sequence (ISYGNDALMP), termed APP-derived peptide inhibitor (APP-IP), which specifically inhibits MMP-2 ($IC_{5n} = 30$ nM), while displaying much weaker activity toward MT1-MMP, MMP-3, MMP-7 and MMP-9, with respective IC_{5n} values of 2 μM and >10 μM. The APP-IP sequence spans amino acids 579 to 601 (APP770 numbering) and is present in sAPPα and sAPPβ [89,90]. This suggests that MT1-MMP/MMP-2 regulation and APP processing are intimately connected. Further evidence for this functional interplay came from the observation of Higashi and Miyazaki showing that MT1-MMP could process APP770 outside the Aβ sequence at the Asn579 and Met580 cleavage site, releasing a soluble truncated APP fragment that lacks the MMP-2 inhibitory decapeptide [90]. This finding suggested that MT1-MMP could indirectly regulate MMP-2 activity by removing its endogenous inhibitor from APP. It is noteworthy that the KPI (kunitz-protease inhibitor) domain of APP, which is present in APP751 and APP770 but absent from the neuronal APP695 isoform, also possesses serine protease inhibitory activity, emphasizing the complex reciprocal regulation between APP and proteases.

Subsequent studies expanded this concept by demonstrating the capacity of several MT-MMPs to directly process APP at multiple sites. Ahmad et al. demonstrated that, when co-expressed with APP770, MT1-, MT3- and MT5-MMP could process it into different fragments. MT1-MMP cleaves at Asn579-Met580 (VLAN579-M580ISEPR), as well as at Ala463-Met464, His622-Ser623, and His685-Gln686 (one residue upstream of the α-cleavage site), but this does not affect Aβ [91].

Among the MT-MMPs, MT1-MMP is the most widely studied because its dual role as the physiological activator of MMP-2 and as a direct regulator of APP metabolism [16,89,90]. MT1-MMP is widely distributed throughout the body, including the brain. We found that MT1-MMP expression is upregulated at symptomatic stages of the pathology in the brains of 5xFAD mice. We have also shown that its overexpression in HEK293 cells expressing the Swedish mutation of APP (HEKswe), one of the familial mutations expressed in 5xFAD mice, stimulates C99 and Aβ production. We have also demonstrated that MT1-MMP interacts with APP to generate sAPP95, thereby increasing Aβ40 levels. Indeed, MT1-MMP favors endosomal APP sorting [21], a major site for Aβ production. Both APP [108] and MT1-MMP [109] share a clathrin-dependent mechanism of internalization involving specific short sequences of their respective cytoplasmic tails. The release of sAPP95 by MT1-MMP is partially dependent on MMP-2, which is activated by MT1-MMP. Indeed, in the absence of MT1-MMP, MMP-2 loses its ability to generate sAPP95, indicating that MT1-MMP is the rate-limiting step [21]. Furthermore, by using C3, a commercial BACE-1 inhibitor, we demonstrated that MT1-MMP exhibits "surrogate β-secretase activity", thereby maintaining Aβ despite BACE-1 inhibition [21]. This activity is dependent on its catalytic function as an inactive MT1-MMP fails to promote Aβ generation. These data highlight a new function for MMP-2, acting in tandem with MT1-MMP, in addition to its known Aβ degradation properties. This provides another example of the multifunctional nature of MMPs.

The relevance of MT1-MMP extends beyond APP metabolism. Recent work has shown that neuroinflammation associated with aging and obesity increases the levels and activation of MT1-MMP in the hippocampus in both mice and humans. Elevated MT1-MMP is associated with synaptic plasticity impairment, independent of neuroinflammation and cognitive decline. Genetic reduction or pharmacological inhibition of MT1-MMP (using a chemical compound or blocking antibody) restores age-associated deficits in synaptic function and cognitive performance, independently of MMP-2 and MMP-9 [110]. Taken together, these findings identify MT1-MMP as a promising therapeutic target with functions extending beyond amyloid processing, highlighting the need for further research into its role in NDs.

Whereas MT1-MMP is broadly expressed and regulates APP metabolism in multiple tissues, MT5-MMP, discovered in 1999 [111–113], stands out due to its predominantly neuronal expression, making it particularly relevant to NDs. As the only MMP primarily expressed in the nervous system [114], MT5-MMP shares close structural and functional similarities (*e.g.*, MMP-2 activation [113]) with MT1-MMP and contributes to several aspects of neural development and plasticity. It has been implicated in axonal outgrowth and reorganization following allodynia [115,116] and traumatic brain injury, notably through astrocytes [117]. Additionally, MT5-MMP participates in neural cell differentiation and morphogenesis [118,119], highlighting its physiological importance in the nervous system. At the synaptic level, MT5-MMP regulates proteins involved in excitatory transmission. It interacts with AMPA receptor binding protein (ABP) and glutamate receptor interacting protein (GRIP) [120], which are involved in targeting AMPA receptors to the cell membrane. Furthermore, MT5-MMP also cleaves synaptic adhesion proteins such as N–Cadherin in concert with γ-secretase, a process that may contribute to synaptic remodeling and, under pathological conditions, synaptic dysfunction [121].

These physiological functions provide a relevant framework for better understanding the potential contribution of MT5-MMP to NDs. It is widely accepted that neuroinflammation is a major driver of AD pathogenesis [122], with pro-inflammatory cytokines like TNF-α and IL-1β playing central roles [123,124]. Interestingly, MT5-MMP may modulate inflammatory signaling, as MT5-MMP knockout mice display reduced IL-1β-mediated inflammatory responses in the peripheral nervous system, probably due to impaired N–Cadherin processing [125].

More recent studies have further expanded the biological relevance of η-processing by demonstrating that η-CTF is not merely an inactive intermediate but can also serve as a substrate for amyloidogenic processing, generating C99 and subsequently Aβ [96]. In the same vein, the proprotein convertase PCSK6, significantly upregulated in the brains of AD patients, activates MT5-MMP by cleaving its pro-peptide, thereby promoting the generation of amyloidogenic fragments (CTFη, CTFβ) and increased Aβ production [126].

Beyond generating η-CTF and Aη peptides, η-secretase cleavage also releases the soluble ectodomain sAPPη (~95 kDa; also referred to as sAPP95), further highlighting the complexity of MT5-MMP-dependent APP processing. Like sAPPα and sAPPβ, sAPPη retains the extension domain (ExD), which mediates binding to sushi domain 1 of the metabotropic GABA$_B$R1a receptor. This interaction negatively regulates neurotransmitter release by inhibiting synaptic vesicle exocytosis, thereby modulating synaptic transmission and plasticity [127]. Although the physiological effects of sAPPη have been less extensively characterized than those of sAPPα and sAPPβ, its ability to bind GABA$_B$R1a suggests that all three soluble APP fragments may share similar neuromodulatory functions. Together, these findings indicate that MT5-MMP-mediated APP processing generates both neuroactive soluble ectodomains and potentially pathogenic membrane-bound and peptide fragments, underscoring the multifaceted consequences of η-secretase activity.

Given the broad spectrum of APP-derived products generated by MT5-MMP, we investigated its contribution to AD pathogenesis *in vivo* by breeding 5xFAD mice (Tg) with MT5-MMP knockout mice (MT5$^{-/-}$) to generate TgMT5$^{-/-}$ [93]. The resulting TgMT5$^{-/-}$ strain exhibits dramatic reductions in hippocampal and cortical Aβ40 and Aβ42 plaque burden, accompanied by lower C99 levels, reduced levels of IL-1β and TNF-α, and attenuated glial reactivity in different brain areas. Importantly, these molecular changes were associated with preserved LTP and learning and memory, compared to Tg controls [93,128]. Complementary *in vitro* work further supported a direct role for MT5-MMP in APP processing. Specifically, overexpression of MT5-MMP in HEK293swe cells increased both Aβ40 production and sAPP95 release, corroborating our observations of reduced sAPP95 levels in TgMT5$^{-/-}$ mice, strongly supporting that APP is a physiological substrate of MT5-MMP *in vivo* [93].

To further define the cellular mechanisms underlying these protective effects, we examined the impact of MT5-MMP deficiency in primary neural cultures. In 11-day-old primary neural cultures, mainly composed of relatively immature neurons and astrocytes, Tg cells exhibited higher levels of MT5-MMP than WT cultures, alongside higher basal levels of inflammatory mediators. Under these conditions, MT5-MMP deficiency limits both basal and IL-1β-mediated inflammation, while also reducing the stability of recombinant IL-1β following incubation with cells lacking MT5-MMP. Notably, Aβ levels were similar in Tg and WT cultures and were not affected by IL-1β treatment. However, MT5$^{-/-}$ and TgMT5$^{-/-}$

cultures showed more than a 40 percent reduction in Aβ levels and no impact on APP metabolism. The absence of MT5-MMP strongly downregulates AAV-induced neuronal accumulation of C99, the precursor of Aβ, and subsequently Aβ levels. Finally, while MT5-MMP deficiency prevented the basal hyperexcitability observed in Tg neurons, it did not lower the hyperexcitability induced by IL-1β treatment, suggesting that in these young neurons, MT5-MMP selectively contributes to basal, but not cytokine-driven, alterations in neuronal excitability [129].

Building on these findings, a subsequent study investigated the role of MT5-MMP in more mature, highly interconnected primary neural cultures [130]. We demonstrated that the absence of MT5-MMP impairs IL-1β-mediated induction of inflammatory genes in TgMT5$^{-/-}$ cells, compared to Tg cells, in 21–24-day-old neural cultures *in vitro*, further supporting its role in neuroinflammatory signaling. In parallel, we observed a reduced density of dendritic spines in Tg neurons, a phenomenon that was prevented in TgMT5$^{-/-}$ neurons. Interestingly, under basal conditions, MT5$^{-/-}$ neurons exhibited fewer spines than WT neurons, yet they were resistant to the spine loss induced by IL-1β. In this context, functional analysis revealed additional effects on network activity: spontaneous rhythmic firing frequency increased in MT5$^{-/-}$ neurons but not in TgMT5$^{-/-}$ neurons, whereas IL-1β selectively enhanced this parameter in Tg neurons. At the molecular level, MT5-MMP deficiency decreased endogenous and overexpressed C83 and C99 levels but did not affect Aβ levels. C99 appears to be cleared by several pathways, including γ-secretase, the autophagolysosomal system, and α-secretase-mediated conversion to C83. Taken together, these findings highlight the pivotal role of MT5-MMP in regulating neuroinflammation, APP metabolism, synaptogenesis, and neural network activity in the early stages of AD pathology [130]. Overall, the two studies are complementary. Together, they support a model in which MT5-MMP plays temporally distinct roles during disease progression: early in development or pathology, it promotes APP amyloidogenic processing and inflammatory signaling, whereas at later stages it primarily modulates neuroinflammation, synaptic maintenance, and neuronal network function, with compensatory mechanisms limiting its impact on steady-state Aβ production.

Mechanistically, these effects extend beyond the proteolytic activity of MT5-MMP. Our work has demonstrated that MT5-MMP, like MT1-MMP, interacts with APP and C99 and promotes their trafficking to endosomes, a major site of Aβ formation, through a mechanism that is independent of its enzymatic activity [22,93]. MT5-MMP and APP share common internalization and recycling pathways, notably through their interaction with Mint3, a type III PDZ domain-containing adaptor protein that promotes APP export from the Golgi complex [131] and interacts with the last three residues from the cytoplasmic tail of MT5-MMP [132]. These observations suggest that MT5-MMP could promote the endosomal targeting of APP, η-CTF and C99 for more efficient amyloidogenic processing by β- and γ-secretases. This trafficking-based mechanism parallels that of asparagine endopeptidase (δ-secretase), which cleaves APP upstream of the β-cleavage site and promotes subsequent processing of the remaining APP fragment by β-secretase [133]. Together, the mechanisms of neurotoxicity derived from APP cleavage cannot be attributed exclusively to Aβ accumulation. Instead, multiple APP-derived metabolites generated by different proteases (*e.g.*, MT-MMPs and δ-secretase) are increasingly recognized as contributors to AD pathology (reviewed in [4]). This broader view of APP proteolysis highlights additional pathogenic pathways and identifies APP metabolites and the involved enzymatic systems as promising potential new therapeutic targets.

Overall, MT1-MMP and MT5-MMP share several key features that support their pro-amyloidogenic functions, including direct interactions with APP, coordinated interdependent intracellular trafficking, and the promotion of APP targeting to endosomal compartments where C99 and Aβ production occurs. However, important mechanistic differences distinguish the two MT-MMPs: i) APP processing by MT1-MMP is partially dependent on MMP-2, whereas MT5-MMP processing is not, ii) MT5-MMP's pro-amyloidogenic activity is, at least in part, independent of its proteolytic activity, whereas that of MT1-MMP is not, iii) In response to BACE-1 inhibition, MT1-MMP continues to promote Aβ production by a mechanism that is still elusive, whereas MT5-MMP preferentially generates η-CTF which can give rise to either Aη-peptides or C99, and subsequently, Aβ. Therefore, both MT-MMPs display distinct contributions to APP processing and suggest that they participate in complementary, rather than redundant, pathogenic pathways.

4. MT5-MMP—an emerging CNS target

Among, MT-MMPs, MT5-MMP has emerged as particularly attractive therapeutic target owing to its predominant expression in the nervous system and its growing implication in neurodegenerative diseases (see [5,9] for review). Currently, the strongest evidence supports the implication of

MT5-MMP in AD, but its contribution to PD, ALS, HD and other neurological disorders such as multiple sclerosis is still being investigated. In AD, MT5-MMP promotes disease progression through multiple mechanisms, including the η-secretase cleavage of APP to generate synaptotoxic η-fragments, as well as by promoting the production and accumulation of C99 and Aβ, through both catalytic- and trafficking-dependent mechanisms [4,22,129,130,134].

These mechanistic insights have opened several potential complementary therapeutic avenues. One strategy involves targeting upstream regulators of MT5-MMP activation. In this context, the PCSK6-mediated activation of MT5-MMP enhances the amyloidogenic processing of APP [126]. Accordingly, genetic silencing of PCSK6 using siRNA or AAV-mediated shRNA reduces MT5-MMP activation, lowers Aβ levels, and attenuates amyloid plaque deposition in AD mice, without altering the expression of APP or the major APP-processing enzymes BACE-1, ADAM10, and presenilin 1. These molecular effects are accompanied by significant improvements in hippocampal LTP and spatial memory, while locomotor activity and anxiety-related behaviors remain unaffected. Collectively, these findings identify PCSK6 as a critical upstream modulator of MT5-MMP-mediated APP processing and AD pathology, highlighting the PCSK6–MT5-MMP axis as a promising therapeutic target for limiting amyloidogenesis and cognitive decline.

A second strategy aims to interfere with the intracellular functions of MT5-MMP that regulate APP trafficking. Recent work has identified several intracellular partners, including N4BP2L1, EIG121, BIN1 and TMX3, which enable MT5-MMP processing of APP across different cellular compartments such as the Golgi apparatus, endosomes and the plasma membrane [135]. Consistent with this concept, our previous studies revealed the role of the C-terminal domains of MT5-MMP (TM and IC) in processing APP and determining the fate of its major toxic metabolites, C99 and Aβ [22]. More recently, we identified specific amino acid residues within the IC domain that influence MT5-MMP interaction with C99 and its subcellular trafficking, leading to the promotion of C99 degradation and, consequently, a decrease in Aβ levels cells [136]. Remarkably, a synthetic peptide mimicking the MT5-MMP IC domain carrying selective mutations in the N-terminus, efficiently decreased C99 levels when transfected into HEK cells [136], providing proof of concept that disrupting MT5-MMP trafficking functions may represent an alternative therapeutic strategy independent of catalytic inhibition.

Direct pharmacological inhibition of MT5-MMP remains attractive but technically challenging due to the high structural conservation of the catalytic domain among MMP family members. Nevertheless, recent structure-guided drug design has led to the identification of the first non-peptidic MT5-MMP inhibitor (compound 17), which achieved an IC_{5n} of 6 μM by substituting a hydroxamic acid with a carboxylic acid zinc-binding group in a homology model. Although further optimization will be required to improve potency, selectivity, and pharmacokinetic properties, this work provides a starting point for the development of selective MT5-MMP inhibitors [137]. Beyond AD, the therapeutic potential of MT5-MMP remains less well defined but is progressively expanding.

MT5-MMP plays a role in neuronal sensitization and neuropathic pain through an FTO (obesity-associated protein)-dependent role in neuropathic pain and neuronal sensitization [138], highlighting its potential relevance to ALS, where spinal motor neurons undergo progressive degeneration. In addition, MT5-MMP deficiency alters neurite branching and astrocyte morphology in iPSC-derived models, suggesting a crucial function in preserving neuronal and glial architecture that is pertinent to neurodegenerative processes in ALS [118,138]. However, direct evidence for MT5-MMP participation in PD, ALS and HD remains considerably less extensive than that available for AD or for other MMP members such as MMP-9.

Overall, the current body of evidence positions MT5-MMP as one of the most promising emerging CNS metalloproteinase targets. While significant challenges remain, notably the need for selective inhibitors and a more comprehensive understanding of MT5-MMPs functions across different neurological disorders, recent advances have substantially broadened the therapeutic landscape. Strategies targeting MT5-MMP activation, intracellular trafficking, or catalytic activity now provide complementary avenues that extend beyond the classical amyloid cascade and may ultimately enable disease-modifying interventions in AD and, potentially, other NDs.

5. Conclusion

MT-MMPs are emerging as multifaceted regulators of APP metabolism, contributing not only to Aβ degradation or its production/accumulation through catalytic and trafficking-based mechanisms, but also to the generation of new truncated APP fragments whose functions are still poorly understood. Similar challenges lie ahead in considering the physiological and pathological roles of breakdown products resulting from the processing of Tau, α-Syn or Htt by MMPs, underscoring the broader relevance of MMP-mediated activities across NDs. Although proteolysis is undoubtedly the primary functional characteristic of metalloproteases, research is now expanding to highlight the importance of their proteolysis-independent functions. Studies on ADAM30 [29] and, more recently, MT5-MMP [22,93,128,129,130], demonstrate that protein-protein interactions may significantly impact AD pathogenesis. Consequently, it is plausible that new therapeutic strategies will emerge to target non-proteolytic interactions between metalloproteinases and their partner proteins rather than their catalytic activity. This could offer an alternative to overcome the specificity limitations associated with the highly conserved catalytic domains shared by metalloproteinases.

Given the central role of inflammation in NDs, a better understanding of how MMPs produced by resident brain cells and infiltrating immune cells shape disease progression is necessary. Defining their spatiotemporal functions and their regulation of key immune receptors and cytokines, including CD44, TREM2, TNF-α and IL-1β, will be essential for clarifying their contribution to disease.

Clearly, the complexity of MMP biology calls for a more global approach that considers both their enzymatic and their non-enzymatic ("moonlighting") activities (*e.g.*, protein network, transcription factor, shuttle...). This integrated approach should combine advanced cellular models such as iPSC-derived systems with animal models, proteomics and interactome techniques to provide a more comprehensive view of MMP roles in neurodegeneration. Such approaches should inspire future research into NDs not only to deepen our understanding of the underlying molecular mechanisms, but also to help identify more precise biomarkers and therapeutic targets for AD and related neurodegenerative disorders. Ultimately, a more complete understanding of MMP biology may pave the way for the development of more effective diagnostic tools and disease-modifying therapeutic strategies.

Conflicts of interest

The authors have no conflicts of interest to declare.

Author contributions

All authors contributed to the literature review, writing, critical discussion, revision, and final approval of the manuscript.

Acknowledgments

This work was supported by funding from the CNRS, La Rochelle Université, Ligue Contre le Cancer CD17 and CD86, and by the French National Research Agency (ANR) HeparIn-fSkin (ANR-21-CE44-0004) project to KB. MF was recipient of a doctoral fellowship from La Rochelle Université (ANR-21-EXES-0010) and (101179124—EU-DOCs for SmUCS). This work was also supported by funding from the CNRS and Aix-Marseille Université and by public grants overseen by the French National Agency for Research (ANR) (MT5-AD, ANR-22-C18-0024–01) and by Fondation Alzheimer (Innovative Project Grant 1315738) to SR. The figure was created using BioRender.

References

[1] Y. Hou, X. Dan, M. Babbar, Y. Wei, S.G. Hasselbalch, D.L. Croteau, et al., Ageing as a risk factor for neurodegenerative disease, Nat. Rev. Neurol. 15 (2019) 565–581, doi:10.1038/s41582-019-0244-7.
[2] S. Wang, Y. Jiang, A. Yang, F. Meng, J. Zhang, The Expanding burden of neurodegenerative diseases: an unmet medical and social need, Aging Dis. 16 (2024) 2937–2952, doi:10.14336/AD.2024.1071.
[3] P.H. Nguyen, P. Derreumaux, Status of Alzheimer's and Parkinson's disease detections and treatments, ACS Chem. Neurosci. 17 (2026) 2441–2453, doi:10.1021/acschemneuro.6c00096.
[4] L. García-González, D. Pilat, K. Baranger, S. Rivera, Emerging alternative proteinases in APP metabolism and Alzheimer's disease pathogenesis: a focus on MT1-MMP and MT5-MMP, Front. Aging Neurosci. 11 (2019) 244, doi:10.3389/fnagi.2019.00244.
[5] S. Rivera, L. García-González, M. Khrestchatisky, K. Baranger, Metalloproteinases and their tissue inhibitors in Alzheimer's disease and other neurodegenerative disorders, Cell. Mol. Life Sci. 76 (2019) 3167–3191, doi:10.1007/s00018-019-03178-2.
[6] L.G.N. de Almeida, H. Thode, Y. Eslambolchi, S. Chopra, D. Young, S. Gill, et al., Matrix metalloproteinases: from molecular mechanisms to physiology, pathophysiology, and pharmacology, Pharmacol. Rev. 74 (2022) 712–768, doi:10.1124/pharmrev.121.000349.
[7] H.-E. Hsia, J. Tüshaus, T. Brummer, Y. Zheng, S.D. Scilabra, S.F. Lichtenthaler, Functions of « A disintegrin and metalloproteases (ADAMs) » in the mammalian nervous system, Cell. Mol. Life Sci. 76 (2019) 3055–3081, doi:10.1007/s00018-019-03173-7.
[8] V.R. Krishnaswamy, A. Benbenishty, P. Blinder, I. Sagi, Demystifying the extracellular matrix and its proteolytic remodeling in the brain: structural and functional insights, Cell. Mol. Life Sci. 76 (2019) 3229–3248, doi:10.1007/s00018-019-03182-6.
[9] P. Zipfel, C. Rochais, K. Baranger, S. Rivera, P. Dallemagne, Matrix metalloproteinases as new targets in Alzheimer's disease: opportunities and challenges, J. Med. Chem. 63 (2020) 10705–10725, doi:10.1021/acs.jmedchem.0c00352.
[10] T. Klein, R. Bischoff, Active metalloproteases of the A Disintegrin and Metalloprotease (ADAM) family: biological function and structure, J. Proteome Res. 10 (2011) 17–33, doi:10.1021/pr100556z.
[11] K. Nakayama, Furin: a mammalian subtilisin/Kex2p-like endoprotease involved in processing of a wide variety of precursor proteins, Biochem. J. 327 (Pt 3) (1997) 625–635, doi:10.1042/bj3270625.
[12] Z. He, A.-M. Khatib, J.W.M. Creemers, The proprotein convertase furin in cancer: more than an oncogene, Oncogene 41 (2022) 1252–1262, doi:10.1038/s41388-021-02175-9.

[13] K.J. Leco, S.S. Apte, G.T. Taniguchi, S.P. Hawkes, R. Khokha, G.A. Schultz, et al., Murine tissue inhibitor of metalloproteinases-4 (Timp-4): cDNA isolation and expression in adult mouse tissues, FEBS Lett. 401 (1997) 213–217, doi:10.1016/s0014-5793(96)01474-3.

[14] N. Pohar, T.A. Godenschwege, E. Buchner, Invertebrate tissue inhibitor of metalloproteinase: structure and nested gene organization within the synapsin locus is conserved from Drosophilato to human, Genomics 57 (1999) 293–296, doi:10.1006/geno.1999.5776.

[15] F.J. Mirza, S. Zahid, The role of synapsins in neurological disorders, Neurosci. Bull. 34 (2018) 349–358, doi:10.1007/s12264-017-0201-7.

[16] K. Shofuda, K. Moriyama, A. Nishihashi, S. Higashi, H. Mizushima, H. Yasumitsu, et al., Role of tissue inhibitor of metalloproteinases-2 (TIMP-2) in regulation of pro-gelatinase A activation catalyzed by membrane-type matrix metalloproteinase-1 (MT1-MMP) in human cancer cells, J. Biochem. 124 (1998) 462–470, doi:10.1093/oxfordjournals.jbchem.a022136.

[17] B. Mroczko, M. Groblewska, M. Barcikowska, The role of matrix metalloproteinases and tissue inhibitors of metalloproteinases in the pathophysiology of neurodegeneration: a literature study, J. Alzheimers Dis. 37 (2013) 273–283, doi:10.3233/JAD-130647.

[18] J. Montaner, L. Ramiro, A. Simats, M. Hernández-Guillamon, P. Delgado, A. Bustamante, et al., Matrix metalloproteinases and ADAMs in stroke, Cell. Mol. Life Sci. 76 (2019) 3117–3140, doi:10.1007/s00018-019-03175-5.

[19] J. Levin, A. Giese, K. Boetzel, L. Israel, T. Högen, G. Nübling, et al., Increased α-synuclein aggregation following limited cleavage by certain matrix metalloproteinases, Exp. Neurol. 215 (2009) 201–208, doi:10.1016/j.expneurol.2008.10.010.

[20] G. Nübling, J. Levin, B. Bader, L. Israel, K. Bötzel, S. Lorenzl, et al., Limited cleavage of tau with matrix-metalloproteinase MMP-9, but not MMP-3, enhances tau oligomer formation, Exp. Neurol. 237 (2012) 470–476, doi:10.1016/j.expneurol.2012.07.018.

[21] J.M. Paumier, N.A. Py, L. Garcia-Gonzalez, A. Bernard, D. Stephan, L. Louis, et al., Proamyloidogenic effects of membrane type 1 matrix metalloproteinase involve MMP-2 and BACE-1 activities, and the modulation of APP trafficking, FASEB J. 33 (2019) 2910–2927, doi:10.1096/fj.201801076R.

[22] L. García-González, J.-M. Paumier, L. Louis, D. Pilat, A. Bernard, D. Stephan, et al., MT5-MMP controls APP and β-CTF/C99 metabolism through proteolytic-dependent and -independent mechanisms relevant for Alzheimer's disease, FASEB J. 35 (2021) e21727, doi:10.1096/fj.202100593R.

[23] A. Dufour, N.S. Sampson, S. Zucker, J. Cao, Role of the hemopexin domain of matrix metalloproteinases in cell migration, J. Cell. Physiol. 217 (2008) 643–651, doi:10.1002/jcp.21535.

[24] A. Dufour, S. Zucker, N.S. Sampson, C. Kuscu, J. Cao, Role of matrix metalloproteinase-9 dimers in cell migration: design of inhibitory peptides, J. Biol. Chem. 285 (2010) 35944–35956, doi:10.1074/jbc.M109.091769.

[25] A. Dufour, N.S. Sampson, J. Li, C. Kuscu, R.C. Rizzo, J.L. Deleon, et al., Small-molecule anticancer compounds selectively target the hemopexin domain of matrix metalloproteinase-9, Cancer Res. 71 (2011) 4977–4988, doi:10.1158/0008-5472.CAN-10-4552.

[26] M. Pavlaki, S. Zucker, A. Dufour, N. Calabrese, W. Bahou, J. Cao, Furin functions as a nonproteolytic chaperone for matrix metalloproteinase-28: MMP-28 propeptide sequence requirement, Biochem. Res. Int. 2011 (2011) 630319, doi:10.1155/2011/630319.

[27] K. Zarrabi, A. Dufour, J. Li, C. Kuscu, A. Pulkoski-Gross, J. Zhi, et al., Inhibition of matrix metalloproteinase 14 (MMP-14)-mediated cancer cell migration, J. Biol. Chem. 286 (2011) 33167–33177, doi:10.1074/jbc.M111.256644.

[28] D.J. Marchant, C.L. Bellac, T.J. Moraes, S.J. Wadsworth, A. Dufour, G.S. Butler, et al., A new transcriptional role for matrix metalloproteinase-12 in antiviral immunity, Nat. Med. 20 (2014) 493–502, doi:10.1038/nm.3508.

[29] F. Letronne, G. Laumet, A.M. Ayral, J. Chapuis, F. Demiautte, M. Laga, et al., ADAM30 downregulates APP-linked defects through cathepsin D activation in Alzheimer's disease, EBioMedicine 9 (2016) 278–292, doi:10.1016/j.ebiom.2016.06.002.

[30] Y. Fu, G.M. Halliday, Dementia with Lewy bodies and Parkinson disease dementia—the same or different and is it important? Nat. Rev. Neurol. 21 (2025) 394–403, doi:10.1038/s41582-025-01090-x.

[31] M.H. Polymeropoulos, C. Lavedan, E. Leroy, S.E. Ide, A. Dehejia, A. Dutra, et al., Mutation in the alpha-synuclein gene identified in families with Parkinson's disease, Science 276 (1997) 2045–2047, doi:10.1126/science.276.5321.2045.

[32] A.B. Singleton, M. Farrer, J. Johnson, A. Singleton, S. Hague, J. Kachergus, et al., alpha-Synuclein locus triplication causes Parkinson's disease, Science 302 (2003) 841, doi:10.1126/science.1090278.

[33] J.B. Leverenz, I. Umar, Q. Wang, T.J. Montine, P.J. McMillan, D.W. Tsuang, et al., Proteomic identification of novel proteins in cortical Lewy bodies, Brain Pathol. 17 (2007) 139–145, doi:10.1111/j.1750-3639.2007.00048.x.

[34] K. Wakabayashi, K. Tanji, S. Odagiri, Y. Miki, F. Mori, H. Takahashi, The Lewy body in Parkinson's disease and related neurodegenerative disorders, Mol. Neurobiol. 47 (2013) 495–508, doi:10.1007/s12035-012-8280-y.

[35] E.-J. Lee, M.-S. Woo, P.-G. Moon, M.-C. Baek, I.-Y. Choi, W.-K. Kim, et al., α-Synuclein activates microglia by inducing the expressions of matrix metalloproteinases and the subsequent activation of protease-activated receptor-1, J. Immunol. 185 (2010) 615–623, doi:10.4049/jimmunol.0903480.

[36] J.Y. Sung, S.M. Park, C.-H. Lee, J.W. Um, H.J. Lee, J. Kim, et al., Proteolytic cleavage of extracellular secreted α-synuclein via matrix metalloproteinases, J. Biol. Chem. 280 (2005) 25216–25224, doi:10.1074/jbc.M503341200.

[37] D.H. Choi, E. Kim, H.J. Son, T.H. Joh, Y.S. Kim, D. Kim, et al., A novel intracellular role of matrix metalloproteinase-3 during apoptosis of dopaminergic cells, J. Neurochem. 106 (2008) 405–415, doi:10.1111/j.1471-4159.2008.05399.x.

[38] Y.S. Kim, D.H. Choi, M.L. Block, S. Lorenzl, L. Yang, Y.J. Kim, et al., A pivotal role of matrix metalloproteinase-3 activity in dopaminergic neuronal degeneration via microglial activation, FASEB J. 21 (2007) 179–187, doi:10.1096/fj.06-5865com.

[39] Y.C. Chung, Y.-S. Kim, E. Bok, T.Y. Yune, S. Maeng, B.K. Jin, MMP-3 contributes to nigrostriatal dopaminergic neuronal loss, BBB damage, and neuroinflammation in an MPTP mouse model of Parkinson's disease, Mediat. Inflamm. 2013 (2013) 1–11, doi:10.1155/2013/370526.

[40] D.-H. Choi, Y.-J. Kim, Y.-G. Kim, T.H. Joh, M.F. Beal, Y.-S. Kim, Role of matrix metalloproteinase 3-mediated α-synuclein cleavage in dopaminergic cell death, J. Biol. Chem. 286 (2011) 14168–14177, doi:10.1074/jbc.M111.222430.

[41] S. Lorenzl, D.S. Albers, S. Narr, J. Chirichigno, M.F. Beal, Expression of MMP-2, MMP-9, and MMP-1 and their endogenous counterregulators TIMP-1 and TIMP-2 in postmortem brain tissue of Parkinson's disease, Exp. Neurol. 178 (2002) 13–20, doi:10.1006/exnr.2002.8019.

[42] S. Lorenzl, N. Calingasan, L. Yang, D.S. Albers, S. Shugama, J. Gregorio, et al., Matrix metalloproteinase-9 is elevated in 1-methyl-4-phenyl-1,2,3,6-tetrahydropyridine-induced parkinsonism in mice, Neuromolecular Med. 5 (2004) 119–132, doi:10.1385/NMM:5:2:119.

[43] V. Bonifati, P. Rizzu, M.J. Van Baren, O. Schaap, G.J. Breedveld, E. Krieger, et al., Mutations in the *DJ-1* gene associated with autosomal recessive early-onset Parkinsonism, Science 299 (2003) 256–259, doi:10.1126/science.1077209.

[44] G.S. Butler, R.A. Dean, E.M. Tam, C.M. Overall, Pharmacoproteomics of a metalloproteinase hydroxamate inhibitor in breast cancer cells: dynamics of membrane type 1 matrix metalloproteinase-mediated membrane protein shedding, Mol. Cell. Biol. 28 (2008) 4896–4914, doi:10.1128/MCB.01775-07.
[45] S. Kim, S. Cho, K.Y. Kim, K.Y. Shin, H. Kim, C. Park, et al., α-Synuclein induces migration of BV-2 microglial cells by up-regulation of CD44 and MT1-MMP, J. Neurochem. 109 (2009) 1483–1496, doi:10.1111/j.1471-4159.2009.06075.x.
[46] M. Benatar, J. Wuu, E.D. Huey, C.T. McMillan, R.C. Petersen, R. Postuma, et al., The Miami framework for ALS and related neurodegenerative disorders: an integrated view of phenotype and biology, Nat. Rev. Neurol. 20 (2024) 364–376, doi:10.1038/s41582-024-00961-z.
[47] G. Logroscino, R. Tortelli, G. Rizzo, B. Marin, P.M. Preux, A. Malaspina, Amyotrophic lateral sclerosis: an aging-related disease, Curr. Geri. Rep. 4 (2015) 142–153, doi:10.1007/s13670-015-0127-8.
[48] M. Łukaszewicz-Zając, B. Mroczko, A. Słowik, Matrix metalloproteinases (MMPs) and their tissue inhibitors (TIMPs) in amyotrophic lateral sclerosis (ALS), J. Neural. Transm. 121 (2014) 1387–1397, doi:10.1007/s00702-014-1205-3.
[49] J.L. Sánchez-Torres, P. Yescas-Gómez, J. Torres-Romero, O.R. Espinosa, L.L. Canovas, ÁC. Tecalco-Cruz, et al., Matrix metalloproteinases deregulation in amyotrophic lateral sclerosis, J. Neurol. Sci. 419 (2020) 117175, doi:10.1016/j.jns.2020.117175.
[50] K.J. Spiller, T. Khan, M.A. Dominique, C.R. Restrepo, D. Cotton-Samuel, M. Levitan, et al., Reduction of matrix metalloproteinase 9 (MMP-9) protects motor neurons from TDP-43-triggered death in rNLS8 mice, Neurobiol. Dis. 124 (2019) 133–140, doi:10.1016/j.nbd.2018.11.013.
[51] B.F. Vahsen, S. Nalluru, G.R. Morgan, L. Farrimond, E. Carroll, Y. Xu, et al., C9orf72-ALS human iPSC microglia are pro-inflammatory and toxic to co-cultured motor neurons via MMP9, Nat. Commun. 14 (2023) 5898, doi:10.1038/s41467-023-41603-0.
[52] S.W. Cheung, E. Bhavnani, D.G. Simmons, M.C. Bellingham, P.G. Noakes, Perineuronal nets are phagocytosed by MMP-9 expressing microglia and astrocytes in the $SOD1^{G93A}$ ALS mouse model, Neuropathol. Appl. Neurobio. 50 (2024) e12982, doi:10.1111/nan.12982.
[53] S. Lorenzl, S. Narr, B. Angele, H.W. Krell, J. Gregorio, M. Kiaei, et al., The matrix metalloproteinases inhibitor Ro 26-2853 extends survival in transgenic ALS mice, Exp. Neurol. 200 (2006) 166–171, doi:10.1016/j.expneurol.2006.01.026.
[54] A.R. Vaz, C. Cunha, C. Gomes, N. Schmucki, M. Barbosa, D. Brites, Glycoursodeoxycholic acid reduces matrix metalloproteinase-9 and caspase-9 activation in a cellular model of superoxide dismutase-1 neurodegeneration, Mol. Neurobiol. 51 (2015) 864–877, doi:10.1007/s12035-014-8731-8.
[55] W. Beuche, M. Yushchenko, M. Mäder, M. Maliszewska, K. Felgenhauer, F. Weber, Matrix metalloproteinase-9 is elevated in serum of patients with amyotrophic lateral sclerosis, Neuroreport 11 (2000) 3419–3422, doi:10.1097/00001756-200011090-00003.
[56] L. Fang, F. Huber-Abel, M. Teuchert, C. Hendrich, J. Dorst, D. Schattauer, et al., Linking neuron and skin: matrix metalloproteinases in amyotrophic lateral sclerosis (ALS), J. Neurol. Sci. 285 (2009) 62–66, doi:10.1016/j.jns.2009.05.025.
[57] I. Niebroj-Dobosz, P. Janik, B. Sokołowska, H. Kwiecinski, Matrix metalloproteinases and their tissue inhibitors in serum and cerebrospinal fluid of patients with amyotrophic lateral sclerosis, Euro. J. Neurol. 17 (2010) 226–231, doi:10.1111/j.1468-1331.2009.02775.x.
[58] R. Bowser, J. An, K. Schwartz, R.L. Sucholeiki, I. Sucholeiki, ALS patients exhibit altered levels of total and active MMP-9 and several other biomarkers in serum and CSF compared to healthy controls and other neurologic diseases, IJMS 26 (2025) 8900, doi:10.3390/ijms26188900.

[59] E. Berrone, G. Chiorino, F. Guana, V. Benedetti, C. Palmitessa, M. Gallo, et al., SOMAscan proteomics identifies novel plasma proteins in amyotrophic lateral sclerosis patients, IJMS 24 (2023) 1899, doi:10.3390/ijms24031899.
[60] J.K. Lee, J.H. Shin, J. Suh, I.S. Choi, K.S. Ryu, B.J. Gwag, Tissue inhibitor of metalloproteinases-3 (TIMP-3) expression is increased during serum deprivation-induced neuronal apoptosis in vitro and in the G93A mouse model of amyotrophic lateral sclerosis: a potential modulator of Fas-mediated apoptosis, Neurobiol. Dis. 30 (2008) 174–185, doi:10.1016/j.nbd.2008.01.004.
[61] M. DiFiglia, E. Sapp, K.O. Chase, S.W. Davies, G.P. Bates, J.P. Vonsattel, et al., Aggregation of huntingtin in neuronal intranuclear inclusions and dystrophic neurites in brain, Science 277 (1997) 1990–1993, doi:10.1126/science.277.5334.1990.
[62] E. Sapp, C. Schwarz, K. Chase, P.G. Bhide, A.B. Young, J. Penney, et al., Huntingtin localization in brains of normal and Huntington's disease patients, Ann. Neurol. 42 (1997) 604–612, doi:10.1002/ana.410420411.
[63] J. Donaldson, D. Hensman Moss, M. Ciosi, K. Usdin, G. Balmus, D.G. Monckton, et al., Huntington disease: somatic expansion, pathobiology and therapeutics, Nat. Rev. Neurol. 22 (2026) 5–21, doi:10.1038/s41582-025-01159-7.
[64] C.L. Wellington, R. Singaraja, L. Ellerby, J. Savill, S. Roy, B. Leavitt, et al., Inhibiting caspase cleavage of huntingtin reduces toxicity and aggregate formation in neuronal and nonneuronal cells, J. Biol. Chem. 275 (2000) 19831–19838, doi:10.1074/jbc.M001475200.
[65] J. Gafni, E. Hermel, J.E. Young, C.L. Wellington, M.R. Hayden, L.M. Ellerby, Inhibition of calpain cleavage of huntingtin reduces toxicity, J. Biol. Chem. 279 (2004) 20211–20220, doi:10.1074/jbc.M401267200.
[66] E. Hermel, J. Gafni, S.S. Propp, B.R. Leavitt, C.L. Wellington, J.E. Young, et al., Specific caspase interactions and amplification are involved in selective neuronal vulnerability in Huntington's disease, Cell Death Differ. 11 (2004) 424–438, doi:10.1038/sj.cdd.4401358.
[67] J.P. Miller, J. Holcomb, I. Al-Ramahi, M. De Haro, J. Gafni, N. Zhang, et al., Matrix metalloproteinases are modifiers of huntingtin proteolysis and toxicity in Huntington's disease, Neuron 67 (2010) 199–212, doi:10.1016/j.neuron.2010.06.021.
[68] D. Hu, X. Sun, A. Magpusao, Y. Fedorov, M. Thompson, B. Wang, et al., Small-molecule suppression of calpastatin degradation reduces neuropathology in models of Huntington's disease, Nat. Commun. 12 (2021) 5305, doi:10.1038/s41467-021-25651-y.
[69] Q. Jia, S. Li, X.-J. Li, P. Yin, Neuroinflammation in Huntington's disease: from animal models to clinical therapeutics, Front. Immunol. 13 (2022) 1088124, doi:10.3389/fimmu.2022.1088124.
[70] K.-H. Chang, Y.-R. Wu, Y.-C. Chen, C.-M. Chen, Plasma inflammatory biomarkers for Huntington's disease patients and mouse model, Brain Behav. Immun. 44 (2015) 121–127, doi:10.1016/j.bbi.2014.09.011.
[71] R.T. Abdel-Latif, H.S. El-Abhar, D.M. Abdallah, I.M. Fawzy, S.M. Mansour, Dual inhibition of canonical and noncanonical PAR-1 by SCH79797 mitigates neurodegeneration in 3-NP-induced Huntington's disease: an in vivo and in silico approach, Arch. Pharm. 358 (2025) e2400846, doi:10.1002/ardp.202400846.
[72] M. Iemmolo, G. Ghersi, G. Bivona, The cytokine CX3CL1 and ADAMs/MMPs in concerted cross-talk influencing neurodegenerative diseases, IJMS 24 (2023) 8026, doi:10.3390/ijms24098026.
[73] B. De Strooper, E. Karran, The cellular phase of Alzheimer's disease, Cell 164 (2016) 603–615, doi:10.1016/j.cell.2015.12.056.
[74] D.J. Selkoe, J. Hardy, The amyloid hypothesis of Alzheimer's disease at 25 years, EMBO Mol. Med. 8 (2016) 595–608, doi:10.15252/emmm.201606210.

[75] M.T. Heneka, W.M. van der Flier, F. Jessen, J. Hoozemanns, D.R. Thal, D. Boche, et al., Neuroinflammation in Alzheimer disease, Nat. Rev. Immunol. 25 (2025) 321–352, doi:10.1038/s41577-024-01104-7.

[76] B. De Strooper, Proteases and proteolysis in Alzheimer disease: a multifactorial view on the disease process, Physiol. Rev. 90 (2010) 465–494, doi:10.1152/physrev.00023.2009.

[77] R.A. Black, C.T. Rauch, C.J. Kozlosky, J.J. Peschon, J.L. Slack, M.F. Wolfson, et al., A metalloproteinase disintegrin that releases tumour-necrosis factor-α from cells, Nature 385 (1997) 729–733, doi:10.1038/385729a0.

[78] B. Decourt, D.K. Lahiri, M.N. Sabbagh, Targeting tumor necrosis factor alpha for Alzheimer's disease, CAR 14 (2017) 412–425, doi:10.2174/1567205013666160930110551.

[79] R. Guerreiro, A. Wojtas, J. Bras, M. Carrasquillo, E. Rogaeva, E. Majounie, et al., TREM2 variants in Alzheimer's disease, N. Engl. J. Med. 368 (2013) 117–127, doi:10.1056/NEJMoa1211851.

[80] T. Jonsson, H. Stefansson, S. Steinberg, I. Jonsdottir, P.V. Jonsson, J. Snaedal, et al., Variant of TREM2 associated with the risk of Alzheimer's disease, N. Engl. J. Med. 368 (2013) 107–116, doi:10.1056/NEJMoa1211103.

[81] D.J. Pilat, H. Le, D. Prokopenko, C.-C.J. Lin, W.A. Eimer, L. Quinti, et al., The gain-of-function TREM2-T96K mutation increases risk for Alzheimer's disease by impairing microglial function, Neuron 114 (2026) 46–66.e13, doi:10.1016/j.neuron.2025.09.032.

[82] G. Kleinberger, Y. Yamanishi, M. Suárez-Calvet, E. Czirr, E. Lohmann, E. Cuyvers, et al., TREM2 mutations implicated in neurodegeneration impair cell surface transport and phagocytosis, Sci. Transl. Med. 6 (2014), doi:10.1126/scitranslmed.3009093.

[83] D. Feuerbach, P. Schindler, C. Barske, S. Joller, E. Beng-Louka, K.A. Worringer, et al., ADAM17 is the main sheddase for the generation of human triggering receptor expressed in myeloid cells (hTREM2) ectodomain and cleaves TREM2 after Histidine 157, Neurosci. Lett. 660 (2017) 109–114, doi:10.1016/j.neulet.2017.09.034.

[84] C. Selvais, L. D'Auria, D. Tyteca, G. Perrot, P. Lemoine, L. Troeberg, et al., Cell cholesterol modulates metalloproteinase-dependent shedding of low-density lipoprotein receptor-related protein-1 (LRP-1) and clearance function, FASEB J. 25 (2011) 2770–2781, doi:10.1096/fj.10-169508.

[85] R.D. Moir, R. Lathe, R.E. Tanzi, The antimicrobial protection hypothesis of Alzheimer's disease, Alzheimers Dement. 14 (2018) 1602–1614, doi:10.1016/j.jalz.2018.06.3040.

[86] W.A. Eimer, A.S. Rodriguez, M.T. DeFao, S. Ehricke, J. Park, D.K. Vijaya Kumar, et al., Phosphorylated tau exhibits antimicrobial activity capable of neutralizing herpes simplex virus 1 infectivity in human neurons, Nat. Neurosci. 29 (2026) 604–616, doi:10.1038/s41593-025-02157-0.

[87] D. Puzzo, L. Privitera, A. Palmeri, Hormetic effect of amyloid-beta peptide in synaptic plasticity and memory, Neurobiol. Aging 33 (2012) 1484.e15–1484.e24, doi:10.1016/j.neurobiolaging.2011.12.020.

[88] A. Palmeri, R. Ricciarelli, W. Gulisano, D. Rivera, C. Rebosio, E. Calcagno, et al., Amyloid-β peptide is needed for cGMP-induced long-term potentiation and memory, J. Neurosci. 37 (2017) 6926–6937, doi:10.1523/JNEUROSCI.3607-16.2017.

[89] S. Higashi, K. Miyazaki, Identification of a region of beta-amyloid precursor protein essential for its gelatinase A inhibitory activity, J. Biol. Chem. 278 (2003) 14020–14028, doi:10.1074/jbc.M212264200.

[90] S. Higashi, K. Miyazaki, Novel processing of beta-amyloid precursor protein catalyzed by membrane type 1 matrix metalloproteinase releases a fragment lacking the inhibitor domain against gelatinase A, Biochemistry 42 (2003) 6514–6526, doi:10.1021/bi020643m.

[91] M. Ahmad, T. Takino, H. Miyamori, T. Yoshizaki, M. Furukawa, H. Sato, Cleavage of amyloid-beta precursor protein (APP) by membrane-type matrix metalloproteinases, J. Biochem. 139 (2006) 517–526, doi:10.1093/jb/mvj054.
[92] M. Willem, S. Tahirovic, M.A. Busche, S.V. Ovsepian, M. Chafai, S. Kootar, et al., η-Secretase processing of APP inhibits neuronal activity in the hippocampus, Nature 526 (2015) 443–447, doi:10.1038/nature14864.
[93] K. Baranger, Y. Marchalant, A.E. Bonnet, N. Crouzin, A. Carrete, J.-M. Paumier, et al., MT5-MMP is a new pro-amyloidogenic proteinase that promotes amyloid pathology and cognitive decline in a transgenic mouse model of Alzheimer's disease, Cell. Mol. Life Sci. 73 (2016) 217–236, doi:10.1007/s00018-015-1992-1.
[94] Y. Sekine-Aizawa, E. Hama, K. Watanabe, S. Tsubuki, M. Kanai-Azuma, Y. Kanai, et al., Matrix metalloproteinase (MMP) system in brain: identification and characterization of brain-specific MMP highly expressed in cerebellum, Eur. J. Neurosci. 13 (2001) 935–948.
[95] M. Mensch, J. Dunot, S.M. Yishan, S.S. Harris, A. Blistein, A. Avdiu, et al., Aη-α and Aη-β peptides impair LTP ex vivo within the low nanomolar range and impact neuronal activity in vivo, Alzheimers Res. Ther. 13 (2021) 125, doi:10.1186/s13195-021-00860-1.
[96] E. Afram, I. Lauritzen, A. Bourgeois, W. El Manaa, E. Duplan, M. Chami, et al., The η-secretase-derived APP fragment ηCTF is localized in Golgi, endosomes and extracellular vesicles and contributes to Aβ production, Cell. Mol. Life Sci. 80 (2023) 97, doi:10.1007/s00018-023-04737-4.
[97] J.R. Backstrom, G.P. Lim, M.J. Cullen, Z.A. Tökés, Matrix metalloproteinase-9 (MMP-9) is synthesized in neurons of the human hippocampus and is capable of degrading the amyloid-beta peptide (1-40), J. Neurosci. 16 (1996) 7910–7919, doi:10.1523/JNEUROSCI.16-24-07910.1996.
[98] A. Fragkouli, E.C. Tsilibary, A.K. Tzinia, Neuroprotective role of MMP-9 overexpression in the brain of Alzheimer's 5xFAD mice, Neurobiol. Dis. 70 (2014) 179–189, doi:10.1016/j.nbd.2014.06.021.
[99] P. Yan, X. Hu, H. Song, K. Yin, R.J. Bateman, J.R. Cirrito, et al., Matrix metalloproteinase-9 degrades amyloid-β fibrils in vitro and compact plaques in situ, J. Biol. Chem. 281 (2006) 24566–24574, doi:10.1074/jbc.M602440200.
[100] K.-J. Yin, J.R. Cirrito, P. Yan, X. Hu, Q. Xiao, X. Pan, et al., Matrix metalloproteinases expressed by astrocytes mediate extracellular amyloid-β peptide catabolism, J. Neurosci. 26 (2006) 10939–10948, doi:10.1523/JNEUROSCI.2085-06.2006.
[101] M. Hernandez-Guillamon, S. Mawhirt, S. Blais, J. Montaner, T.A. Neubert, A. Rostagno, et al., Sequential amyloid-β degradation by the matrix metalloproteases MMP-2 and MMP-9, J. Biol. Chem. 290 (2015) 15078–15091, doi:10.1074/jbc.M114.610931.
[102] M.-C. Liao, W.E. Van Nostrand, Degradation of soluble and fibrillar amyloid β-protein by matrix metalloproteinase (MT1-MMP) *in vitro*, Biochemistry 49 (2010) 1127–1136, doi:10.1021/bi901994d.
[103] B. Terni, I. Ferrer, Abnormal expression and distribution of MMP2 at initial stages of Alzheimer's disease-related pathology, J. Alzheimers Dis. 46 (2015) 461–469, doi:10.3233/JAD-142460.
[104] M. Bjerke, H. Zetterberg, Å Edman, K. Blennow, A. Wallin, U. Andreasson, Cerebrospinal fluid matrix metalloproteinases and tissue inhibitor of metalloproteinases in combination with subcortical and cortical biomarkers in vascular dementia and Alzheimer's disease, JAD 27 (2011) 665–676, doi:10.3233/JAD-2011-110566.
[105] C.E. Hanzel, M.F. Iulita, H. Eyjolfsdottir, E. Hjorth, M. Schultzberg, M. Eriksdotter, et al., Analysis of matrix metallo-proteases and the plasminogen system in mild cognitive impairment and Alzheimer's disease cerebrospinal fluid, JAD 40 (2014) 667–678, doi:10.3233/JAD-132282.

[106] M. Vervuurt, A.M. de Kort, L. Jäkel, I. Kersten, W.F. Abdo, F.H.B.M. Schreuder, et al., Decreased ratios of matrix metalloproteinases to tissue-type inhibitors in cerebrospinal fluid in sporadic and hereditary cerebral amyloid angiopathy, Alzheimers Res. Ther. 15 (2023) 26, doi:10.1186/s13195-023-01171-3.
[107] K. Miyazaki, M. Hasegawa, K. Funahashi, M. Umeda, A metalloproteinase inhibitor domain in Alzheimer amyloid protein precursor, Nature 362 (1993) 839–841, doi:10.1038/362839a0.
[108] R.G. Perez, S. Soriano, J.D. Hayes, B. Ostaszewski, W. Xia, D.J. Selkoe, et al., Mutagenesis identifies new signals for beta-amyloid precursor protein endocytosis, turnover, and the generation of secreted fragments, including Abeta42, J. Biol. Chem. 274 (1999) 18851–18856, doi:10.1074/jbc.274.27.18851.
[109] T. Uekita, Y. Itoh, I. Yana, H. Ohno, M. Seiki, Cytoplasmic tail-dependent internalization of membrane-type 1 matrix metalloproteinase is important for its invasion-promoting activity, J. Cell. Biol. 155 (2001) 1345–1356, doi:10.1083/jcb.200108112.
[110] P. Asthana, L. Li, L. Lu, J. Wu, S. Zhang, N. Li, et al., MT1-MMP inhibition rejuvenates ageing brain and rescues cognitive deficits in obesity, Cell. Discov. 11 (2025) 76, doi:10.1038/s41421-025-00825-w.
[111] E. Llano, A.M. Pendás, J.P. Freije, A. Nakano, V. Knäuper, G. Murphy, et al., Identification and characterization of human MT5-MMP, a new membrane-bound activator of progelatinase a overexpressed in brain tumors, Cancer Res. 59 (1999) 2570–2576.
[112] D. Pei, Identification and characterization of the fifth membrane-type matrix metalloproteinase MT5-MMP, J. Biol. Chem. 274 (1999) 8925–8932, doi:10.1074/jbc.274.13.8925.
[113] X. Wang, J. Yi, J. Lei, D. Pei, Expression, purification and characterization of recombinant mouse MT5-MMP protein products, FEBS Lett. 462 (1999) 261–266, doi:10.1016/s0014-5793(99)01534-3.
[114] D.M. Jaworski, Developmental regulation of membrane type-5 matrix metalloproteinase (MT5-MMP) expression in the rat nervous system, Brain Res. 860 (2000) 174–177.
[115] H. Hayashita-Kinoh, H. Kinoh, A. Okada, K. Komori, Y. Itoh, T. Chiba, et al., Membrane-type 5 matrix metalloproteinase is expressed in differentiated neurons and regulates axonal growth, Cell Growth Differ. 12 (2001) 573–580.
[116] K. Komori, T. Nonaka, A. Okada, H. Kinoh, H. Hayashita-Kinoh, N. Yoshida, et al., Absence of mechanical allodynia and Aβ-fiber sprouting after sciatic nerve injury in mice lacking membrane-type 5 matrix metalloproteinase, FEBS Lett. 557 (2004) 125–128, doi:10.1016/S0014-5793(03)01458-3.
[117] K.M. Warren, T.M. Reeves, L.L. Phillips, MT5-MMP, ADAM-10, and N-cadherin act in concert to facilitate synapse reorganization after traumatic brain injury, J. Neurotrauma 29 (2012) 1922–1940, doi:10.1089/neu.2012.2383.
[118] N. Arnst, P. Belio-Mairal, L. García-González, L. Arnaud, L. Greetham, E. Nivet, et al., Deficiency in MT5-MMP supports branching of human iPSCs-derived neurons and reduces expression of GLAST/S100 in iPSCs-derived astrocytes, Cells 10 (2021) 1705, doi:10.3390/cells10071705.
[119] E. Porlan, B. Martí-Prado, J.M. Morante-Redolat, A. Consiglio, A.C. Delgado, R. Kypta, et al., MT5-MMP regulates adult neural stem cell functional quiescence through the cleavage of N-cadherin, Nat. Cell Biol. 16 (2014) 629–638, doi:10.1038/ncb2993.
[120] S. Monea, B.A. Jordan, S. Srivastava, S. DeSouza, E.B. Ziff, Membrane localization of membrane type 5 matrix metalloproteinase by AMPA receptor binding protein and cleavage of cadherins, J. Neurosci. 26 (2006) 2300–2312, doi:10.1523/JNEUROSCI.3521-05.2006.

[121] S. Restituito, L. Khatri, I. Ninan, P.M. Mathews, X. Liu, R.J. Weinberg, et al., Synaptic autoregulation by metalloproteases and γ-secretase, J. Neurosci. 31 (2011) 12083–12093, doi:10.1523/JNEUROSCI.2513-11.2011.

[122] A. Webers, M.T. Heneka, P.A. Gleeson, The role of innate immune responses and neuroinflammation in amyloid accumulation and progression of Alzheimer's disease, Immunol. Cell Biol. 98 (2020) 28–41, doi:10.1111/imcb.12301.

[123] C. Domingues, O.A.B. da Cruz e Silva, A.G. Henriques, Impact of cytokines and chemokines on Alzheimer's disease neuropathological hallmarks, CAR 14 (2017), doi:10.2174/1567205014666170317113606.

[124] M.T. Heneka, R.M. McManus, E. Latz, Inflammasome signalling in brain function and neurodegenerative disease, Nat. Rev. Neurosci. 19 (2018) 610–621, doi:10.1038/s41583-018-0055-7.

[125] A.R. Folgueras, T. Valdes-Sanchez, E. Llano, L. Menendez, A. Baamonde, B.L. Denlinger, et al., Metalloproteinase MT5-MMP is an essential modulator of neuro-immune interactions in thermal pain stimulation, Proc. Natl. Acad. Sci. 106 (2009) 16451–16456, doi:10.1073/pnas.0908507106.

[126] M. Xu, J. Li, L. Xia, Y. Du, B. Wu, X. Shi, et al., PCSK6 exacerbates Alzheimer's disease pathogenesis by promoting MT5-MMP maturation, Exp. Neurol. 374 (2024) 114688, doi:10.1016/j.expneurol.2024.114688.

[127] H.C. Rice, D. de Malmazet, A. Schreurs, S. Frere, I. Van Molle, A.N. Volkov, et al., Secreted amyloid-β precursor protein functions as a GABABR1a ligand to modulate synaptic transmission, Science 363 (2019) eaao4827, doi:10.1126/science.aao4827.

[128] K. Baranger, A.E. Bonnet, S.D. Girard, J.-M. Paumier, L. García-González, W. Elmanaa, et al., MT5-MMP promotes Alzheimer's pathogenesis in the frontal cortex of 5xFAD Mice and APP trafficking in vitro, Front. Mol. Neurosci. 9 (2017), doi:10.3389/fnmol.2016.00163.

[129] D. Pilat, J.-M. Paumier, L. García-González, L. Louis, D. Stephan, C. Manrique, et al., MT5-MMP promotes neuroinflammation, neuronal excitability and Aβ production in primary neuron/astrocyte cultures from the 5xFAD mouse model of Alzheimer's disease, J. Neuroinflammation 19 (2022) 65, doi:10.1186/s12974-022-02407-z.

[130] D. Pilat, J.-M. Paumier, L. Louis, C. Manrique, L. García-González, D. Stephan, et al., Suppression of MT5-MMP reveals early modulation of Alzheimer's pathogenic events in primary neuronal cultures of 5xFAD mice, Biomolecules 14 (2024) 1645, doi:10.3390/biom14121645.

[131] A.H. Caster, R.A. Kahn, Recruitment of the Mint3 adaptor is necessary for export of the amyloid precursor protein (APP) from the Golgi complex, J. Biol. Chem. 288 (2013) 28567–28580, doi:10.1074/jbc.M113.481101.

[132] P. Wang, X. Wang, D. Pei, Mint-3 regulates the retrieval of the internalized membrane-type matrix metalloproteinase, MT5-MMP, to the plasma membrane by binding to its carboxyl end motif EWV, J. Biol. Chem. 279 (2004) 20461–20470, doi:10.1074/jbc.M400264200.

[133] Z. Zhang, M. Song, X. Liu, S. Su Kang, D.M. Duong, N.T. Seyfried, et al., Delta-secretase cleaves amyloid precursor protein and regulates the pathogenesis in Alzheimer's disease, Nat. Commun. 6 (2015) 8762, doi:10.1038/ncomms9762.

[134] D. Jadhav, A.M. Knapinska, H. Wang, G.B. Fields, Membrane-type 5 matrix metalloproteinase (MT5-MMP): background and proposed roles in normal physiology and disease, Biomolecules 15 (2025) 1114, doi:10.3390/biom15081114.

[135] H. Wang, M.K. Lakshmana, G.B. Fields, Identification of binding partners that facilitate membrane-type 5 matrix metalloproteinase (MT5-MMP) processing of amyloid precursor protein, J. Cell. Physiol. 239 (2024) e31218, doi:10.1002/jcp.31218.

[136] P. Belio-Mairal, A. Kamitsou, D. Stephan, N. Jullien, D. Thiane, M. Ramos, et al., MT5-MMP C-terminal mutations reduce C99 and β accumulation: a foundation for peptide-based Alzheimer's interventions, Cell Biol. (2025), doi:10.1101/2025.05.22.655451.
[137] A. Magniez, J. Giovannini, P. Zipfel, C. Rémondin, H. Saïdoun, A. Davis, et al., Structure-guided discovery of a nonpeptidic MT5-MMP inhibitor, ACS Med. Chem. Lett. 17 (2026) 1407–1415, doi:10.1021/acsmedchemlett.6c00219.
[138] L. Ma, Y. Huang, F. Zhang, D.S. Gao, N. Sun, J. Ren, et al., MMP24 contributes to neuropathic pain in an FTO-dependent manner in the spinal cord neurons, Front. Pharmacol. 12 (2021) 673831, doi:10.3389/fphar.2021.673831.

CHAPTER EIGHT

Role of MMPs in neutrophil biology

Jennifer Vandooren*
Department of Microbiology, Immunology and Transplantation, KU Leuven, Kortrijk, Belgium
*Corresponding author. e-mail address: jennifer.vandooren@kuleuven.be.

Contents

Abstract

Neutrophils are central effectors of the innate immune system, providing rapid responses to harmful stimuli and contributing to both host defense and resolution of inflammation. Among their effector mechanisms is the production and secretion of matrix metalloproteinases (MMPs), particularly MMP-8, MMP-9, and MMP-25. This chapter examines the roles of neutrophil-derived MMPs across neutrophil biology, from their development and maturation in the bone marrow to their functional activities in peripheral tissues. We discuss how these proteases regulate immune responses, tissue remodeling and inflammatory resolution. In addition, we explore the involvement of neutrophil MMPs in a range of pathological conditions, including cardiovascular disease, pulmonary disease, infection, trauma, and cancer. Collectively, these insights highlight the dual roles of neutrophil MMPs in host protection and disease progression.

Fungal and Protozoan Carbonic Anhydrases, Volume 60
ISSN 1874-6047
https://doi.org/10.1016/bs.enz.2026.06.004

Keywords: Inflammation, Innate immunity, Matrix metalloproteinases (MMPs), MMP-8, MMP-9, MMP-25, Neutrophils, Neutrophil recruitment, Neutrophil-derived proteases, Proteolysis

1. Introduction

Events such as trauma or infection require a rapid and coordinated immune response to eliminate harmful agents and restore tissue homeostasis. Neutrophils, also referred to as polymorphonuclear leukocytes (PMNs), are among the first responders to such challenges and play a central role in pathogen clearance and the initiation of inflammation. Patients with neutropenia, defined as an absolute neutrophil count below 1500 cells per microliter of blood, exhibit increased susceptibility to infections, particularly of the respiratory tract, as well as periodontal disease, oral ulcers, and skin abscesses [1,2]. Similarly, inborn errors affecting genes critical for neutrophil function, such as defects in the nicotinamide adenine dinucleotide phosphate (NADPH) oxidase complex leading to chronic granulomatous disease, result in recurrent bacterial and fungal infections, granuloma formation, and other inflammatory complications [3]. Thus, neutrophils are essential for host defense and for protective inflammatory responses.

Neutrophils originate from hematopoietic stem cells (HSC) in the bone marrow, where they are generated through granulopoiesis, the process by which granulocytes are formed and mature. During this process, neutrophils acquire specialized granules containing a range of bioactive molecules, as well as surface receptors that enable rapid responses upon activation [4]. Given their short lifespan ($\pm$12 h for mouse cells and up to 5 days for human neutrophils) [5,6], continuous production is required, with daily neutrophil output reaching $1–2 \times 10^{11}$ cells in adult humans [4,5]. Studies in mice have demonstrated that circulating neutrophil levels follow a circadian rhythm, peaking during the active phase and declining during the rest phase. Aging neutrophils are cleared by returning to the bone marrow, where they are engulfed by macrophages and contribute to the regulation of the hematopoietic niche. This represents a mechanism by which neutrophils influence the homeostatic trafficking of hematopoietic precursors [7]. Under non-homeostatic conditions such as stress or inflammation, neutrophil production is markedly increased through a process known as emergency granulopoiesis. This response results in enhanced release of neutrophils into the circulation, including the mobilization of immature forms, thereby contributing to the growing complexity of neutrophil phenotypes and functional states [8–12]. In this context, granulocyte colony-stimulating factor (G-CSF) plays a central role in meeting increased neutrophil demand [4,5].

Mature neutrophils are equipped with a diverse repertoire of receptors, including pattern recognition receptors and cytokine receptors, which enable rapid detection of, and response to, external stimuli. In addition, their granules contain effector molecules, including MMPs such as MMP-8, MMP-9 and MMP-25 [4]. Upon activation, neutrophils exit the circulation through a process known as extravasation and migrate into tissues, where they stand against invading pathogens through the release of antimicrobial granule contents, phagocytosis, modulation of cytokine and chemokine responses, and the formation of neutrophil extracellular traps (NETs) [4,5,9]. While these effector functions are essential for host defense and tissue repair, they also render neutrophils potentially harmful when dysregulated. Indeed, aberrant neutrophil activity has been implicated in a wide range of diseases [9], including inflammatory lung diseases [13] and cancer [14,15].

In this chapter, we discuss the role of neutrophil-derived MMPs across the different stages of neutrophil biology, including neutrophil maturation and mobilization (Section 2), extravasation and interstitial migration (Section 3), activation, degranulation, and the release of MMPs (Section 4), proteolytic regulation of the inflammatory milieu (Section 5), antimicrobial effector functions (Section 6), and emerging roles for neutrophil MMPs in resolution and tissue repair (Section 7). Finally, we examine the role of neutrophil-derived MMPs in pathological conditions (Section 8), including cardiovascular disease and thrombosis, pulmonary diseases, infection and sepsis, ischemia and trauma, and cancer.

2. Neutrophil maturation and mobilization

Before entering the peripheral blood, neutrophils undergo a maturation process in the bone marrow. This process proceeds through five main stages: myeloblasts, (pro)myelocytes, metamyelocytes, band neutrophils, and finally mature segmented neutrophils. The latter constitute a reservoir of cells that can be released into the circulation.

During maturation, neutrophils assemble distinct granule populations containing molecules essential for their effector functions (Fig. 1). This stepwise accumulation enables neutrophils to respond rapidly upon activation, without the need for *de novo* synthesis. Granule formation occurs in a defined sequence, beginning with primary (azurophilic) granules, followed

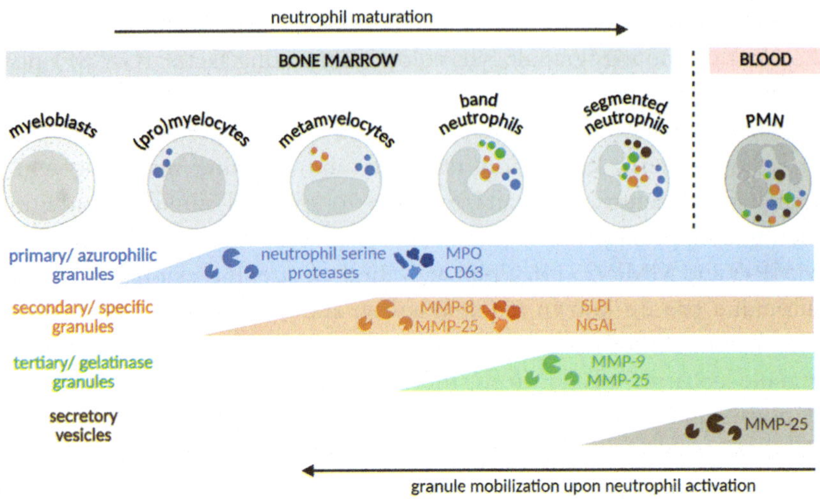

Fig. 1 Neutrophil maturation and sequential accumulation of MMPs.Neutrophils undergo stepwise maturation in the bone marrow, progressing from myeloblasts through (pro)myelocytes, metamyelocytes, and band neutrophils to mature segmented neutrophils, which are released into the circulation as PMNs. During this process, distinct granule subsets are formed, each characterized by the synthesis and storage of specific proteins. Primary (azurophilic) granules contain serine proteases and MPO, secondary (specific) granules are enriched in MMP-8, SLPI and NGAL, tertiary (gelatinase) granules contain MMP-9 and MMP-25, secretory vesicles contain MMP-25. This ordered granule biogenesis is reflected by a reverse hierarchy of vesicle mobilization upon neutrophil activation. *Created in BioRender. Vandooren, J. (2026) https://BioRender.com/63cxcby*

by secondary (specific) granules, tertiary (gelatinase) granules, and finally secretory vesicles [16]. This process also defines the temporal expression and storage of distinct MMP family members in neutrophils.

Primary (azurophilic) granule formation is characterized by the synthesis of serine proteases such as neutrophil elastase (NE), cathepsin G, and proteinase 3 [16–20]. However, the machinery required for their regulated release only becomes fully functional from the band neutrophil stage onward [18]. This stage also coincides with the production of key antimicrobial components such as myeloperoxidase (MPO), as well as markers associated with neutrophil activation, including CD63 [16–19]. The formation of secondary (specific) granules is associated with the production of proteins such as the serine protease inhibitor secretory leukocyte protease inhibitor (SLPI) [16], MMP-8 [16,17] and neutrophil gelatinase-associated lipocalin (NGAL) [20]. Tertiary granules, also referred to as gelatinase granules, mark the appearance of MMP-9 [17,18,20]. Finally, MMP-25, also known as leukolysin or MT6-MMP, has a distinct subcellular distribution compared to MMP-8 and MMP-9, being found across specific granules (±10 percent), gelatinase granules (±40 percent), secretory vesicles (±30 percent), and the plasma membrane (±20 percent) of resting neutrophils [21]. It is anchored to the membrane via a glycosylphosphatidylinositol (GPI) anchor and is present within lipid rafts, where it forms a disulfide-linked homodimer of ±120 kDa [22,23].

Neutrophil-derived proteases, such as MMP-9, have been proposed to play an important role in the mobilization of hematopoietic stem and progenitor cells (HSPCs), a process that is relevant for stem cell transplantation [24]. This mobilization can be induced by various agents, including G-CSF, chemotherapy and interleukin-8 (IL-8). Notably, stem cell mobilization has been associated with systemic release of MMP-9 [25]. Supporting this concept, inhibition of MMP-9 blocks IL-8-induced mobilization [26], and HSC recruitment in response to chemotherapy (5-fluorouracil) or G-CSF is impaired in MMP-9-deficient mice [27]. Collectively, it is suggested that, following treatment, bone marrow myeloid cells release a broad range of proteases, including MMP-9, thereby creating a more proteolytic microenvironment. This shift is thought to facilitate the cleavage of key proteins involved in HSPC mobilization, such as Kit ligand [27] or C–X–C motif chemokine ligand 12 (CXCL12; also referred to as stromal cell–derived factor–1, SDF–1) [28,29]. Nevertheless, these findings remain debated, as other studies have failed to confirm a central role for MMP-9 in HSPC mobilization in knockout models, warranting caution in interpretation [19,24,30,31].

In addition to MMPs, their endogenous inhibitors may also influence hematopoiesis. Interestingly, Kobuch et al. have shown that systemic increases in the MMP inhibitor tissue inhibitor of metalloproteinases–1 (TIMP–1), as observed during infection and inflammatory diseases, can induce neutrophilia in mice. Mechanistically, TIMP-1 promotes granulopoiesis in the bone marrow through its interaction with CD63 [32].

3. Neutrophil extravasation and interstitial migration

Neutrophils are recruited from the circulation into tissues through a process known as extravasation. This process is initiated by activation of the endothelium in response to inflammatory stimuli. In most vascular beds, extravasation proceeds as a coordinated sequence of events. Initially, selectin-mediated interactions (e.g., involving L-selectin/CD62L) slow down neutrophils within the bloodstream and promote rolling along the endothelial surface [33]. Subsequently, chemokines presented on the endothelium activate neutrophil integrins, such as Mac1 (CD11b/CD18), resulting in firm adhesion to endothelial ligands. Neutrophils then undergo intravascular crawling along the endothelial surface to locate permissive sites for transmigration, after which they cross the endothelial barrier [5]. Following transendothelial migration, neutrophils must traverse the perivenular basement membrane and pericyte sheaths before entering the interstitial space, where they migrate directionally toward the site of inflammation [34]. For a comprehensive overview of these processes, the reader is referred to other literature reviews [5,35,36]. Neutrophil-derived proteases, including MMPs, can contribute to multiple stages of this process (Fig. 2), which will be discussed in the following two parts.

3.1 Neutrophil extravasation

Most evidence linking MMPs to neutrophil extravasation relates to the later stages, particularly adhesion, transmigration, and interstitial migration (Fig. 2). Big endothelin-1 (ET-1), an inactive precursor of a potent vasoconstrictor produced by endothelial cells, is found together with gelatinase activity at sites of tissue injury and inflammation. MMP-2 cleaves ET-1 to generate a truncated and biologically active peptide, ET-1(1-32), which we will now refer to as ET* [37]. This fragment directly modulates neutrophil function via binding to the endothelin A receptor (ET_AR), resulting in downregulation of L-selectin and upregulation of CD11b expression, thereby promoting a pro-adhesive phenotype. In addition, ET-1* induces the release of MMP-9 from neutrophils, which in turn can further cleave ET-1 to generate additional ET-1*, establishing a positive feedback loop. Functionally, ET-1* enhances neutrophil adhesion to endothelial cells, an effect that can be largely inhibited by combined blockade of selectins and CD18 integrins [38]. Together, these findings suggest that MMP-mediated processing of ET-1 establishes a feed-forward mechanism that amplifies neutrophil activation and adhesion.

Following transmigration, neutrophils must cross the basement membrane before entering the interstitial space. Although MMPs are well known for their ability to degrade basement membrane components, their contribution to neutrophil migration across these barriers remains debated. In a transwell system coated with Matrigel, migration of human neutrophils toward IL-8 was not impaired by MMP inhibition [39]. Likewise, neutrophil

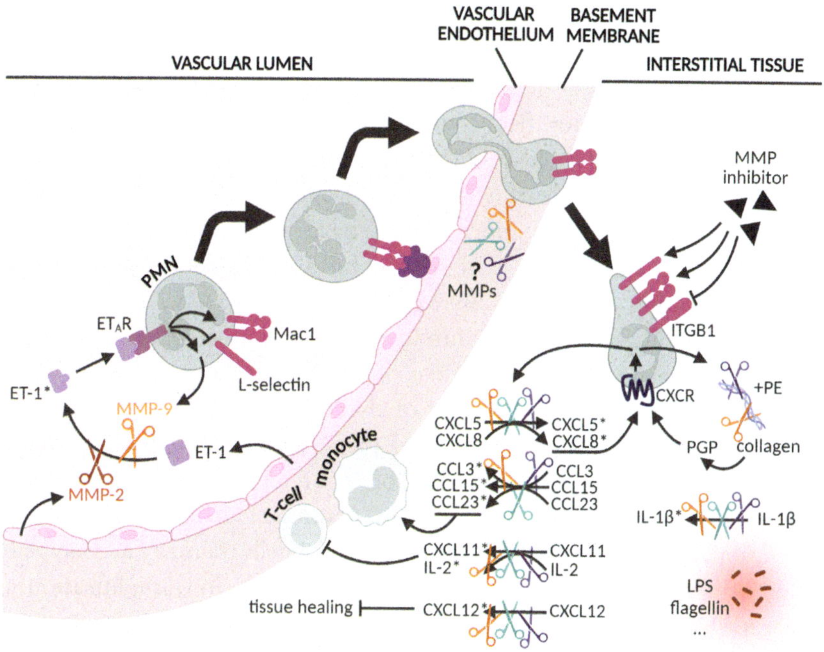

Fig. 2 Neutrophil-derived MMPs regulate extravasation, interstitial migration and inflammatory gradients.Schematic representation of how neutrophil-derived MMPs regulate neutrophil recruitment and function. During extravasation, endothelial-derived endothelin-1 (ET-1) is proteolytically processed by MMP-2 into the active fragment ET-1(1-32)(shown as ET*), which induces L-selectin shedding and Mac-1 upregulation, thereby promoting neutrophil adhesion and activation. MMP-9 release further amplifies this feed-forward loop. Following transmigration, MMPs modulate adhesion molecule expression and support interstitial migration, in part through non-proteolytic mechanisms including integrin switching ($\beta2$ to $\beta1$). Within the tissue, MMPs reshape the inflammatory milieu by proteolytic processing of chemokines and cytokines, leading to modified forms (indicated by an asterisk*) of CXCL5, CXCL8, CXCL11, IL-2, CCL3, CCL15, CCL23 and IL-1β, thereby regulating leukocyte recruitment. In parallel, MMP-8 and MMP-9 generate the matrikine PGP from collagen in cooperation with prolyl endopeptidase, reinforcing neutrophil recruitment through CXCR1/2 signaling. *Created in BioRender. Vandooren, J. (2026) https://BioRender.com/63cxcby*

migration across human pulmonary endothelial cells and associated basal lamina could not be inhibited by natural nor synthetic MMP or serine proteases inhibitors [40]. Similarly, studies in MMP-9-deficient mice showed that neutrophil extravasation is not uniformly impaired in models of acute lung, peritoneal, or skin inflammation [41]. In a mouse cremaster muscle model, the role of MMP-9 in neutrophil recruitment was stimulus-dependent: inhibition of MMP-2/−9 reduced leukocyte adhesion and transmigration in response to ischemia–reperfusion and C–C motif chemokine ligand 3 (CCL3), also known as CCL3/ macrophage inflammatory protein–1 alpha (MIP–1α), but only affected transmigration under platelet–activating factor (PAF) stimulation [42]. Moreover, in a murine peritonitis model, absence of MMP-9 resulted in increased NE activity, compensating for the loss of MMP-9 and leading to delayed but enhanced neutrophil infiltration. In murine models of acute lung injury, MMP–8 deficiency leads to increased neutrophil and macrophage infiltration, exacerbated lung injury and increased mortality [43,44]. These findings highlight potential functional redundancy between protease classes and may explain discrepancies between studies on neutrophil migration.

Nevertheless, the consequences of dysregulated MMP activity during extravasation are illustrated in several pathological contexts. In psoriasis, neutrophil infiltration is a hallmark of lesion development, and MMP-9 is overexpressed compared to healthy controls. During transendothelial migration, neutrophils and vascular endothelial cells mutually activate each other, resulting in increased vascular permeability and amplification of inflammatory infiltration. In this setting, MMP-9 promotes activation of ERK1/2 and p38 MAPK signaling pathways in endothelial cells, thereby enhancing inflammatory cell recruitment [45]. Similarly, in experimental acute pancreatitis, neutrophils are a major source of increased pulmonary MMP-9. Pharmacological inhibition of MMP-9 reduces tumor necrosis factor–alpha (TNF–α)-induced neutrophil transmigration across basement membrane components in vitro, as well as neutrophil infiltration and alveolar–capillary leakage in vivo [46].

3.2 Interstitial migration

Within the interstitium, increasing collagen density during inflammation provides structural guidance for migrating neutrophils, which traverse the interstitial space without extensive pericellular matrix degradation. This amoeboid mode of locomotion depends on actin polymerization at the leading edge rather than actomyosin contraction, and is impaired by MMP inhibition but not by serine protease inhibition, indicating that MMPs support interstitial migration through a mechanism likely involving signaling and adhesion molecule modulation rather than matrix remodeling. Consistent with this, MMP inhibition alters the surface expression of adhesion molecules on neutrophils in response to PAF, namely, it blocks L-selectin/CD62L shedding and reduces the PAF-elicited upregulation of β1 integrins (ITGB1), and further enhances the already elevated surface expression of $\beta 2$ integrins Mac-1 and LFA-1 [34]. This suggests that MMPs are required not just for physical passage through the interstitium, but for completing the adhesion molecule switch initiated at the endothelial surface, driving the transition from the selectin- and $\beta 2$ integrin-dependent endothelial phenotype toward the $\beta 1$ integrin-dependent tissue-navigation program. By contrast, serine proteases such as NE appear more important during earlier transendothelial steps.

Interestingly, it was shown that during neutrophil migration, fragments of extracellular matrix proteins, such as collagen and elastin, play an important role in inflammatory cell recruitment. MMP-8 and MMP-9 participate in a sequential proteolytic cascade alongside prolyl endopeptidase (PE) to generate the matrikine proline-glycine-proline (PGP) from interstitial collagen. PGP acts as a neutrophil chemoattractant via C–X–C motif chemokine receptor 1 (CXCR1) and CXCR2 [47,48], and its N-acetylated form (AcPGP), a form induced by cigarette smoke, is resistant to degradation and drives a self-sustaining feed-forward loop in which neutrophil-derived MMP-9 release generates further AcPGP from collagen, a mechanism particularly implicated in chronic neutrophilic inflammation [49,50].

3.3 Lymphatic trafficking

In addition to trafficking from blood to tissues, neutrophils can also migrate via lymphatic vessels to draining lymph nodes, where they contribute to shaping adaptive immune responses. This process similarly depends on endothelial activation and adhesion, initially mediated by E-selectin, followed by integrin-dependent interactions with intercellular adhesion molecule–1 (ICAM–1) and vascular cell adhesion molecule–1 (VCAM–1). Importantly, neutrophils localize MMP activity at sites of endothelial contact, in conjunction with lipid mediators such as 12-Hydroxyeicosatetraenoic acid, to induce transient junctional retraction and facilitate transmigration. Consistent with this, MMP inhibitors reduce neutrophil transmigration in both a transendothelial migration assay and mouse *ex vivo* neutrophil migration assays with mouse dermal tissue explants [51].

Taken together, the direct contribution of neutrophil-derived MMPs to cell migration is complex. While MMPs are not universally required for transendothelial migration, they can modulate key aspects of neutrophil recruitment, particularly under acute inflammatory conditions. Factors such as the nature of the stimulus, tissue context, timing and protease redundancy are critical determinants of their functional relevance.

4. Neutrophil activation, degranulation, and release of MMPs

Neutrophils are capable of mounting a rapid response to inflammatory stimuli, owing to the pre-storage of granule proteins during their maturation process. For example, Pugin et al. [52] demonstrated that PMN in whole blood rapidly release MMP-9 (within ±20 min) upon stimulation with Lipopolysaccharide (LPS), TNF, IL-8, or G-CSF. In comparison, the monocytic fraction required more than 7 h of LPS stimulation to initiate secretion of MMP-9, with maximal levels observed only after 24–48 h of treatment [52]. The sequential build-up of neutrophil granules is also mirrored by the order of granule mobilization upon activation (degranulation), albeit in reverse (Fig. 1): secretory vesicles are released first, followed by tertiary, secondary, and primary granules, depending on the strength and duration of stimulation [53]. Consequently, secretory vesicles (containing MMP-25) are most readily exocytosed, followed by gelatinase granules (MMP-9/ MMP-25), secondary granules (MMP-8/ MMP-25), and primary granules (containing neutrophil serine proteases). Interestingly, neutrophils are also able to release extracellular vesicles such as ectosomes upon stimulation. These are rightside-out oriented vesicles that expose phosphatidylserine on their outer layer and carry a selectively sorted subset of membrane proteins. Neutrophil-derived ectosomes were shown to contain enzymatically active granule-derived proteins including NE, proteinase 3, MPO and MMP-9, reflecting incorporation of contents from azurophilic and gelatinase granules [54].

Upon neutrophil stimulation (with e.g., IL-8), active soluble MMP-25 is secreted into the supernatant, while the lipid raft-associated membrane-bound pool remains associated with these structures and is not displaced by stimulation [21,22]. Importantly, surface exposure of MMP-25 was shown to only occurs upon induction of apoptosis, at which point membrane-bound MMP-25 localizes to the extracellular face [22]. Overall, the functional role of MMP-25 during neutrophil maturation and priming remains incompletely defined. Interestingly, serum MMP-25 has been proposed as a potential metabolic marker for atopy-associated inflammation in the context of asthma and chronic obstructive pulmonary disease (COPD), with serum leukolysin levels found to correlate with elevated total IgE and with anti-IL-1 antibody levels. Interestingly, serum MMP-25 has been proposed as a potential marker of atopy-associated inflammation and correlated with elevated total IgE, a marker known to be increased in human atopic asthma [55].

Whereas many neutrophil responses to TNF depend on β2-integrin- and Src family kinase–mediated signaling, TNF-α has been shown to induce the secretion of MMP-9 from tertiary (gelatinase) granules via TNF receptor I through a pathway that is only partially reliant on this classical signaling cascade. Instead, protein kinase C (PKC) plays a dominant role, as broad PKC inhibition reduces MMP-9 release by more than 85 percent. Furthermore, PKCα and PKCδ appear to be required for maximal MMP-9 secretion [56].

Release of neutrophil granules can also be enhanced through neutrophil priming. For example, pretreatment with bacterial LPS [57] enhances the exocytosis of gelatinase and azurophilic granules, while cytokines such as TNFα [58], GM-CSF [58] and IL-1β [59] promote the release of specific and azurophilic granules. In addition, TNFα induces exocytosis of secretory vesicles and gelatinase granules and primes neutrophils for enhanced release of specific and azurophilic granules in response to subsequent stimulation with N–formyl–methionyl–leucyl–phenylalanine (fMLF) [60].

In contrast, certain mediators can suppress the release and activity of neutrophil-derived MMPs from granules. For example, adenosine, a nucleoside with cardioprotective effects, has been shown to inhibit the secretion of MMP-9 from neutrophils stimulated with agents such as N–formyl–methionyl–leucyl–phenylalanine (fMLP), LPS, or Hydrogen peroxide (H_2O_2) [61]. Clusterin, an abundant extracellular chaperone protein present in tissue fluids, has been identified as an endogenous inhibitor of MMP-25. It forms a stable complex with the soluble form of MMP-25, rendering the enzyme inactive. This complex has also been detected in human neutrophils [62].

5. Proteolytic regulation of inflammation by neutrophil MMPs

5.1 Shaping the neutrophil gradient by chemokine processing

One of the best-characterized functions of neutrophil-derived MMPs is their ability to proteolytically process chemokines, modulating their biological activity ranging from no effect to potentiation of inflammatory signaling or complete inactivation. Here we focus specifically on processes mediated by neutrophil MMPs (MMP-8, MMP-9 and MMP-25). Comprehensive overviews of chemokine processing by MMPs can be found in several literature reviews [63–65].

Neutrophil chemotaxis is primarily orchestrated by CXC chemokines with glutamate-leucine-arginine (ELR) motif [66]. These chemokines act largely through the G protein–coupled receptors CXCR1 and CXCR2. The most notable member regulating neutrophil chemotaxis in humans being CXCL8/IL-8 [67]. Neutrophil MMP-9 is able to cleave CXCL8 into forms with increased signaling through CXCR1 and results in a 10- to 27-fold increased potency of IL-8 to activate neutrophils (increased intracellular calcium, additional MMP-9 secretion and chemotaxis) [68–70]. MMP-9 degrades several CXC chemokines, including connective tissue–activating peptide III (CTAP–III), CXCL4/ platelet factor 4 (PF–4), and CXCL1/growth–regulated oncogene–alpha (GRO–α) [70]. CXCL6/ Granulocyte chemotactic protein–2 (GCP-2) is cleaved by both MMP-9 and MMP-8, although at a lower rate by MMP-8, which does not result in any altered biological activity [68].

CXCL5 (also known as ENA–78 in humans or LIX in mice) presents a more complex picture: MMP-9 cleaves it at up to eight positions, with sequential processing likely initially potentiating and ultimately inactivating the chemokine [68], suggesting a built-in kinetic mechanism for temporal regulation of neutrophil recruitment. In contrast, MMP-25 cleaves CXCL5 in an activating manner, further reinforcing the feed-forward model of neutrophil recruitment [71]. MMP-8 also cleaves and activates CXCL5, and neutrophils lacking MMP-8 show reduced chemotaxis toward intact LIX in vivo, demonstrating that MMP-8-mediated processing also potentiates LIX activity through a feed-forward mechanism [72,73]. In vivo evidence for the physiological relevance of MMP-mediated CXCL5 processing was provided by Song et al., who demonstrated that MMP-2 and MMP-9 act synergistically to potentiate CXCL5 activity during IL-1β-induced peritonitis, with neutrophil-derived MMP-9 and resident-cell-derived MMP-2 representing distinct but cooperative sources [74], highlighting that neutrophil MMP activity operates within a broader proteolytic microenvironment rather than in isolation.

Also ELR-negative CXCR3 ligands (CXCL9/MIG, CXCL10/IP-10 and CXCL11/I-TAC) which regulate the migration of Th1 lymphocytes, are potential substrates for MMP-8 and MMP-9 [75,76]. CXCL11/I-TAC was shown to be cleaved at both the N-terminus and C-terminus, generating three forms: CXCL11-(5–73), -(5–63), and -(5–58). Amino-terminal processing resulted in loss of agonist activity (calcium flux and chemotactic migration of human T lymphocytes). However, CXCL11-(5–73) retained receptor affinity and hence serving as a potential antagonist for full length CXCL11, as well as displaying enhanced affinity to heparin. COOH-terminal processing caused a reduction in receptor binding and loss of affinity for heparin [76]. Together, cleavage of CXCL11 by neutrophil MMPs dampens T cell migration by loss of agonist activity.

Notably, MMP-9 cleaves and inactivates SDF-1/CXCL12, the chemokine responsible for CXCR4-mediated neutrophil retention in the bone marrow [77], potentially mirroring at the tissue level the disruption of this axis that drives neutrophil mobilization. Furthermore, the biological relevance of this cleavage has been confirmed in vivo, where protection of SDF-1 from MMP-9-mediated inactivation improved skin wound healing [78].

Although MMP-9 has been reported to leave certain CC chemokines such as RANTES and MCP-2 intact [70], a broader family-wide analysis revealed that MMPs collectively cleave the full set of monocyte-directed CC chemokines, in some cases generating more potent monocyte chemoattractants [79]. Notably, MMP-25 participates in this broader regulatory landscape, with cleavage of CCL15 and CCL23, likely generating more potent monocyte chemoattractants [71]. Also within the neutrophil-specific context, MMP-8—particularly in its membrane-bound form on lung neutrophils—was shown to cleave and inactivate CCL3/MIP-1α, thereby attenuating lung inflammation and injury in murine models of acute lung injury [43], identifying MMP-8 as capable of both amplifying and resolving inflammatory chemokine gradients depending on substrate and context.

5.2 Beyond chemokines

During systemic inflammation, cytokines such as IL–1β require proteolytic processing to generate soluble, bioactive forms. This step is primarily mediated by dedicated converting enzymes such as caspase–1 [80]. During local inflammation, however, neutrophil-derived proteases can provide additional processing capacity. In this context, in vitro studies have demonstrated that incubation of pro–IL–1β with MMP–9 results in efficient proteolysis and the generation of stable intermediate products [81]. Subsequent processing by other MMPs, such as MMP–3, is thought to promote further degradation, ultimately leading to reduced IL–1β bioactivity [81,82]. Together, these findings suggest that MMPs contribute to a context-dependent regulation of IL–1β activity, acting both as alternative activators and as enzymes that terminate cytokine signaling.

Neutrophil-derived MMP-9 has been shown to proteolytically process both IL-2 [83] and IL-7 [84], two cytokines critical for T cell development and homeostasis, thereby potentially modulating adaptive immune signaling at sites of innate inflammation. MMP-9 cleavage of IL-2 generates fragments with markedly reduced affinity for all IL-2 receptor subunits, resulting in significantly impaired T cell proliferation. Complete inactivation of IL-2 was further achieved by the combined action of the broader proteolytic milieu generated by activated neutrophils, including serine proteases [83]. In contrast, IL-7 demonstrated partial resistance to MMP-9-mediated inactivation: although cleavage into stable fragments was observed, glycosylation of IL-7 and the retention of inter-fragment interactions preserved much of its biological activity [84]. Collectively, these findings suggest that in neutrophil-rich inflammatory conditions, neutrophil MMPs may attenuate IL-mediated adaptive immune signaling, an emerging counter-regulatory function that extends beyond innate immunity. Furthermore, MMP-9-mediated cytokine inactivation extends to the antiviral mediator IFN-β, whose proteolytic cleavage by MMP-9 abolishes its biological activity, a finding with particular relevance in multiple sclerosis where elevated MMP-9 may counteract therapeutically administered IFN-β [85].

Beyond direct cytokine processing, MMP-9 has also been shown to cleave cytokine receptors. In a murine model of desiccating stress-induced ocular surface inflammation, proteolytic shedding of the IL-2 receptor alpha chain (CD25) was demonstrated in a concentration-dependent manner, generating a soluble form of this receptor subunit [86]. Together with MMP-9-mediated IL-2 inactivation discussed above, this raises the possibility that neutrophil-derived MMP-9 may dampen T-cell activation at inflammatory sites through dual targeting of both the cytokine and its receptor. A similar mechanism controls the IL-1 system. MMP activity causes the shedding of soluble IL-1RII. The shed ectodomain acts as a decoy receptor, sequestering free IL-1 and preventing its interaction with the signaling IL-1RI [87].

Finally, MMP-9 has also been shown to proteolytically activate latent transforming growth factor–beta (TGF-β), a cytokine with potent immunosuppressive and pro-resolution functions. By cleaving the latency-associated complex, MMP-9 converts latent TGF-β into its bioactive form, a mechanism that may link the acute neutrophil response to downstream resolution and tissue remodeling processes, though this has been primarily characterized in non-neutrophil contexts [88,89].

6. Antimicrobial effector functions

One of the most recognized functions of neutrophils is the attack and destruction of invading pathogens. To achieve this, neutrophils employ several microbicidal effector mechanisms. These include the phagocytosis of pathogens into phagosomes, where intracellular killing is mediated by the production of NADPH oxidase–derived reactive oxygen species (ROS), as well as by the fusion of phagosomes with cytotoxic azurophilic granules containing antimicrobial proteins such as NE [90–93]. In addition to these intracellular mechanisms, neutrophil degranulation results in the release of various antimicrobial proteins into the extracellular environment (e.g., lysozyme, lactoferrin, and cathelicidins) [93]. Finally, neutrophils are characterized by their ability to form NETs, which are extracellular chromatin structures decorated with antimicrobial proteins that trap and restrict pathogens within localized areas of high microbicidal activity [5,12,93]. These mechanisms enable neutrophils to efficiently eliminate pathogens in both intracellular and extracellular environments. Although these processes are core functions of neutrophils, direct involvement of MMPs in their execution appears to be limited, with current evidence suggesting predominantly indirect or regulatory roles. Consistent with this, Betsuyaku et al. demonstrated that neutrophils from *Mmp9* KO mice exhibit antibacterial activity against *Staphylococcus aureus* and *Klebsiella pneumoniae* comparable to wild-type cells [41], indicating that MMP–9 is not essential for direct bacterial killing. However, MMPs may contribute to antimicrobial responses in a context-dependent manner. In *Plasmodium falciparum* malaria, inhibition of parasite growth was associated with increased levels of neutrophil proteases, including MMP–9 and cathepsin G, both of which were able to inhibit parasite growth and erythrocyte invasion in vitro [94]. Similarly, in tuberculosis (TB), elevated levels of MMP–8 have been observed, where it colocalizes with NETs [95].

More direct roles for MMPs have been described in the regulation of NET formation. IL–1β production and NET release in both murine and human neutrophils were shown to depend not only on NOD–like receptor family pyrin domain containing 3 (NLRP3) inflammasome activation and caspase–1, but also on MMP–9 [96]. Supporting this, MMP–9^{high} neutrophil subsets have been identified as key drivers of NET formation during myocardial ischemia–reperfusion injury, correlating with disease severity [11]. In addition, NETs may function as proteolytic scaffolds. For example, laminin–111 is degraded upon incubation with NET-containing conditioned media, a process that is prevented by DNase I–mediated NET degradation. This proteolysis involves sequential cleavage by NE followed by MMP–9. Notably, the resulting laminin fragments were shown to promote proliferation of dormant cancer cells via activation of integrin $\alpha3\beta1$ signaling [97].

A central component of neutrophil antimicrobial activity is the generation of reactive oxygen species (ROS), a process known as the oxidative burst. In neutrophils, H_2O_2 is utilized by MPO to generate hypochlorous acid (HOCl), a highly potent antimicrobial oxidant [12,92]. Although no direct role for neutrophil-derived MMPs in the oxidative burst has been demonstrated, reactive species such as HOCl and H_2O_2 can regulate MMP activity by promoting the activation of proMMPs, thereby linking oxidative antimicrobial pathways to protease-dependent processes [98,99].

7. Neutrophil MMPs in inflammation resolution and tissue repair

Neutrophil-derived MMPs may also contribute to the termination of inflammation and the restoration of tissue homeostasis. Vimentin has been identified as a substrate of MMP–25 at the neutrophil cell surface. Interestingly, intact vimentin acts as a chemoattractant for THP–1 monocytic cells, whereas cleavage by MMP–25 abolishes this chemotactic activity and instead promotes phagocytic uptake by these cells [71]. In resting and activated neutrophils, MMP–25 is predominantly localized to intracellular granules and membranes. During apoptosis, however, membrane-anchored MMP–25 becomes externalized and is prominently displayed on the neutrophil surface [22]. Together, these findings suggest that MMP–25-mediated processing of surface-exposed vimentin on apoptotic neutrophils may function as an "eat-me" signal promoting their recognition and clearance by phagocytes. This mechanism may facilitate the efficient removal of dying neutrophils and thereby contribute to the resolution of inflammation.

In contrast to this pro-resolving role of MMP-25, MMP-9 activity may impair the resolution of inflammation. Following myocardial infarction (MI), the ventricular wall undergoes a coordinated healing process characterized by the infiltration of neutrophils and macrophages, which remove dead cardiomyocytes and promote extracellular matrix (ECM) remodeling. In a murine model of MI, MMP–9 was shown to cleave the macrophage scavenger receptor CD36, a membrane glycoprotein required for the phagocytic uptake of apoptotic cells [100,101]. This process, known as efferocytosis, is essential for initiating anti-inflammatory responses through the release of IL–10 and TGF–β [102]. Consequently, MMP–9-mediated cleavage of

CD36 impaired macrophage efferocytosis and delayed the resolution of inflammation following MI [100].

These observations illustrate that the impact of neutrophil-derived MMPs on inflammation resolution is highly context- and substrate-dependent. While MMP–25-mediated processing of vimentin promotes phagocytic clearance of apoptotic neutrophils, MMP–9-mediated cleavage of the scavenger receptor CD36 impairs macrophage efferocytosis [100], thereby delaying resolution [71].

Beyond their roles in modulating specific cellular interactions during inflammation resolution, neutrophil-derived MMPs are also implicated in broader physiological processes of tissue breakdown and repair. Menstruation represents a unique physiological model of regulated inflammatory tissue breakdown and repair, in which neutrophils play a central role. Menstruation is an inflammatory process, characterized by the influx of various myeloid cells into the endometrium. These inflammatory cells produce a variety of cytokines and proteases, including MMPs, thereby contributing to degradation of the endometrial extracellular matrix [103]. Neutrophils are found particularly during the immediate premenstrual and menstrual phases. In an in vitro co-culture system, interaction between endometrial stromal fibroblasts and neutrophils purified from either peripheral blood or endometrial tissue resulted in activation of proMMP-2, proMMP-3 and proMMP-9, as well as degradation of TIMP-1 and TIMP-2 [104]. However, in a mouse model for endometrial breakdown and repair, the broad MMP inhibitors doxycycline and batimastat did not have significant effects [105]. Hence, the precise contribution of neutrophil-derived MMPs to physiological endometrial breakdown and repair remains inconclusive. Nevertheless, given their potent proteolytic activity and role in inflammatory tissue remodeling, it would be of considerable interest to further investigate their involvement in menstrual pathologies characterized by aberrant tissue breakdown and repair, such as heavy menstrual bleeding, endometriosis, adenomyosis, and other disorders associated with dysregulated endometrial remodeling.

8. Neutrophil MMPs in pathology

Although beneficial in the short term, many of the pro-inflammatory and microbicidal mechanisms employed by neutrophils can also have cytotoxic and tissue-damaging effects on host tissues. Consequently, excessive activation or persistent recruitment of neutrophils contributes to the pathogenesis of a wide range of diseases, in which neutrophil-derived MMPs play an important role. In this section, we explore the context-dependent contributions of neutrophil MMPs to a selection of pathological conditions.

8.1 Cardiovascular disease and thrombosis

In an in vitro system, 30 min of exposure of the human iliac artery vessel wall to activated neutrophils or neutrophil-derived proteases (MMP-8, MMP-9 and NE) led to degradation of the proteoglycan meshwork. This, in turn, exposed collagen fibers and increased platelet adhesion by up to 10-fold. These findings suggest that neutrophils may contribute to the early stages of thrombus formation by structurally modifying the thromboresistant vessel wall and enhancing its platelet-adhesive properties [106]. Indeed, neutrophils are a prominent component of thrombi [107], and NETs can provide a scaffold for platelet adhesion and aggregation, as well as promote red blood cell recruitment, fibrin deposition and thrombus formation [108]. Consequently, neutrophil-derived MMPs are likely to contribute to this process. Histological analysis of thrombectomy samples from patients with acute myocardial infarction demonstrated that MMPs and their tissue inhibitors (TIMPs) are present across all stages of thrombus development. Notably, high numbers of neutrophils were observed in early-stage thrombi, which correlated with strong immunostaining for MMP-8 and MMP-9. As thrombi progressed to more organized stages, both granulocyte abundance and MMP-8/−9 signals were markedly reduced or absent. A similar temporal pattern was observed for MMP-1, TIMP-2 and TIMP-3, whereas MMP-2 and MMP-14 staining remained consistently high across different stages of thrombus maturation [109].

In the microvasculature, fluid shear stress, sensed by the formyl peptide receptor (FPR) acting as a mechanosensor, induces neutrophils to adopt a spherical shape, thereby facilitating their passage through small vessels [110]. In contrast, under hypertensive conditions, circulating neutrophils display an activated morphology with pseudopod formation, which increases the risk of capillary obstruction. Studies in spontaneously hypertensive rats have shown that elevated plasma levels of MMP-9 promote proteolytic cleavage of FPR, thereby impairing the neutrophil response to fluid shear stress [111]. Furthermore, single-cell transcriptomic analysis of neutrophils from patients with idiopathic pulmonary arterial hypertension, compared to matched controls, identified a neutrophil subset characterized by high MMP-9 expression. The abundance of this subset was associated with worse survival, also suggesting a potential role for neutrophil-derived MMP-9 in the pathogenesis and progression of pulmonary arterial hypertension [112].

In contrast, studies using an elastase-induced abdominal aortic aneurysm (AAA) model in mice have shown that aortic degeneration is reduced by treatment with the broad-spectrum MMP inhibitor doxycycline as well as by MMP-9 deficiency. This effect was attributed to MMP-9 from bone marrow–derived inflammatory cells [113]. In a later study, neutrophil depletion experiments further demonstrated that neutrophils play an important role in AAA formation and are a major source of MMP-8 in this context, whereas no increase in MMP-2 or MMP-9 expression was observed. However, MMP-8 deficiency did not confer protection against AAA formation, suggesting that MMP-8 primarily reflects neutrophil infiltration rather than playing a critical functional role in aneurysm development [114].

8.2 Pulmonary diseases

Neutrophil abundance and associated MMP levels and their activity correlate with disease severity in several pulmonary pathologies. In sputum from patients with COPD and idiopathic pulmonary fibrosis (IPF)—both characterized by dysregulated matrix remodeling and fibrosis—neutrophil counts are markedly increased. MMP-9 levels are elevated in both conditions, with higher levels observed in COPD. Although TIMP-1 levels are increased in both COPD and IPF compared to healthy controls, the MMP-9/TIMP-1 ratio is significantly elevated only in COPD. Notably, MMP-9 concentrations negatively correlate with lung function in COPD patients [115]. In addition, increased levels and activity of both MMP-8 and MMP-9 correlate with the degree of airflow limitation in COPD [116]. Antigen challenge in allergic individuals significantly increased proMMP-9 and TIMP-1 levels in bronchoalveolar lavage fluid (BALF) at 48 h, but not after saline challenge. MMP-9 immunoreactivity localized predominantly to BALF neutrophils, and its concentration correlated with airway neutrophil counts [117]. In non-cystic fibrosis bronchiectasis (NCFB), a disease characterized by recurrent cycles of airway infection, inflammation, and structural damage, neutrophil counts as well as gelatinase activity and NE levels are significantly increased compared to healthy controls. Levels of NE and MMP-9 correlate with sputum purulence, a marker of inflammation and disease severity [118,119].

In the diseased lung, neutrophil serine proteases and MMPs form an amplifying proteolytic cascade in which reciprocal activation and inhibitor degradation enhance overall protease activity, contributing to tissue damage and inflammation. For example, SLPI—a key protector of mucosal surfaces against serine protease–mediated damage with additional anti-inflammatory and antimicrobial functions—is susceptible to proteolytic degradation by multiple MMPs, including MMP-8 and MMP-9, as well as by neutrophil serine proteases. This degradation results in loss of its antiproteolytic and anti-inflammatory functions. Consistent with this, proteolytically degraded SLPI fragments have been detected in sputum from patients with NCFB [119]. Similarly, in cystic fibrosis (CF), sputum levels of MMP-9, NE, the MMP-9/TIMP-1 ratio, and overall gelatinolytic activity are increased. Mechanistically, NE has been shown to cleave and inactivate TIMP-1 while also converting proMMP-9 into an active form. Importantly, these proteolytic fragments have been identified in sputum from CF patients, supporting the presence of this protease cascade in vivo [120].

Ventilator-associated pneumonia (VAP) occurs in 15–30 percent of intubated, mechanically ventilated patients. Neutrophil proteases (NE, MMP-8, MMP-9) are significantly elevated in VAP, whereas levels of MMP inhibitors (TIMP-1, TIMP-2) and NE inhibitors (elafin, alpha–1 proteinase inhibitor, SLPI) remain unchanged [121]. Notably, in a mouse model of ventilator-induced lung injury (VILI), neutrophil depletion resulted in delayed lung repair, accompanied by increases in TNFα, IFNγ, and MIP-2, as well as absence of MMP-9. Administration of exogenous MMP-9 to neutropenic mice following VILI reduced tissue damage without altering cytokine levels, suggesting that neutrophil-derived MMP-9 contributes to matrix remodeling and tissue repair [122]. MMP-9 has also been implicated in bronchopulmonary dysplasia (BPD), the most common chronic lung disease in infants. However, in a transgenic mouse model of IL-1β–induced BPD, MMP-9 deficiency reduced neutrophil accumulation but exacerbated alveolar hypoplasia and increased cell death [123]. Collectively, these findings indicate that MMP-9 activity may also protect against inflammatory lung injury and have a positive role for neutrophils lung repair [124].

8.3 Infection and sepsis

Microbial components, collectively referred to as pathogen-associated molecular patterns (PAMPs), are key triggers of neutrophil activation. Neutrophils express a broad array of pattern recognition receptors, including Toll-like receptors (TLRs), such as TLR2, TLR4, and TLR5, which recognize bacterial lipoproteins, LPS from Gram-negative bacteria, and flagellin, respectively. Engagement of these receptors initiates intracellular signaling cascades that promote neutrophil activation and the release of pro-inflammatory mediators, including MMPs [125,126]. Consequently, infection is characterized by rapid recruitment and activation of neutrophils at the affected site, accompanied by the release of neutrophil-derived MMPs.

In humans, this response can be observed systemically: intravenous administration of LPS in healthy volunteers induces a rapid increase in circulating proMMP-9 levels, peaking within 1.5–3 h after exposure [52]. Notably, these responses are not limited to bacterial infections. In parasitic diseases such as Chagas disease, neutrophils from infected individuals exhibit elevated intracellular levels of MMP-2 and MMP-9 compared to non-infected controls, further supporting the role of neutrophils as a major source of MMPs across diverse infectious contexts [127]. In influenza virus infection, neutrophils are also a major source of MMP-9. Moreover, in a mouse model of influenza virus infection, MMP-9 was required for efficient neutrophil migration into the respiratory tract and contributed to the control of viral replication. However, this same mechanism may also drive excessive neutrophil infiltration, thereby contributing to lung pathology during influenza virus infection [128].

Recent transcriptomic and proteomic analyses have provided additional insight into neutrophil heterogeneity during infection. In acute bacterial infection, MMP-9 has been identified as a major component of a pro-inflammatory neutrophil subpopulation characterized by increased expression of S100 proteins and genes associated with NET formation [129]. In line with this, transcriptomic profiling of whole blood from patients with sepsis complicated by acute respiratory distress syndrome revealed upregulation of neutrophil-associated genes, including MMP-8 and CD24 [130]. CD24 functions as a ligand for platelet P-selectin, a key mediator of neutrophil–platelet interactions, thereby also linking neutrophil activation and MMP expression to platelet-dependent inflammatory processes.

Platelets have also been proposed to play a central role in septic lung injury by promoting neutrophil infiltration into the pulmonary compartment, thereby contributing to tissue damage. In this context, platelet-derived CD40 ligand (CD40L) has been identified as a key mediator. CD40L enhances neutrophil activation by upregulating Mac1 expression and promoting neutrophil migration into the bronchoalveolar space [131]. Upon platelet activation, CD40L can be cleaved from the platelet surface to generate soluble CD40L (sCD40L), which exerts potent pro-inflammatory effects, including activation of neutrophils and endothelial cells. Using

MMP-9-deficient mice in the cecal ligation and puncture model for polymicrobial sepsis, it was demonstrated that MMP-9 regulates the shedding of CD40L from platelets. MMP-9 deficiency resulted in reduced neutrophil infiltration into the lungs, decreased edema formation, and improved histopathological outcomes. In line with this, recombinant MMP-9 was shown to directly cleave surface-expressed CD40L on activated platelets [132]. Furthermore, ras-related C3 botulinum toxin substrate 1 (Rac1) has been identified as a key regulator of this pathway in sepsis. Rac1 signaling not only controls the surface mobilization of CD40L on activated platelets but also regulates MMP-9 secretion from neutrophils, thereby coordinating the release of sCD40L into the circulation [133]. Collectively, these findings highlight an important role for neutrophil-derived MMP-9 in modulating platelet-dependent neutrophil recruitment and tissue injury during sepsis, through the regulation of CD40L shedding from activated platelets. Beneficial effects of targeting neutrophil-derived MMPs have also been demonstrated in an LPS-induced endotoxemia mouse model. Genetic deletion of MMP-9 confers protection against LPS-induced mortality. Similarly, treatment with a peptidomimetic inhibitor targeting MMP-8, MMP-9, and a disintegrin and metalloproteinase 17 (ADAM–17) reduces mortality in this model [134].

Neutrophil-derived MMP-9 is an important factor in bacterial urinary tract infection. Infiltration of neutrophils into the uroepithelium, required for effective bacterial clearance, depends on coordinated interactions with two macrophage subsets, Ly6C^{-} and Ly6C^{+} macrophages. Tissue-resident Ly6C^{-} macrophages initiate the recruitment of Ly6C^{+} macrophages and neutrophils. At the site of infection, Ly6C^{+} macrophages produce TNF, which in turn stimulates Ly6C^{-} macrophages to secrete CXCL2. This chemokine promotes activation of MMP-9 in neutrophils, thereby enabling their entry into the uroepithelium [135].

In TB, caused by *Mycobacterium tuberculosis*, neutrophil influx is associated with poor clinical outcomes. TB infection upregulates MMP expression and disrupts the balance between MMPs and their tissue inhibitors [136]. In particular, neutrophil-derived MMP-8 is increased in TB, where it is associated with NET formation and contributes to extracellular matrix degradation, both in vitro and in patient samples [95].

Microbial virulence factors can also directly modulate neutrophil MMP release and activation. Pneumolysin, a pore-forming toxin produced by *Streptococcus pneumoniae*, facilitates bacterial dissemination from the lungs into the vascular system. It induces release of MMP-8 and MMP-9 from neutrophils, both independently and synergistically with other stimuli such as fMLP [137]. In addition, neutrophil-derived proMMPs, including proMMP-8 and proMMP-9, can be proteolytically activated by bacterial proteinases. Thermolysin family enzymes (M4), such as *Pseudomonas aeruginosa* elastase, *Vibrio cholerae* proteinase, and thermolysin, are potent activators of these MMPs [138].

8.4 Ischemia and trauma

Permanent middle cerebral artery occlusion (MCAO) in rats, a well-established model of focal cerebral ischemia (stroke), has demonstrated that early increases in MMP–9 levels (within 24 h) in both the circulation and infarcted brain tissue are primarily derived from neutrophils and endothelial cells. Notably, systemic administration of an MMP–9–neutralizing monoclonal antibody 1 h prior to MCAO significantly reduced infarct size [139]. In a reperfusion model using MCAO followed by restoration of blood flow in hypertensive rats, MMP–9 immunostaining was observed at later time points (48 h) in endothelial cells, infiltrating neutrophils, and neurons, suggesting a broader cellular contribution during injury progression [140]. However, findings from a model of transient focal cerebral ischemia in superoxide dismutase 2 (SOD2)–deficient mice indicate that neutrophils may not represent the primary source of MMP–9. Although MMP–9 expression spatially correlated with blood–brain barrier (BBB) disruption, the temporal and spatial patterns of neutrophil accumulation did not parallel MMP–9 levels. This suggests that other cellular sources, including resident brain cells, contribute substantially to MMP–9 production in this context [141]. Thus, while the involvement of MMP–9 in ischemic brain injury is well established, its temporal dynamics and cellular origin appear to be strongly context- and model-dependent.

In a model of repeated intraperitoneal (i.p.) LPS administration inducing NLRP3-dependent BBB permeabilization, microglia were shown to upregulate CXCL chemokines and MMPs, including MMP–8 and MMP–9, via NLRP3 and gasdermin D signaling. This response was mediated in part through the production of growth differentiation factor 15 (GDF–15), which promotes recruitment of CXCR2-expressing neutrophils to the BBB. Pharmacological inhibition of neutrophil infiltration or MMP activity (e.g., using ilomastat) attenuated NLRP3-mediated BBB disruption [142]. In a mouse model of moderate spinal cord contusion, MMP-9 levels increased rapidly in the injured tissue, peaking at 24 h post-injury, and co-localized with glial cells, macrophages, neutrophils, and the vasculature. Both genetic deletion of MMP-9 and pharmacological inhibition using a hydroxamic acid–based MMP inhibitor resulted in reduced neutrophil infiltration, decreased blood–spinal cord barrier disruption, and improved functional recovery compared to wild-type controls [143].

Beyond the central nervous system, neutrophil-derived MMP–9 also contributes to ischemia–reperfusion injury in peripheral tissues. In canine myocardial ischemia–reperfusion models, infiltrating neutrophils were identified as an early source of MMP–9, with in situ zymography confirming active MMP–9 localization in proximity to these cells [144]. Similarly, in a rat model of lung transplantation–associated ischemia–reperfusion injury, MMP–2 and MMP–9 levels increased during ischemia and further during reperfusion. Inhibition of MMP activity with the broad-spectrum inhibitor batimastat reduced alveolar–capillary permeability, attenuated neutrophil transmigration, and improved pulmonary gas exchange [145]. In hepatic ischemia–reperfusion injury, pharmacological inhibition of MMP–2 and MMP–9 prevented the ischemia-induced increase in hepatic MMP–9 levels. This intervention also reduced leukocyte rolling, adhesion, and sinusoidal accumulation, resulting in decreased infiltration of neutrophils and T cells into the liver [146].

8.5 Cancer

MMPs have a long-established and extensively reviewed role in cancer progression, including extracellular matrix remodeling, invasion and metastasis [147–149]. Rather than revisiting this broad literature, this section focuses specifically on neutrophil-derived MMPs and highlights selected examples of their context-dependent functions in the tumor microenvironment.

An abundancy of studies exist, showing infiltration of MMP-9-bearing neutrophils into primary tumors [150]. For example, in glioblastoma patients, plasma MMP–9 levels are elevated compared to patients with cerebral aneurysm or healthy controls and decrease following tumor resection. Baseline MMP–9 levels also predict response to bevacizumab (anti–vascular endothelial growth factor A (anti–VEGFA) therapy), as patients with low MMP–9 show a significant 5.2–month improvement in overall survival. Notably, MMP–9 levels do not correlate with tumor volume, invasion, or angiogenesis, suggesting a role in alternative tumor microenvironment processes. Immunohistochemical analyses indicate that over 80 percent of MMP–9 signal originates from $CD45^{+}$ immune cells, predominantly neutrophils (alongside macrophages and dendritic cells). These findings suggest that tumor-infiltrating neutrophils are a major source of MMP–9 in glioblastoma and may contribute to resistance to bevacizumab therapy [151].

Tumor angiogenesis has long been studied as a means to halt cancer progression [152] and neutrophils play a key role in the induction of angiogenesis [153,154]. MMP-9 delivered by tumor-infiltrating myeloid cells induces tumor angiogenesis through both VEGF-dependent and VEGF-independent pathways. However, MMP-9 originates from several cellular sources, depending on the tumor type and stage of tumor development, findings which are well documented in multiple studies [154]. A particularly interesting concept is that, whereas most cell types produce both MMP-9 and its endogenous inhibitor TIMP-1, neutrophils lack TIMP-1 production, rendering them especially effective at promoting angiogenesis [155].

Single-cell analysis of leukocytes in a stem cell leukemia/lymphoma model revealed that PMN-myeloid-derived suppressor cells constitute over 50 percent of circulating leukocytes in leukemic mice, with neutrophils displaying the greatest phenotypic diversity. In leukemia-associated neutrophil subsets, MMP-related pathways were consistently upregulated, with particularly high expression of *Mmp8, Mmp9*, and *Mmp25* in Ly6G^{+} neutrophils. Treatment with the MMP inhibitor ilomastat improved survival in leukemic mice, accompanied by a reduction in leukocyte counts, including leukemia-associated neutrophils. In clinical samples from leukemia patients, elevated MMP–8, but not MMP–9, expression was associated with poor survival [156].

In a murine model of epithelial carcinogenesis, MMP–9 supplied by bone marrow–derived cells was shown to contribute to tumor development [157]. In contrast, MMP–8–deficient mice display an increased incidence of skin tumors, particularly in males and ovariectomized females, highlighting a sex-dependent effect on cancer susceptibility. This phenotype can be rescued by bone marrow transplantation, indicating that neutrophil-derived MMP–8 acts as a protective factor in this context. Mechanistically, MMP–8 deficiency is associated with a prolonged inflammatory response following carcinogen exposure [72]. Consistent with these findings, mutational analyses of MMPs in human melanoma have identified *MMP8* as one of the most frequently mutated genes, with several recurrent mutations resulting in reduced enzymatic activity [158]. Taken together, these observations indicate that the contribution of neutrophil-derived MMPs to tumorigenesis is complex and highly context-dependent.

Neutrophils also play a critical role in cancer progression through their interactions with circulating tumor cells, which detach from primary tumors, enter the bloodstream, and seed metastatic lesions [159]. Neutrophil-derived MMPs may contribute to this process at multiple levels. As discussed above, NETs can function as proteolytic scaffolds that concentrate neutrophil-derived proteases, including MMPs, thereby promoting the degradation of extracellular matrix components such as laminin–111 [97]. In addition to their proteolytic activity, neutrophils can exert immunosuppressive effects within the tumor microenvironment. In a murine model of inducible intestinal adenoma formation, neutrophils were shown to suppress T-cell activity, thereby promoting tumor growth. This effect was mediated by the activation of latent transforming growth factor–β (TGF–β) by neutrophil-derived MMP–9 [160].

9. Emerging concepts and therapeutic targeting

Neutrophils are highly responsive cells that require both rapid action and adaptability to diverse stimuli, including tissue damage and invading microorganisms. These defining features are reflected in the diverse roles of neutrophil-derived MMPs discussed throughout this chapter. During neutrophil maturation, MMPs are synthesized and stored in distinct granule subsets (Fig. 1), with the exception of azurophilic granules, which predominantly contain serine proteases. Consistent with their ordered granule mobilization, MMP activity becomes most prominent during later stages of neutrophil activation, particularly during adhesion and interstitial migration (Fig. 2).

Following their initial identification as ECM–remodeling enzymes, early studies focused primarily on the role of MMPs in matrix degradation, such as during basement membrane traversal and interstitial migration, as well as in pathological ECM remodeling. However, a key concept emerging is that neutrophil-derived MMPs act as important regulators of the inflammatory milieu. Through the proteolytic processing of chemokines, cytokines, growth factors as well as ECM components, MMPs generate bioactive fragments that modulate immune cell recruitment, activation, and resolution of inflammation.

Increasing evidence indicates that neutrophils exist as heterogeneous populations with distinct functional phenotypes, as revealed by single-cell transcriptomic analyses. This heterogeneity is reflected in the differential expression and regulation of MMPs, contributing to their functional diversity.

Consequently, the roles of neutrophil-derived MMPs are highly context-dependent and may even appear paradoxical. This is particularly evident in processes such as extravasation and basement membrane traversal, as well as in pathological conditions such as cancer, where the same protease may exert tumor-promoting or tumor-suppressive effects depending on tissue context, cellular source, and disease stage. These insights have important implications for therapeutic targeting, suggesting that strategies aimed at modulating specific MMP activities within defined cellular and temporal contexts may be more promising than previously developed broad-spectrum inhibitors. A deeper understanding of neutrophil MMP biology across distinct microenvironments will be essential for harnessing their beneficial functions while limiting tissue damage in disease.

References

[1] D.C. Dale, A.A. Bolyard, An update on the diagnosis and treatment of chronic idiopathic neutropenia, Curr. Opin. Hematol. 24 (2017) 46–53, doi:10.1097/MOH.0000000000000305.

[2] W. Dobrewa, M. Bielska, K. Bąbol-Pokora, S. Janczar, W. Młynarski, Congenital neutropenia: from lab bench to clinic bedside and back, Mutat. Res. Rev. Mutat. Res. 793 (2024) 108476, doi:10.1016/J.MRREV.2023.108476.

[3] M.C. Dinauer, Inflammatory consequences of inherited disorders affecting neutrophil function, Blood 133 (2019) 2130–2139, doi:10.1182/BLOOD-2018-11-844563.

[4] N. Borregaard, Neutrophils, from marrow to microbes, Immunity 33 (2010) 657–670, doi:10.1016/j.immuni.2010.11.011.

[5] E. Kolaczkowska, P. Kubes, Neutrophil recruitment and function in health and inflammation, Nat. Rev. Immunol. 13 (2013) 159–175, doi:10.1038/nri3399.

[6] J. Pillay, I. Den Braber, N. Vrisekoop, L.M. Kwast, R.J. De Boer, J.A.M. Borghans, et al., In vivo labeling with 2H2O reveals a human neutrophil lifespan of 5.4 days, Blood 116 (2010) 625–627, doi:10.1182/BLOOD-2010-01-259028.

[7] M. Casanova-Acebes, C. Pitaval, L.A. Weiss, C. Nombela-Arrieta, R. Chèvre, N. A-González, et al., Rhythmic modulation of the hematopoietic niche through neutrophil clearance, Cell 153 (2013) 1025, doi:10.1016/j.cell.2013.04.040.

[8] L.G. Ng, I. Ballesteros, M.A. Cassatella, M. Egeblad, Z.G. Fridlender, D. Gabrilovich, et al., From complexity to consensus: a roadmap for neutrophil classification, Immunity 58 (2025) 1890–1903, doi:10.1016/j.immuni.2025.07.011.

[9] W. He, L. Yan, D. Hu, J. Hao, Y.-C. Liou, G. Luo, Neutrophil heterogeneity and plasticity: unveiling the multifaceted roles in health and disease, MedComm (Beijing) 6 (2025) e70063, doi:10.1002/mco2.70063.

[10] G. Christoffersson, E. Vågesjö, J. Vandooren, M. Lidén, S. Massena, R.B. Reinert, et al., VEGF-A recruits a proangiogenic MMP-9–delivering neutrophil subset that induces angiogenesis in transplanted hypoxic tissue, Blood 120 (2012) 4653, doi:10.1182/BLOOD-2012-04-421040.

[11] S. Hu, F. Zhang, J. Wang, J. Zhang, C. Li, Y. Lyu, et al., MMP9High neutrophils are critical mediators of neutrophil extracellular traps formation and myocardial ischemia/reperfusion injury, Adv. Sci. 12 (2025), doi:10.1002/ADVS.202415205.

[12] A.H. Mahmoud, M.A.C. Olivieri, J.P. Bannantine, G.S. Abdellrazeq, Neutrophil biology and fate reprogramming in bacterial infections: mechanisms, plasticity, and therapeutic targeting, J. Leukoc. Biol. 118 (2026) 1–19, doi:10.1093/jleuko/qiag033.
[13] X. Lai, J. Zhong, C. Liao, B. Chen, W. Zhang, R. Liao, miRNAs-neutrophil axis: novel insights into acute lung injury and chronic inflammatory lung diseases, Front. Immunol. 17 (2026) 1803762, doi:10.3389/FIMMU.2026.1803762/PDF.
[14] H. Luyang, F. Zeng, Y. Lei, Q. He, Y. Zhou, J. Xu, Bidirectional role of neutrophils in tumor development, Mol. Cancer 24 (2025), doi:10.1186/S12943-025-02228-7.
[15] Y. Zhou, H. Shan, Y. Pei, W. Zhang, M. Xu, Neutrophil extracellular traps: the hidden driver of gastrointestinal cancer and emerging therapeutic strategies, Front. Immunol. 17 (2026) 1793013, doi:10.3389/FIMMU.2026.1793013/PDF.
[16] J.B. Cowland, N. Borregaard, Granulopoiesis and granules of human neutrophils, Immunol. Rev. 273 (2016) 11–28, doi:10.1111/IMR.12440.
[17] K. Theilgaard-Mönch, L.C. Jacobsen, R. Borup, T. Rasmussen, M.D. Bjerregaard, F.C. Nielsen, et al., The transcriptional program of terminal granulocytic differentiation, Blood 105 (2005) 1785–1796, doi:10.1182/BLOOD-2004-08-3346.
[18] L. Grassi, F. Pourfarzad, S. Ullrich, A. Merkel, F. Were, E. Carrillo-de-Santa-Pau, et al., Dynamics of transcription regulation in human bone marrow myeloid differentiation to mature blood neutrophils, Cell Rep. 24 (2018) 2784–2794, doi:10.1016/j.celrep.2018.08.018.
[19] P.E. Van Den Steen, B. Dubois, I. Nelissen, P.M. Rudd, R.A. Dwek, G. Opdenakker, Critical reviews in biochemistry and molecular biology biochemistry and molecular biology of gelatinase B or matrix metalloproteinase-9 (MMP-9), Crit. Rev. Biochem. Mol. Biol. 37 (2002) 375–536, doi:10.1080/10409230290771546.
[20] J.B. Cowland, N. Borregaard, The individual regulation of granule protein mRNA levels during neutrophil maturation explains the heterogeneity of neutrophil granules, J. Leukoc. Biol. 66 (1999) 989–995, doi:10.1002/jlb.66.6.989.
[21] T. Kang, J. Yi, A. Guo, X. Wang, C.M. Overall, W. Jiang, et al., Subcellular distribution and cytokine- and chemokine-regulated secretion of leukolysin/MT6-MMP/MMP-25 in neutrophils, J. Biol. Chem. 276 (2001) 21960–21968, doi:10.1074/jbc.M007997200.
[22] C.F. Fortin, A. Sohail, Q. Sun, P.P. McDonald, R. Fridman, T. Fülöp, MT6-MMP is present in lipid rafts and faces inward in living human PMNs but translocates to the cell surface during neutrophil apoptosis, Int. Immunol. 22 (2010) 637–649, doi:10.1093/INTIMM/DXQ048.
[23] H. Zhao, A. Sohail, Q. Sun, Q. Shi, S. Kim, S. Mobashery, et al., Identification and role of the homodimerization interface of the glycosylphosphatidylinositol-anchored membrane type 6 matrix metalloproteinase (MMP25), J. Biol. Chem. 283 (2008) 35023–35032, doi:10.1074/jbc.M806553200.
[24] A.M. Greenbaum, D.C. Link, Mechanisms of G-CSF-mediated hematopoietic stem and progenitor mobilization, Leukemia 25 (2010) 211–217, doi:10.1038/leu.2010.248.
[25] L.M. Pelus, H. Bian, A.G. King, S. Fukuda, Neutrophil-derived MMP-9 mediates synergistic mobilization of hematopoietic stem and progenitor cells by the combination of G-CSF and the chemokines GROβ/CXCL2 and GROβT /CXCL2Δ4, Blood 103 (2004) 110–119, doi:10.1182/BLOOD-2003-04-1115.
[26] J.F.M. Pruijt, W.E. Fibbe, L. Laterveer, R.A. Pieters, I.J.D. Lindley, L. Paemen, et al., Prevention of interleukin-8-induced mobilization of hematopoietic progenitor cells in rhesus monkeys by inhibitory antibodies against the metalloproteinase gelatinase B (MMP-9), Proc. Natl. Acad. Sci. U.S.A. 96 (1999) 10863–10868, doi:10.1073/PNAS.96.19.10863;SUBPAGE:STRING:FULL.
[27] B. Heissig, K. Hattori, S. Dias, M. Friedrich, B. Ferris, N.R. Hackett, et al., Recruitment of stem and progenitor cells from the bone marrow niche requires MMP-9 mediated release of Kit-ligand, Cell 109 (2002) 625–637, doi:10.1016/S0092-8674(02)00754-7.

[28] K.R. Georgiou, M.A. Scherer, T.J. King, B.K. Foster, C.J. Xian, Deregulation of the CXCL12/CXCR4 axis in methotrexate chemotherapy-induced damage and recovery of the bone marrow microenvironment, Int. J. Exp. Pathol. 93 (2012) 104–114, doi:10.1111/j.1365-2613.2011.00800.x.
[29] C. Steinl, M. Essl, T.D. Schreiber, K. Geiger, L. Prokop, S. Stevanović, et al., Release of matrix metalloproteinase-8 during physiological trafficking and induced mobilization of human hematopoietic stem cells, Stem Cells Dev. 22 (2013) 1307–1318, doi:10.1089/scd.2012.0063.
[30] J.P. Levesque, F. Liu, P.J. Simmons, T. Betsuyaku, R.M. Senior, C. Pham, et al., Characterization of hematopoietic progenitor mobilization in protease-deficient mice, Blood 104 (2004) 65–72, doi:10.1182/BLOOD-2003-05-1589.
[31] J.F.M. Pruijt, P. Verzaal, R. Van Os, E. De Kruijf, M.L.J. Van Schie, A. Mantovani, et al., Neutrophils are indispensable for hematopoietic stem cell mobilization induced by interleukin-8 in mice, Proc. Natl. Acad. Sci. U.S.A. 99 (2002) 6228, doi:10.1073/PNAS.092112999.
[32] J. Kobuch, H. Cui, B. Grünwald, P. Saftig, P.A. Knolle, A. Krüger, TIMP-1 signaling via CD63 triggers granulopoiesis and neutrophilia in mice, Haematologica 100 (2015) 1005–1013, doi:10.3324/HAEMATOL.2014.121590.
[33] M.B. Lawrence, D.F. Bainton, T.A. Springer, Neutrophil tethering to and rolling on E-selectin are separable by requirement for L-selectin, Immunity 1 (1994) 137–145, doi:10.1016/1074-7613(94)90107-4.
[34] M. Lerchenberger, B. Uhl, K. Stark, G. Zuchtriegel, A. Eckart, M. Miller, et al., Matrix metalloproteinases modulate ameboid-like migration of neutrophils through inflamed interstitial tissue, Blood 122 (2013) 770–780, doi:10.1182/BLOOD-2012-12-472944.
[35] M. Phillipson, P. Kubes, The neutrophil in vascular inflammation, Nat. Med. 17 (2011) 1381–1390, doi:10.1038/nm.2514.
[36] S. Nourshargh, R. Alon, Leukocyte migration into inflamed tissues, Immunity 41 (2014) 694–707, doi:10.1016/j.immuni.2014.10.008.
[37] C. Fernandez-Patron, M.W. Radomski, S.T. Davidge, Vascular matrix metalloproteinase-2 cleaves big endothelin-1 yielding a novel vasoconstrictor, Circ. Res. 85 (1999) 906–911, doi:10.1161/01.RES.85.10.906.
[38] C. Fernandez-Patron, C. Zouki, R. Whittal, J.S.D. Chan, S.T. Davidge, J.G. Filep, Matrix metalloproteinases regulate neutrophil-endothelial cell adhesion through generation of endothelin-1[1-32], FASEB J. 15 (2001) 2230–2240, doi:10.1096/FJ.01-0178COM.
[39] C. Felkel, U. Schöll, M. Mäder, P. Schwartz, K. Felgenhauer, R. Hardeland, et al., Migration of human granulocytes through reconstituted basement membrane is not dependent on matrix metalloproteinase-9 (MMP-9), J. Neuroimmunol. 116 (2001) 49–55, doi:10.1016/S0165-5728(01)00294-6.
[40] A.J. Mackarel, D.C. Cottell, K.J. Russell, M.X. FitzGerald, C.M. O'Connor, Migration of neutrophils across human pulmonary endothelial cells is not blocked by matrix metalloproteinase or serine protease inhibitors, Am. J. Respir. Cell Mol. Biol. 20 (1999) 1209–1219, doi:10.1165/AJRCMB.20.6.3539.
[41] T. Betsuyaku, J.M. Shipley, Z. Liu, R.M. Senior, Neutrophil emigration in the lungs, peritoneum, and skin does not require gelatinase B, J. Respir. Cell Mol. Biol. 20 (1999) 1303–1309.
[42] C.A. Reichel, M. Rehberg, P. Bihari, C.M. Moser, S. Linder, A. Khandoga, et al., Gelatinases mediate neutrophil recruitment in vivo: evidence for stimulus specificity and a critical role in collagen IV remodeling, J. Leukoc. Biol. 83 (2008) 864–874, doi:10.1189/JLB.1007666.
[43] P.A. Quintero, M.D. Knolle, L.F. Cala, Y. Zhuang, C.A. Owen, Matrix metalloproteinase-8 inactivates macrophage inflammatory protein-1α to reduce acute lung inflammation and injury in mice, J. Immunol. 184 (2010) 1575–1588, doi:10.4049/JIMMUNOL.0900290.

[44] E. Kolaczkowska, W. Grzybek, N. van Rooijen, H. Piccard, B. Plytycz, B. Arnold, et al., Neutrophil elastase activity compensates for a genetic lack of matrix metalloproteinase-9 (MMP-9) in leukocyte infiltration in a model of experimental peritonitis, J. Leukoc. Biol. 85 (2009) 374–381, doi:10.1189/jlb.0808460.

[45] J. Chen, Z. Zhu, Q. Li, Y. Lin, E. Dang, H. Meng, et al., Neutrophils enhance cutaneous vascular dilation and permeability to aggravate psoriasis by releasing matrix metallopeptidase 9, J. Invest. Dermatol. 141 (2021) 787–799, doi:10.1016/J.JID.2020.07.028.

[46] T. Keck, J.H. Balcom, C. Fernández-del Castillo, B.A. Antoniu, A.L. Warshaw, Matrix metalloproteinase-9 promotes neutrophil migration and alveolar capillary leakage in pancreatitis-associated lung injury in the rat, Gastroenterology 122 (2002) 188–201, doi:10.1053/gast.2002.30348.

[47] A. Gaggar, P.L. Jackson, B.D. Noerager, P.J. O'Reilly, D.B. McQuaid, S.M. Rowe, et al., A novel proteolytic cascade generates an extracellular matrix-derived chemoattractant in chronic neutrophilic inflammation, J. Immunol. 180 (2008) 5662–5669, doi:10.4049/jimmunol.180.8.5662.

[48] P.J. O'Reilly, M.T. Hardison, P.L. Jackson, X. Xu, R.J. Snelgrove, A. Gaggar, et al., Neutrophils contain prolyl endopeptidase and generate the chemotactic peptide, PGP, from collagen, J. Neuroimmunol. 217 (2009) 51–54, doi:10.1016/J.JNEUROIM.2009.09.020.

[49] R.J. Snelgrove, P.L. Jackson, M.T. Hardison, B.D. Noerager, A. Kinloch, A. Gaggar, et al., A critical role for LTA4H in limiting chronic pulmonary neutrophilic inflammation, Science 330 (2010) 90, doi:10.1126/SCIENCE.1190594.

[50] X. Xu, P.L. Jackson, S. Tanner, M.T. Hardison, M.A. Roda, J.E. Blalock, et al., A self-propagating matrix metalloprotease-9 (MMP-9) dependent cycle of chronic neutrophilic inflammation, PLoS One 6 (2011) e15781, doi:10.1371/JOURNAL.PONE.0015781.

[51] D.A. Rigby, D.J.P. Ferguson, L.A. Johnson, D.G. Jackson, Neutrophils rapidly transit inflamed lymphatic vessel endothelium via integrin-dependent proteolysis and lipoxin-induced junctional retraction, J. Leukoc. Biol. 98 (2015) 897–912, doi:10.1189/JLB.1HI0415-149R.

[52] J. Pugin, M.C. Widmer, S. Kossodo, C.M. Liang, H.L. Preas, A.F. Suffredini, Human neutrophils secrete gelatinase B in vitro and in vivo in response to endotoxin and proinflammatory mediators, Am. J. Respir. Cell Mol. Biol. 20 (1999) 458–464, doi:10.1165/AJRCMB.20.3.3311.

[53] H. Sengeløv, L. Kjeldsen, N. Borregaard, Control of exocytosis in early neutrophil activation, J. Immunol. 150 (1993) 1535–1543, doi:10.4049/JIMMUNOL.150.4.1535.

[54] O. Gasser, C. Hess, S. Miot, C. Deon, J.C. Sanchez, J.A. Schifferli, Characterisation and properties of ectosomes released by human polymorphonuclear neutrophils, Exp. Cell. Res. 285 (2003) 243–257, doi:10.1016/S0014-4827(03)00055-7.

[55] M.N. Blumenthal, W. Zhong, M. Miller, C. Wendt, J.E. Connett, D. Pei, Serum metalloproteinase leukolysin (MMP-25/MT-6): a potential metabolic marker for atopy-associated inflammation, Clin. Exp. Allergy 40 (2010) 859–866, doi:10.1111/J.1365-2222.2010.03475.X.

[56] S. Chakrabarti, J.M. Zee, K.D. Patel, Regulation of matrix metalloproteinase-9 (MMP-9) in TNF-stimulated neutrophils: novel pathways for tertiary granule release, J. Leukoc. Biol. 79 (2006) 214–222, doi:10.1189/JLB.0605353.

[57] J.L. Johnson, H. Hong, J. Monfregola, W.B. Kiosses, S.D. Catz, Munc13-4 restricts motility of Rab27a-expressing vesicles to facilitate lipopolysaccharide-induced priming of exocytosis in neutrophils, J. Biol. Chem. 286 (2011) 5647–5656, doi:10.1074/jbc.M110.184762.

[58] H.M. Burt, J.K. Jackson, The priming action of tumour necrosis factor-alpha (TNF-α) and granulocyte-macrophage colony-stimulating factor (GM-CSF) on neutrophils activated by inflammatory microcrystals, Clin. Exp. Immunol. 108 (2003) 432–437, doi:10.1046/J.1365-2249.1997.4011298.X.
[59] L. Brandolini, R. Bertini, C. Bizzarri, R. Sergi, G. Caselli, D. Zhou, et al., IL-1β primes IL-8-activated human neutrophils for elastase release, phospholipase D activity, and calcium flux, J. Leukoc. Biol. 59 (1996) 427–434, doi:10.1002/JLB.59.3.427.
[60] K.R. McLeish, M.L. Merchant, T. Michael Creed, S. Tandon, M.T. Barati, S.M. Uriarte, et al., Frontline science: tumor necrosis factor-α stimulation and priming of human neutrophil granule exocytosis, J. Leukoc. Biol. 102 (2017) 19–29, doi:10.1189/JLB.3HI0716-293RR.
[61] I. Ernens, D. Rouy, E. Velot, Y. Devaux, D.R. Wagner, Adenosine inhibits matrix metalloproteinase-9 secretion by neutrophils: implication of A2a receptor and cAMP/PKA/Ca2+ pathway, Circ. Res. 99 (2006) 590–597, doi:10.1161/01.RES.0000241428.82502.D4.
[62] A. Matsuda, Y. Itoh, N. Koshikawa, T. Akizawa, I. Yana, M. Seiki, Clusterin, an abundant serum factor, is a possible negative regulator of MT6-MMP/MMP-25 produced by neutrophils, J. Biol. Chem. 278 (2003) 36350–36357, doi:10.1074/jbc.M301509200.
[63] P. Proost, S. Struyf, J. Van Damme, P. Fiten, E. Ugarte-Berzal, G. Opdenakker, Chemokine isoforms and processing in inflammation and immunity, J. Autoimmun. 85 (2017) 45–57, doi:10.1016/j.jaut.2017.06.009.
[64] V. Vanheule, M. Metzemaekers, R. Janssens, S. Struyf, P. Proost, How post-translational modifications influence the biological activity of chemokines, Cytokine 109 (2018) 29–51, doi:10.1016/J.CYTO.2018.02.026.
[65] D. Young, N. Das, A. Anowai, A. Dufour, Matrix metalloproteases as influencers of the cells' social media, Int. J. Mol. Sci. 20 (2019), doi:10.3390/ijms20163847.
[66] M.-J. Sanz, P. Kubes, Neutrophil-active chemokines in in vivo imaging of neutrophil trafficking, Eur. J. Immunol. 42 (2012) 278–283, doi:10.1002/eji.201142231.
[67] S. Cambier, M. Gouwy, P. Proost, The chemokines CXCL8 and CXCL12: molecular and functional properties, role in disease and efforts towards pharmacological intervention, Cell Mol. Immunol. 20 (2023) 217–251, doi:10.1038/s41423-023-00974-6.
[68] P.E. Van den Steen, A. Wuyts, S.J. Husson, P. Proost, J. Van Damme, G. Opdenakker, Gelatinase B/MMP-9 and neutrophil collagenase/MMP-8 process the chemokines human GCP-2/CXCL6, ENA-78/CXCL5 and mouse GCP-2/LIX and modulate their physiological activities, Eur. J. Biochem. 270 (2003) 3739–3749, doi:10.1046/J.1432-1033.2003.03760.X.
[69] A.M. Tester, J.H. Cox, A.R. Connor, A.E. Starr, R.A. Dean, X.S. Puente, et al., LPS responsiveness and neutrophil chemotaxis in vivo require PMN MMP-8 activity, PLoS One 2 (2007) e312, doi:10.1371/JOURNAL.PONE.0000312.
[70] P.E. Van den Steen, P. Proost, A. Wuyts, J. Van Damme, G. Opdenakker, Neutrophil gelatinase B potentiates interleukin-8 tenfold by aminoterminal processing, whereas it degrades CTAP-III, PF-4, and GRO-alpha and leaves RANTES and MCP-2 intact, Blood 96 (2000) 2673–2681.
[71] A.E. Starr, C.L. Bellac, A. Dufour, V. Goebeler, C.M. Overall, Biochemical characterization and N-terminomics analysis of leukolysin, the membrane-type 6 matrix metalloprotease (MMP25): chemokine and vimentin cleavages enhance cell migration and macrophage phagocytic activities, J. Biol. Chem. 287 (2012) 13382–13395, doi:10.1074/jbc.M111.314179.
[72] M. Balbín, A. Fueyo, A.M. Tester, A.M. Pendás, A.S. Pitiot, A. Astudillo, et al., Loss of collagenase-2 confers increased skin tumor susceptibility to male mice, Nat. Genet. 35 (2003) 252–257, doi:10.1038/NG1249.

[73] A.M. Tester, J.H. Cox, A.R. Connor, A.E. Starr, R.A. Dean, X.S. Puente, et al., LPS responsiveness and neutrophil chemotaxis in vivo require PMN MMP-8 activity, PLoS One 2 (2007), doi:10.1371/JOURNAL.PONE.0000312.

[74] J. Song, C. Wu, X. Zhang, L.M. Sorokin, In vivo processing of CXCL5 (LIX) by matrix metalloproteinase (MMP)-2 and MMP-9 promotes early neutrophil recruitment in IL-1β-induced peritonitis, J. Immunol. 190 (2013) 401–410, doi:10.4049/jimmunol.1202286.

[75] P.E. Van den Steen, S.J. Husson, P. Proost, J. Van Damme, G. Opdenakker, Carboxyterminal cleavage of the chemokines MIG and IP-10 by gelatinase B and neutrophil collagenase, Biochem. Biophys. Res. Commun. 310 (2003) 889–896, doi:10.1016/J.BBRC.2003.09.098.

[76] J.H. Cox, R.A. Dean, C.R. Roberts, C.M. Overall, Matrix metalloproteinase processing of CXCL11/I-TAC results in loss of chemoattractant activity and altered glycosaminoglycan binding, J. Biol. Chem. 283 (2008) 19389–19399, doi:10.1074/JBC.M800266200.

[77] G.A. McQuibban, G.S. Butler, J.H. Gong, L. Bendall, C. Power, I. Clark-Lewis, et al., Matrix metalloproteinase activity inactivates the CXC chemokine stromal cell-derived factor-1, J. Biol. Chem. 276 (2001) 43503–43508, doi:10.1074/JBC.M107736200.

[78] R.V.S. Pereira, M. EzEldeen, E. Ugarte-Berzal, J. Vandooren, E. Martens, M. Gouwy, et al., Protection of stromal cell-derived factor-1 SDF-1/CXCL12 against proteases yields improved skin wound healing, Front. Immunol. 15 (2024), doi:10.3389/FIMMU.2024.1359497.

[79] A.E. Starr, A. Dufour, J. Maier, C.M. Overall, Biochemical analysis of matrix metalloproteinase activation of chemokines CCL15 and CCL23 and increased glycosaminoglycan binding of CCL16, J. Biol. Chem. 287 (2012) 5848–5860, doi:10.1074/jbc.M111.314609.

[80] F. Martinon, K. Burns, J. Tschopp, The Inflammasome: a molecular platform triggering activation of inflammatory caspases and processing of proIL-β, Mol. Cell 10 (2002) 417–426, doi:10.1016/S1097-2765(02)00599-3.

[81] U. Schönbeck, F. Mach, P. Libby, Generation of biologically active IL-1 beta by matrix metalloproteinases: a novel caspase-1-independent pathway of IL-1 beta processing, J. Immunol. 161 (1998) 3340–3346.

[82] A. Ito, A. Mukaiyama, Y. Itoh, H. Nagase, I.B. Thogersen, J.J. Enghild, et al., Degradation of interleukin 1beta by matrix metalloproteinases, J. Biol. Chem. 271 (1996) 14657–14660, doi:10.1074/jbc.271.25.14657.

[83] V. Rybakin, M. Stas, E. Ugarte-Berzal, S. Noppen, J. Vandooren, I. Van Aelst, et al., Gelatinase B/matrix metalloproteinase-9 and other neutrophil proteases switch off interleukin-2 activity, Biochem. J. 476 (2019) 2191–2208, doi:10.1042/BCJ20180382.

[84] J. Vandooren, R.V.S. Pereira, E. Ugarte-Berzal, V. Rybakin, S. Noppen, M.R. Stas, et al., Internal disulfide bonding and glycosylation of interleukin-7 protect against proteolytic inactivation by neutrophil metalloproteinases and serine proteases, Front. Immunol. 12 (2021), doi:10.3389/fimmu.2021.701739.

[85] I. Nelissen, E. Martens, P.E. Van den Steen, P. Proost, I. Ronsse, G. Opdenakker, Gelatinase B/matrix metalloproteinase-9 cleaves interferon-beta and is a target for immunotherapy, Brain 126 (2003) 1371–1381, doi:10.1093/BRAIN/AWG129.

[86] C.S. De Paiva, K.C. Yoon, S.B. Pangelinan, S. Pham, L.M. Puthenparambil, E.Y. Chuang, et al., Cleavage of functional IL-2 receptor alpha chain (CD25) from murine corneal and conjunctival epithelia by MMP-9, J. Inflamm. 6 (2009) 31, doi:10.1186/1476-9255-6-31.

[87] S. Orlando, M. Sironi, G. Bianchi, A.H. Drummond, D. Boraschi, D. Yabes, et al., Role of metalloproteases in the release of the IL-1 type II decoy receptor, J. Biol. Chem. 272 (1997) 31764–31769, doi:10.1074/jbc.272.50.31764.

[88] Q. Yu, I. Stamenkovic, Cell surface-localized matrix metalloproteinase-9 proteolytically activates TGF-β and promotes tumor invasion and angiogenesis, Genes Dev. 14 (2000) 163–176, doi:10.1101/GAD.14.2.163.

[89] S.L. Dallas, J.L. Rosser, G.R. Mundy, L.F. Bonewald, Proteolysis of latent transforming growth factor-β (TGF-β)-binding protein-1 by osteoclasts, J. Biol. Chem. 277 (2002) 21352–21360, doi:10.1074/jbc.m111663200.

[90] Y. Weinrauch, D. Drujan, S.D. Shapiro, J. Weiss, A. Zychlinsky, Neutrophil elastase targets virulence factors of enterobacteria, Nature 417 (2002) 91–94, doi:10.1038/417091a.

[91] E. Naish, A.J.T. Wood, A.P. Stewart, M. Routledge, A.C. Morris, E.R. Chilvers, et al., The formation and function of the neutrophil phagosome, Immunol. Rev. 314 (2022) 158, doi:10.1111/IMR.13173.

[92] G.T. Nguyen, E.R. Green, J. Mecsas, Neutrophils to the ROScue: mechanisms of NADPH oxidase activation and bacterial resistance, Front. Cell. Infect. Microbiol. 7 (2017) 264679, doi:10.3389/FCIMB.2017.00373/FULL.

[93] C. Nathan, Neutrophils and immunity: challenges and opportunities, Nat. Rev. Immunol. 6 (2006) 173–182, doi:10.1038/NRI1785;KWRD.

[94] A. Georgiadou, H.J. Lee, M. Walther, A.E. van Beek, F. Fitriani, D. Wouters, et al., Modelling pathogen load dynamics to elucidate mechanistic determinants of host–Plasmodium falciparum interactions, Nat. Microbiol. 4 (2019) 1592–1602, doi:10.1038/s41564-019-0474-x.

[95] C.W.M. Ong, P.T. Elkington, S. Brilha, C. Ugarte-Gil, M.T. Tome-Esteban, L.B. Tezera, et al., Neutrophil-derived MMP-8 drives AMPK-dependent matrix destruction in human pulmonary tuberculosis, PLoS Pathog. 11 (2015) e1004917, doi:10.1371/JOURNAL.PPAT.1004917.

[96] V.N.C. Leal, F. Bork, M. Mateo Tortola, J.C. von Guilleaume, C.L. Greve, S. Bugl, et al., Bruton's tyrosine kinase (BTK) and matrix metalloproteinase-9 (MMP-9) regulate NLRP3 inflammasome-dependent cytokine and neutrophil extracellular trap responses in primary neutrophils, J. Allergy Clin. Immunol. 155 (2025) 569–582, doi:10.1016/j.jaci.2024.10.035.

[97] J. Albrengues, M.A. Shields, D. Ng, C.G. Park, A. Ambrico, M.E. Poindexter, et al., Neutrophil extracellular traps produced during inflammation awaken dormant cancer cells in mice, Science (1979) 361 (2018), doi:10.1126/science.aao4227.

[98] Y. Wang, C.Y. Chuang, C.L. Hawkins, M.J. Davies, Activation and Inhibition of Human Matrix Metalloproteinase-9 (MMP9) by HOCl, Myeloperoxidase and Chloramines, Antioxidants 11 (2022) 1616, doi:10.3390/ANTIOX11081616/S1.

[99] B. Cauwe, G. Opdenakker, Intracellular substrate cleavage: a novel dimension in the biochemistry, biology and pathology of matrix metalloproteinases, Crit. Rev. Biochem. Mol. Biol. 45 (2010) 351–423, doi:10.3109/10409238.2010.501783.

[100] K.Y. Deleon-Pennell, Y. Tian, B. Zhang, C.A. Cates, R.P. Iyer, P. Cannon, et al., CD36 is a matrix metalloproteinase-9 substrate that stimulates neutrophil apoptosis and removal during cardiac remodeling, Circ. Cardiovasc. Genet. 9 (2016) 14–25, doi:10.1161/CIRCGENETICS.115.001249;JOURNAL:JOURNAL:CIRCGENETICS;WGROUP:STRING:PUBLICATION.

[101] V.A. Fadok, M.L. Warner, D.L. Bratton, P.M. Henson, CD36 is required for phagocytosis of apoptotic cells by human macrophages that use either a phosphatidylserine receptor or the vitronectin receptor ($\alpha v\beta 3$), J. Immunol. 161 (1998) 6250–6257, doi:10.4049/JIMMUNOL.161.11.6250.

[102] V.A. Fadok, D.L. Bratton, A. Konowal, P.W. Freed, J.Y. Westcott, P.M. Henson, Macrophages that have ingested apoptotic cells in vitro inhibit proinflammatory cytokine production through autocrine/paracrine mechanisms involving TGF-beta, PGE2, and PAF, J. Clin. Invest. 101 (1998) 890, doi:10.1172/JCI1112.

[103] L.A. Salamonsen, D.E. Woolley, Menstruation: induction by matrix metalloproteinases and inflammatory cells, J. Reprod. Immunol. 44 (1999) 1–27, doi:10.1016/S0165-0378(99)00002-9.

[104] L.J. Lathbury, L.A. Salamonsen, In-vitro studies of the potential role of neutrophils in the process of menstruation, Mol. Hum. Reprod. 6 (2000) 899–906, doi:10.1093/MOLEHR/6.10.899.

[105] T.J. Kaitu'u, J. Shen, J. Zhang, N.B. Morison, L.A. Salamonsen, Matrix metalloproteinases in endometrial breakdown and repair: functional significance in a mouse model, Biol. Reprod. 73 (2005) 672–680, doi:10.1095/BIOLREPROD.105.042473.

[106] N. Wohner, Z. Keresztes, P. Sótonyi, L. Szabó, E. Komorowicz, R. Machovich, et al., Neutrophil granulocyte-dependent proteolysis enhances platelet adhesion to the arterial wall under high-shear flow, J. Thromb. Haemost. 8 (2010) 1624–1631, doi:10.1111/j.1538-7836.2010.03890.x.

[107] E. Laridan, F. Denorme, L. Desender, O. François, T. Andersson, H. Deckmyn, et al., Neutrophil extracellular traps in ischemic stroke thrombi, Ann. Neurol. 82 (2017) 223–232, doi:10.1002/ANA.24993;JOURNAL:JOURNAL:15318249;CTYPE:STRING:JOURNAL.

[108] T.A. Fuchs, A. Brill, D. Duerschmied, D. Schatzberg, M. Monestier, D.D. Myers, et al., Extracellular DNA traps promote thrombosis, Proc. Natl. Acad. Sci. U.S.A. 107 (2010) 15880–15885, doi:10.1073/PNAS.1005743107.

[109] X. Li, O.J. De Boer, H. Ploegmaker, P. Teeling, M.J. Daemen, R.J. De Winter, et al., Granulocytes in coronary thrombus evolution after myocardial infarction—time-dependent changes in expression of matrix metalloproteinases, Cardiovasc. Pathol. 25 (2016) 40–46, doi:10.1016/j.carpath.2015.09.007.

[110] A.G. Harris, T.C. Skalak, Leukocyte cytoskeletal structure determines capillary plugging and network resistance, Am. J. Physiol. 265 (1993) H1670–H1675, doi:10.1152/ajpheart.1993.265.5.H1670.

[111] A.Y. Chen, F.A. DeLano, S.R. Valdez, J.N. Ha, H.Y. Shin, G.W. Schmid-Schönbein, Receptor cleavage reduces the fluid shear response in neutrophils of the spontaneously hypertensive rat, Am. J. Physiol. Cell Physiol. 299 (2010) 1441–1449, doi:10.1152/AJPCELL.00157.2010.

[112] R. Zhang, J. Zhang, Y.L. Zhang, S.G. Gong, Q.H. Zhao, X.J. Wang, et al., Single-cell transcriptome analysis of peripheral neutrophils from patients with idiopathic pulmonary arterial hypertension, Hypertension 80 (2023) 1784–1794, doi:10.1161/HYPERTENSIONAHA.123.21142.

[113] R. Pyo, J.K. Lee, J.M. Shipley, J.A. Curci, D. Mao, S.J. Ziporin, et al., Targeted gene disruption of matrix metalloproteinase-9 (gelatinase B) suppresses development of experimental abdominal aortic aneurysms, J. Clin. Invest. 105 (2000) 1641–1649, doi:10.1172/JCI8931.

[114] J.L. Eliason, K.K. Hannawa, G. Ailawadi, I. Sinha, J.W. Ford, M.P. Deogracias, et al., Neutrophil depletion inhibits experimental abdominal aortic aneurysm formation, Circulation 112 (2005) 232–240, doi:10.1161/CIRCULATIONAHA.104.517391.

[115] K.M. Beeh, J. Beier, O. Kornmann, R. Buhl, Sputum matrix metalloproteinase-9, tissue inhibitor of metalloprotinease-1, and their molar ratio in patients with chronic obstructive pulmonary disease, idiopathic pulmonary fibrosis and healthy subjects, Respir. Med. 97 (2003) 634–639, doi:10.1053/rmed.2003.1493.

[116] J.H.J. Vernooy, J.H.N. Lindeman, J.A. Jacobs, R. Hanemaaijer, E.F.M. Wouters, Increased activity of matrix metalloproteinase-8 and matrix metalloproteinase-9 in induced sputum from patients with COPD, Chest 126 (2004) 1802–1810, doi:10.1378/chest.126.6.1802.

[117] E.A.B. Kelly, W.W. Busse, N.N. Jarjour, Increased matrix metalloproteinase-9 in the airway after allergen challenge, Am. J. Respir. Crit. Care Med. 162 (2000) 1157–1161, doi:10.1164/AJRCCM.162.3.9908016.

[118] P.C. Goeminne, J. Vandooren, E.A. Moelants, A. Decraene, E. Rabaey, A. Pauwels, et al., The Sputum Colour Chart as a predictor of lung inflammation, proteolysis and damage in non-cystic fibrosis bronchiectasis: a case-control analysis, Respirology 19 (2014) 203–210, doi:10.1111/RESP.12219.

[119] J. Vandooren, P. Goeminne, L. Boon, E. Ugarte-Berzal, V. Rybakin, P. Proost, et al., Neutrophils and activated macrophages control mucosal immunity by proteolytic cleavage of antileukoproteinase, Front. Immunol. 9 (2018), doi:10.3389/FIMMU.2018.01154.

[120] P.L. Jackson, X. Xu, L. Wilson, N.M. Weathington, J.P. Clancy, J.E. Blalock, et al., Human neutrophil elastase-mediated cleavage sites of MMP-9 and TIMP-1: implications to cystic fibrosis proteolytic dysfunction, Mol. Med. 16 (2010) 159–166, doi:10.2119/MOLMED.2009.00109.

[121] T.S. Wilkinson, A. Conway Morris, K. Kefala, C.M. O'Kane, N.R. Moore, N.A. Booth, et al., Ventilator-associated pneumonia is characterized by excessive release of neutrophil proteases in the lung, Chest 142 (2012) 1425–1432, doi:10.1378/chest.11-3273.

[122] J. Blázquez-Prieto, I. López-Alonso, L. Amado-Rodríguez, C. Huidobro, A. González-López, W.M. Kuebler, et al., Impaired lung repair during neutropenia can be reverted by matrix metalloproteinase-9, Thorax 73 (2018) 321–330, doi:10.1136/THORAXJNL-2017-210105.

[123] H. Lukkarinen, A. Hogmalm, U. Lappalainen, K. Bry, Matrix metalloproteinase-9 deficiency worsens lung injury in a model of bronchopulmonary dysplasia, Am. J. Respir. Cell Mol. Biol. 41 (2009) 59–68, doi:10.1165/RCMB.2008-0179OC.

[124] J. Blázquez-Prieto, I. López-Alonso, C. Huidobro, G.M. Albaiceta, The emerging role of neutrophils in repair after acute lung injury, Am. J. Respir. Cell Mol. Biol. 59 (2018) 289–294, doi:10.1165/RCMB.2018-0101PS.

[125] T. Duan, Y. Du, C. Xing, H.Y. Wang, R.F. Wang, Toll-like receptor signaling and its role in cell-mediated immunity, Front. Immunol. 13 (2022), doi:10.3389/FIMMU.2022.812774.

[126] K. Vijay, Toll-like receptors in immunity and inflammatory diseases: past, present, and future, Int. Immunopharmacol. 59 (2018) 391–412, doi:10.1016/j.intimp.2018.03.002.

[127] N.I. Medeiros, R.C.G. Fares, E.P. Franco, G.R. Sousa, R.T. Mattos, A.T. Chaves, et al., Differential expression of matrix metalloproteinases 2, 9 and cytokines by neutrophils and monocytes in the clinical forms of chagas disease, PLoS Negl. Trop. Dis. 11 (2017), doi:10.1371/JOURNAL.PNTD.0005284.

[128] L.M. Bradley, M.F. Douglass, D. Chatterjee, S. Akira, B.J.G. Baaten, Matrix metalloprotease 9 mediates neutrophil migration into the airways in response to influenza virus-induced toll-like receptor signaling, PLoS Pathog. 8 (2012) e1002641, doi:10.1371/JOURNAL.PPAT.1002641.

[129] R. Kaiser, C. Gold, M. Joppich, Q. Loew, A. Akhalkatsi, T.T. Mueller, et al., Peripheral priming induces plastic transcriptomic and proteomic responses in circulating neutrophils required for pathogen containment, Sci. Adv. 10 (2024), doi:10.1126/SCIADV.ADL1710.

[130] K.N. Kangelaris, A. Prakash, K.D. Liu, B. Aouizerat, P.G. Woodruff, D.J. Erle, et al., Increased expression of neutrophil-related genes in patients with early sepsis-induced ARDS, Am. J. Physiol. Lung Cell. Mol. Physiol. 308 (2015) L1102–L1113, doi:10.1152/AJPLUNG.00380.2014.

[131] M. Rahman, S. Zhang, M. Chew, A. Ersson, B. Jeppsson, H. Thorlacius, Platelet-derived CD40L (CD154) mediates neutrophil upregulation of Mac-1 and recruitment in septic lung injury, Ann. Surg. 250 (2009) 783–789, doi:10.1097/SLA.0B013E3181BD95B7.

[132] M. Rahman, S. Zhang, M. Chew, I. Syk, B. Jeppsson, H. Thorlacius, Platelet shedding of CD40L is regulated by matrix metalloproteinase-9 in abdominal sepsis, J. Thromb. Haemost. 11 (2013) 1385–1398, doi:10.1111/jth.12273.

[133] R. Hwaiz, M. Rahman, E. Zhang, H. Thorlacius, Rac1 regulates platelet shedding of CD40L in abdominal sepsis, Lab. Invest. 94 (2014) 1054–1063, doi:10.1038/labinvest.2014.92.

[134] J. Hu, P.E. Van Den Steen, C. Dillen, G. Opdenakker, Targeting neutrophil collagenase/matrix metalloproteinase-8 and gelatinase B/matrix metalloproteinase-9 with a peptidomimetic inhibitor protects against endotoxin shock, Biochem. Pharmacol. 70 (2005) 535–544, doi:10.1016/J.BCP.2005.04.047.

[135] M. Schiwon, C. Weisheit, L. Franken, S. Gutweiler, A. Dixit, C. Meyer-Schwesinger, et al., Crosstalk between sentinel and helper macrophages permits neutrophil migration into infected uroepithelium, Cell 156 (2014) 456–468, doi:10.1016/j.cell.2014.01.006.

[136] N. Sabir, T. Hussain, M.H. Mangi, D. Zhao, X. Zhou, Matrix metalloproteinases: expression, regulation and role in the immunopathology of tuberculosis, Cell Prolif. 52 (2019), doi:10.1111/CPR.12649.

[137] R. Cockeran, T.J. Mitchell, C. Feldman, R. Anderson, Pneumolysin induces release of matrix metalloproteinase-8 and -9 from human neutrophils, Eur. Respir. J. 34 (2009) 1167–1170, doi:10.1183/09031936.00007109.

[138] T. Okamoto, T. Akaike, M. Suga, S. Tanase, H. Horie, S. Miyajima, et al., Activation of human matrix metalloproteinases by various bacterial proteinases, J. Biol. Chem. 272 (1997) 6059–6066, doi:10.1074/jbc.272.9.6059.

[139] A.M. Romanic, R.F. White, A.J. Arleth, E.H. Ohlstein, F.C. Barone, Matrix metalloproteinase expression increases after cerebral focal ischemia in rats: inhibition of matrix metalloproteinase-9 reduces infarct size, Stroke 29 (1998) 1020–1030, doi:10.1161/01.STR.29.5.1020.

[140] G.A. Rosenberg, L.A. Cunningham, J. Wallace, S. Alexander, E.Y. Estrada, M. Grossetete, et al., Immunohistochemistry of matrix metalloproteinases in reperfusion injury to rat brain: activation of MMP-9 linked to stromelysin-1 and microglia in cell cultures, Brain Res. 893 (2001) 104–112, doi:10.1016/S0006-8993(00)03294-7.

[141] C.M. Maier, L. Hsieh, F. Yu, P. Bracci, P.H. Chan, Matrix metalloproteinase-9 and myeloperoxidase expression: quantitative analysis by antigen immunohistochemistry in a model of transient focal cerebral ischemia, Stroke 35 (2004) 1169–1174, doi:10.1161/01.STR.0000125861.55804.F2.

[142] S.H. Yoon, L.E. Kim C youn, C. Lee, K.S. Lee, J. Lee, et al., Microglial NLRP3-gasdermin D activation impairs blood-brain barrier integrity through interleukin-1β-independent neutrophil chemotaxis upon peripheral inflammation in mice, Nat. Commun. 16 (2025), doi:10.1038/S41467-025-56097-1.

[143] L.J. Noble, F. Donovan, T. Igarashi, S. Goussev, Z. Werb, Matrix metalloproteinases limit functional recovery after spinal cord injury by modulation of early vascular events, J. Neurosci. 22 (2002) 7526–7535, doi:10.1523/JNEUROSCI.22-17-07526.2002.

[144] M. Lindsey, K. Wedin, M.D. Brown, C. Keller, A.J. Evans, J. Smolen, et al., Matrix-dependent mechanism of neutrophil-mediated release and activation of matrix metalloproteinase 9 in myocardial ischemia/reperfusion, Circulation 103 (2001) 2181–2187, doi:10.1161/01.CIR.103.17.2181.

[145] P.M. Soccala, Y. Gasche, D.N. Miniati, G. Hoyt, G.J. Berry, R.L. Doyle, et al., Matrix metalloproteinase inhibition decreases ischemia-reperfusion injury after lung transplantation, Am. J. Transplant. 4 (2004) 41–50, doi:10.1046/j.1600-6135.2003.00277.x.

[146] A. Khandoga, J.S. Kessler, M. Hanschen, A.G. Khandoga, D. Burggraf, C. Reichel, et al., Matrix metalloproteinase-9 promotes neutrophil and T cell recruitment and migration in the postischemic liver, J. Leukoc. Biol. 79 (2006) 1295–1305, doi:10.1189/JLB.0805468.

[147] K. Kessenbrock, V. Plaks, Z. Werb, Matrix metalloproteinases: regulators of the tumor microenvironment, Cell 141 (2010) 52, doi:10.1016/J.CELL.2010.03.015.
[148] S. Mustafa, S. Koran, L. AlOmair, Insights into the role of matrix metalloproteinases in cancer and its various therapeutic aspects: a review, Front Mol. Biosci. 9 (2022) 896099, doi:10.3389/FMOLB.2022.896099.
[149] C.M. Overall, O. Kleifeld, Validating matrix metalloproteinases as drug targets and anti-targets for cancer therapy, Nat. Rev. Cancer 6 (2006) 227–239, doi:10.1038/NRC1821; KWRD.
[150] A.D. Gregory, A.M.G. Houghton, Tumor-associated neutrophils: new targets for cancer therapy, Cancer Res. 71 (2011) 2411–2416, doi:10.1158/0008-5472.CAN-10-2583.
[151] C. Jiguet-Jiglaire, S. Boissonneau, E. Denicolai, V. Hein, R. Lasseur, J. Garcia, et al., Plasmatic MMP9 released from tumor-infiltrating neutrophils is predictive for bevacizumab efficacy in glioblastoma patients: an AVAglio ancillary study, Acta Neuropathol. Commun. 10 (2022), doi:10.1186/S40478-021-01305-4.
[152] J. Folkman, K. Watson, D. Ingber, D. Hanahan, Induction of angiogenesis during the transition from hyperplasia to neoplasia, Nature 339 (1989) 58–61, doi:10.1038/339058A0.
[153] J.P. Shaw, N. Chuang, H. Yee, P. Shamamian, Polymorphonuclear neutrophils promote rFGF-2-induced angiogenesis in vivo, J. Surg. Res. 109 (2003) 37–42, doi:10.1016/S0022-4804(02)00020-3.
[154] E.I. Deryugina, J.P. Quigley, Tumor angiogenesis: MMP-mediated induction of intravasation- and metastasis-sustaining neovasculature, Matrix Biol. 44–46 (2015) 94–112, doi:10.1016/j.matbio.2015.04.004.
[155] V.C. Ardi, T.A. Kupriyanova, E.I. Deryugina, J.P. Quigley, Human neutrophils uniquely release TIMP-free MMP-9 to provide a potent catalytic stimulator of angiogenesis, Proc. Natl. Acad. Sci. U.S.A. 104 (2007) 20262–20267, doi:10.1073/PNAS.0706438104.
[156] T. Hu, B. Cheng, A. Matsunaga, T. Zhang, X. Lu, H. Fang, et al., Single-cell analysis defines highly specific leukemia-induced neutrophils and links MMP8 expression to recruitment of tumor associated neutrophils during FGFR1 driven leukemogenesis, Exp. Hematol. Oncol. 13 (2024), doi:10.1186/S40164-024-00514-6.
[157] L.M. Coussens, C.L. Tinkle, D. Hanahan, Z. Werb, MMP-9 supplied by bone marrow-derived cells contributes to skin carcinogenesis, Cell 103 (2000) 481–490, doi:10.1016/S0092-8674(00)00139-2.
[158] L.H. Palavalli, T.D. Prickett, J.R. Wunderlich, X. Wei, A.S. Burrell, P. Porter-Gill, et al., Analysis of the matrix metalloproteinase family reveals that MMP8 is often mutated in melanoma, Nat. Genet. 41 (2009) 518–520, doi:10.1038/ng.340.
[159] C. Hu, L. Long, J. Lou, M. Leng, Q. Yang, X. Xu, et al., CTC-neutrophil interaction: a key driver and therapeutic target of cancer metastasis, Biomed. Pharmacother. 180 (2024), doi:10.1016/j.biopha.2024.117474.
[160] M. Germann, N. Zangger, M. Sauvain, C. Sempoux, A.D. Bowler, P. Wirapati, et al., Neutrophils suppress tumor-infiltrating T cells in colon cancer via matrix metalloproteinase-mediated activation of TGFβ, EMBO Mol. Med. 12 (2020), doi:10.15252/EMMM.201910681.

CHAPTER NINE

Remodeling at the intersection of time and tissue: Aging, obesity, and the MT1-MMP proteolytic driver in ovarian cancer

Brighton C. Strasiotto[a,b], **Mary Sharon Stack**[a,b,*]

[a] Harper Cancer Research Institute, University of Notre Dame, South Bend, IN, USA
[b] Department of Chemistry and Biochemistry, University of Notre Dame, Notre Dame, IN, USA
*Corresponding author. e-mail address: Sharon.Stack.11@nd.edu.

Contents

Abstract

Ovarian cancer is the leading cause of gynecological cancer death for women in the United States. Aging and obesity are known risk factors that increase incidence rate, severity, and metastatic progression of ovarian cancer. With aging and obesity, physiological changes occur including increased inflammation, altered metabolism, and, most notably, extracellular matrix (ECM) remodeling. Remodeling of the ECM is critical for

Fungal and Protozoan Carbonic Anhydrases, Volume 60
ISSN 1874-6047
https://doi.org/10.1016/bs.enz.2026.07.002

accommodating adipocyte expansion in obesity and arises from accumulated modifications with aging. The already remodeled collagen under these host factors is driven by remodeling enzymes such as matrix metalloproteinases (MMPs). Of particular interest is membrane-bound MT1-MMP, which is required for invasion in metastatic ovarian cancer. MT1-MMP is responsible for the directed collagenolysis of the ECM, which is a barrier to metastatic growth. Because of metastatic dependence on MT1-MMP, there have been many attempts to inhibit this enzyme, but current therapies lack specificity and cause significant adverse side effects. Nonetheless, novel therapeutic interventions targeting MT1-MMP are promising to improve selectivity. In this chapter, we will highlight key pathways regulating MT1-MMP expression, impact on oncogenic phenotypes, and the mechanisms that aging and obesity use to further promote ECM remodeling and accelerate disease progression, while assessing past and future challenges to study and targeting MMPs for cancer treatment.

Keywords: Aging, Extracellular matrix remodeling, MT1-MMP (MMP14), Obesity, Ovarian cancer metastasis

1. Overview

Ovarian cancer is the leading cause of gynecological cancer death in women in the United States, with a 5-year survival rate of 32 percent for stage III/IV cancers at diagnosis [1]. While the prevalence of ovarian cancer has reduced by 45 percent since 1976, treatment options and early detection methods have remained unchanged. The median age for ovarian cancer diagnosis is 63 years old, and it is rarely diagnosed before the age of 40 [2], highlighting aging as a major risk factor for ovarian cancer. Not only does aging increase the risk of ovarian cancer incidence, it also increases the severity of disease with a reduction in survival at an advanced stage diagnosis after the age of 65 [2]. Furthermore, obesity is an additional host factor linked with greater metastatic progression and reduced survival compared to non-obese patients [3–10]. The mechanisms that contribute to increased ovarian cancer severity in aging and obesity overlap through alterations in metabolism, increased inflammation, and, as the primary focus of this review, extracellular matrix (ECM) remodeling [11–13]. Throughout normal homeostasis, the ECM is being remodeled to accommodate changes in structural reorganization and tissue size changes, and is regulated by proteolytic enzymes. The matrix metalloproteinase (MMP) family heavily contributes to collagen remodeling in the ECM, which enhances permissive tumor growth.

MMPs are a major class of ECM remodeling enzymes that contribute to tissue homeostasis under normal conditions and have been implicated in many disease states [14]. While many enzymes in this class have been shown to play a role in cancer progression, membrane type 1 MMP (MT1-MMP, MMP-14) has been implicated in the progression of ovarian cancer at multiple stages, including proliferation, angiogenesis, invasion, and metastasis [15]. MT1-MMP consists of 5 key elements within a single polypeptide containing: (1) a prodomain, (2) a zinc catalytic domain, (3) a hemopexin-like domain flanked by two linkers, (4) a transmembrane domain, and (5) a cytoplasmic tail [16]. MT1-MMPs unique ability to localize at the point of invasion in cancer cells allows for site-specific collagen remodeling, which is upregulated in cancer and metastasis [17]. Additionally, MT1-MMP has many feedback loops that upregulate other key enzymes and regulators of cancer metastasis such as proMMP-2, epithelial-mesenchymal transition (EMT) pathways, and adhesion pathways to aid in cancer progression [18]. In turn, MT1-MMP expression is induced by integrin-mediated signaling pathways activated by ovarian cancer cell adhesion to collagen after dissemination, highlighting the upregulation of MT1-MMP by cancer cells to assist invasive phenotypes [19]. Clinical data have demonstrated that MT1-MMP expression correlates with worse clinical outcomes in comparison to those with minimal MT1-MMP expression [20]. Additionally, human genetic studies have uncovered polymorphisms in MT1-MMP that correspond with poor prognosis. The +6767 G>A polymorphism has a hazard ratio of 2.09 for recurrence-free survival, highlighting the influential role MT1-MMP plays in the growth success of ovarian cancer [21]. Overall, the dependence of ovarian cancer progression on collagen remodeling heavily relies on the upregulation of MMPs, specifically MT1-MMP. Here, we cover the regulation of collagen remodeling by MT1-MMP in cancer and consider the impact of aging and obesity on MT1-MMP expression.

2. Impact on clinical cancer phenotypes

From the functional networks previously described, it is apparent that MT1-MMP is not simply a workhorse for the degradation of physical barriers but is a crucial translator of environmental cues that play a critical role in the progression of aggressive cancers. Besides catalyzing the cleavage of key ECM components, including type I collagen, MT1-MMP can also activate pro-MMP2, which leads to the remodeling of type IV collagen [22,23]. Together, these proteinases can efficiently break down the primary collagenous barriers to invasion.

Clinical data highlight the importance of MT1-MMP expression as a potential prognostic marker. Immunohistochemical studies have identified an increase in MT1-MMP expression in aggressive tumors, demonstrating a correlation with poor prognosis in ovarian cancer [17,24]. Clinical analyses revealed that heightened expression of MT1-MMP mRNA levels in tumor cells was associated with poor survival, whereas expression in surrounding stromal cells correlated with increased survival [25,26]. These studies corroborated findings that MT1-MMP was expressed at the leading edge of ovarian cancer cell cultures, highlighting its role in invasion and aggressive cellular behavior [27]. Furthermore, additional studies demonstrated that mRNA levels of MT1-MMP correlated with increased tumor size or spread in ovarian cancer pre-clinical murine models and in human tumors, respectively [28]. Together, these data provided the rationale for in depth investigations into the molecular mechanisms that regulate MT1-MMP expression in ovarian cancer and its disease-driving mechanisms.

3. Unique metastatic mechanism of ovarian cancer

Ovarian cancer is the deadliest gynecological cancer, with a lifetime risk of 1 in 87 [1]. Ovarian cancer metastasizes in a unique fashion, in which malignant cells are shed from the primary tumor site and disseminate through the peritoneal cavity, often aided by the buildup of malignant ascites fluid that facilitates metastatic dissemination [29]. Metastasis occurs as either single cells or as multicellular aggregates (MCAs) through integrin-mediated interactions with intra-peritoneal collagen [30]. Initial attachment of metastatic ovarian cancer cells relies on adherence to the mesothelial cell monolayer lining the omentum and the peritoneal cavity, followed by displacement of mesothelial cells, exposing the collagen dense sub-mesothelial matrix [30]. Ovarian cancer cells use MT1-MMP to remodel the extracellular matrix at secondary tumor sites to invade, anchor in the sub-mesothelial collagen matrix, and proliferate to form widely disseminated metastatic lesions [31]. The formation of MCAs early in dissemination allows for ovarian cancer cells to survive in hypoxic and ECM-detached ascites fluid and avoid anoikis, a form of cell death triggered by the absence of cell adhesion [30]. This event is promoted by MT1-MMP-catalyzed shedding of the α3-integrin ectodomain, which allows cells to detach as cell-cell adherent sheets rather than as anoikis-prone single cells [32]. Metastatic tumor cells preferentially bind to the omentum, a collagen-rich fatty tissue that drapes over the peritoneal cavity and acts as an immune filter, the ultrastructure of which is formed by a dense and uniquely fenestrated collagen network [29]. The milky spots of the omentum are composed of immune cells which assist in cancer cell homing to the omentum as well as support immune evasion in cancer progression [33]. Furthermore, the adipose tissue of the omentum forms a nutrient-rich environment that provides metabolic support for cancerous growth [34].

Changes to collagen ultrastructure have been shown to have clinical significance in a variety of contexts. Tumor-associated collagen remodeling can be defined using a tumor-associated collagen signature (TACS) that is based on ultrastructural changes in collagen density (TACS1), collagen alignment (TACS2), and collagen alignment in relation to the tumor (TACS3) [35–37]. These features have prognostic significance in a variety of tumors, including ovarian cancer. As initial events in ovarian cancer metastasis are initiated by interactions between cells or MCAs and sub-mesothelial collagen, exposure or remodeling of collagen promotes metastatic success [38–40]. The dysregulation of extracellular matrix structural integrity in cancer progression has provided the rationale for subsequent investigations into the regulation of remodeling enzymes in the tumor microenvironment.

4. Structure and internalization of MT1-MMP

The discovery of MT1-MMP, the first membrane-bound matrix metalloproteinase, dates back to 1994 when Sato et al. observed a correlation between the membrane-bound enzyme and an invasive cancerous phenotype [41]. Subsequent studies uncovered the structure and molecular interactions of MT1-MMP and demonstrated that MT1-MMP can effectively cleave collagen type 1, but can also process less prominent ECM components such as collagen types II and III, fibronectin, laminin-1,–2/4, and −5, and fibrinogen [42–43]. The discovery of MT1-MMP and its interactions with pro-MMP2 highlighted its potential as a tumor-promoting enzyme [44]. Structurally similar to other MMP family members (Fig. 1), the catalytic domain of MT1-MMP is adjacent to a hemopexin ectodomain that regulates substrate recognition and enzymatic activity [45], activation of proenzymes [46], and binding of ECM components and inhibitors [22,47]. Distinct from soluble MMPs, MT1-MMP contains a transmembrane domain and a cytoplasmic tail domain. The expression of MT1-MMP at specific locations of the cell surface is ensured by anchorage of the enzyme through the transmembrane domain, highlighting the importance of proper localization signals. Furthermore, the cytoplasmic tail has essential roles in the regulation

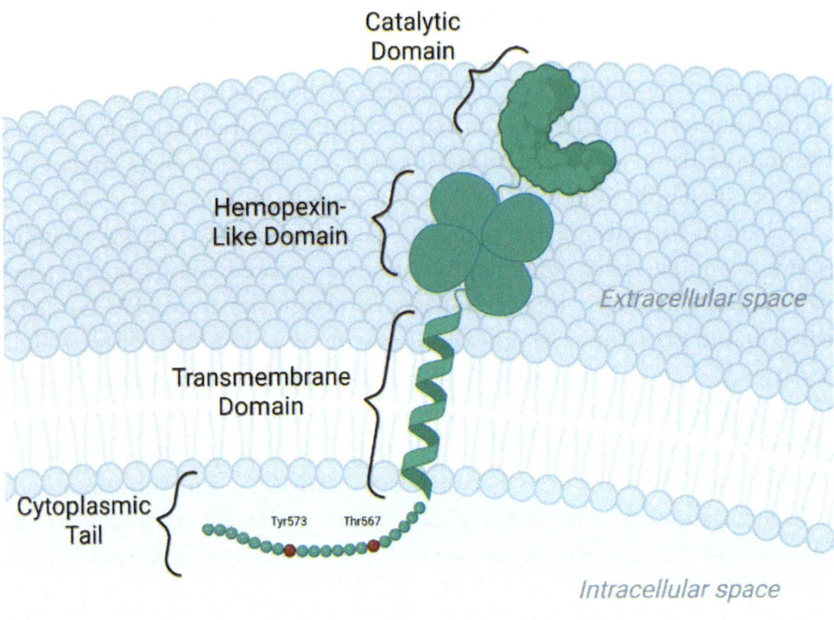

Fig. 1 Embedded Structure of MT1-MMP in the cell membrane.

of MT1-MMP expression on the surface of the cell, with its primary role in the regulation of internalization signals [16,22,32].

Regulation of MT1-MMP quaternary structure is essential for controlling MT1-MMP activity. It has been shown that homodimerization of MT1-MMP is required for effective collagenolysis at the cell surface. Itoh et al. showed that when dimerization of MT1-MMP was blocked by exogenously expressed membrane-anchored or soluble hemopexin domain, collagenolytic activity was inhibited, while general catalytic activity remained. The exact mechanism by which homodimerization of MT1-MMP promotes collagenolysis remains unknown, but there is speculation that this process facilitates the stabilization and exposure of cleavage sites in the collagen trimer [45]. Homodimerization also plays an important role in the activation of proenzymes, specifically pro-MMP2. Formation of a trimolecular complex containing an MT1-MMP homodimer and tissue inhibitor of metalloproteases-2 (TIMP-2) recruits proMMP-2 to the complex, stabilizing proMMP-2 for zymogen activation by the adjacent MT1-MMP [46].

Surface retention of MT1-MMP results in high enzymatic activity, with studies demonstrating that MT1-MMP internalization is a primary pathway for MT1-MMP catalytic activity regulation [48]. Studies have shown that truncation of the cytoplasmic tail inhibits internalization of MT1-MMP, restricting it to the cell surface, effectively promoting its activity. These studies demonstrate the crucial role of the cytoplasmic tail in regulating activity [48]. Phosphorylation sites on the cytoplasmic tail play important roles in cytoplasmic signaling and regulation of surface activity [16]. Additional data suggest that MT1-MMP is necessary for the fragmentation of collagen for phagocytosis, and intracellular co-localization of MT1-MMP with collagen fragments has been observed in invasive cancer cells, including ovarian cancer [49].

The process of MT1-MMP internalization occurs via two distinct pathways: (1) clathrin-mediated internalization and (2) caveolae-mediated internalization. The internalization of MT1-MMP via clathrin-mediated endocytosis is constitutively active to maintain homeostatic levels of proteolysis without excess remodeling. Alternatively, caveolae and recycled MT1-MMP reside near lipid rafts in the cell membrane (Fig. 2), enabling highly regulated caveolae-mediated endocytosis. The caveolae-mediated pathway can be activated by external signals, such as integrin binding to activate RhoA. Upon endocytosis, MT1-MMP can be recycled at the site of invadopodia in migrating cells and allows for directed collagenolysis [22].

5. Regulation of MT1-MMP via cytoplasmic tail phosphorylation

The activation of MT1-MMP to perform its variety of functions not only relies on its localization to the cell surface, but also on its phosphorylation state. The cytoplasmic tail of MT1-MMP has three major phosphorylation sites that potentially modulate its activity: Thr567, Tyr573, and Ser 577 [16]. While Ser577 has not been extensively evaluated, the phosphorylation site at Thr567 was shown to promote cell detachment from the primary tumor, either through the formation of multicellular aggregates or as single cells [16]. Utilizing a Thr567 phosphomimetic mutant (T567E), Yang et al. demonstrated that phosphorylation of Thr567 in MT1-MMP enhanced cellular migration in comparison to phosphodeficient or wild-type MT1-MMP-expressing cells. Additionally, phosphomimetic T567E MT1-MMP expressing cells formed more cohesive ovarian cancer spheroids that also demonstrated enhanced adhesion to peritoneal explants. Furthermore,

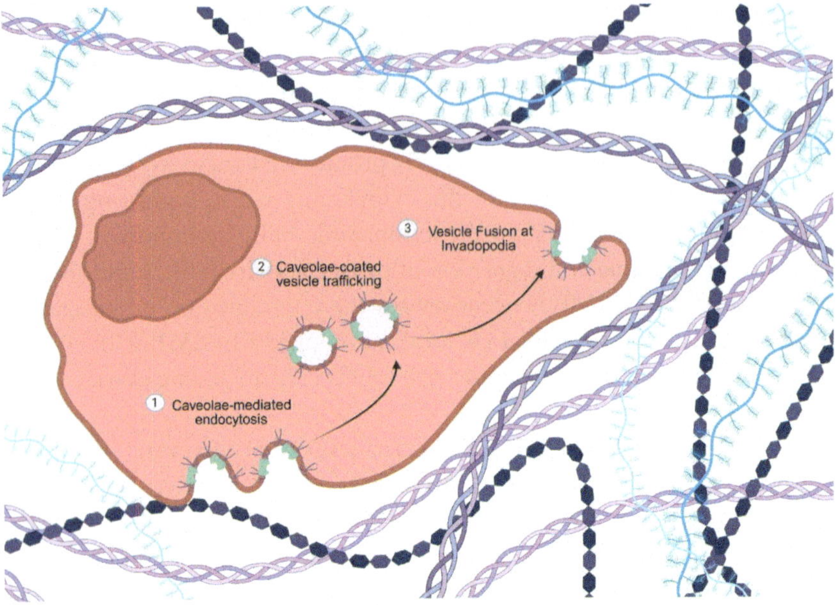

Fig. 2 Caveolae mediated endocytosis of MT1-MMP and subsequent trafficking of MT1-MMP to the invadopodia formation resulting in site-specific collagenolysis at the migratory leading edge to promote metastasis.

T567E MT1-MMP was shown to enhance MCA invasion into collagen in comparison to non-phosphorylated MT1-MMP [16]. Overall, this study demonstrated the role of phosphorylation of Thr567 as a key regulator of ovarian cancer MCA dynamics as a regulatory mechanism to impact ovarian cancer metastatic success.

Regulation of MT1-MMP by epidermal growth factor receptor (EGFR) signaling leading to phosphorylation of Tyr573 was shown to modulate caveolae-mediated internalization of MT1-MMP, removing proteolytic activity from the cell surface [50]. Moss et al. found that a Y573F phosphodeficient MT1-MMP mutant was no longer internalized following EGF treatment, leading to sustained cell surface MMP activity [50] accompanied by a phenotypic EMT and enhanced collagen invasion. In contrast, growth within three dimensional collagen gels was inhibited. These data demonstrate the ability for MT1-MMP Tyr573 phosphorylation to act as a switch regulating the transition between invasive and expansive growth within a three-dimensional collagen microenvironment.

6. Adhesion-mediated regulatory mechanisms

The expression and regulation of MMPs by ovarian cancer cells occurs in constant communication with the microenvironment. At the primary tumor site, epithelial integrity is tightly maintained by E-cadherin-mediated cell-cell junctions between neighboring cells. An early event in EMT is the loss of surface-associated E-cadherin and the gain of a migratory phenotype [51]. Metastatic success is highly dependent on the shedding of cancer cells from the primary tumor and subsequent cellular adhesion to a secondary site(s). The heightened expression of proteolytic enzymes in cancer is responsible for removing extracellular matrix barriers to invasion into or expansive growth within a tissue [52]. Metastatic dissemination of cancer into secondary sites requires cell-matrix adhesion, which is primarily reliant on cell surface integrins, transmembrane heterodimeric receptors that recognize ECM components [53]. Integrin-mediated adhesion can stimulate a complex array of signaling pathways involving FAK, Src, and Rho, ultimately modulating the actin cytoskeleton to control cell motility and survival gene programs [22].

Integrin signaling is hierarchical, enabling a cell to distinguish between integrin occupancy (for example, by a peptide fragment of a matrix protein) and integrin clustering (such as would be incurred when a cell encounters a complex 3-dimensional matrix), and this, in turn, greatly impacts the activation of downstream signal transduction pathways. Monomeric ligand occupancy induces integrin redistribution to focal adhesion sites, but does not induce Tyr phosphorylation or redistribution of cytoskeletal proteins to the cytoplasmic face of the integrin. In contrast, integrin occupancy and aggregation together stimulate activation of FAK, phosphorylation of downstream effectors, and accumulation of actin-containing cytoskeletal molecules to the cytoplasmic face of the integrin [22].

The regulation of MT1-MMP expression and activity is highly dependent on this integrin hierarchy. As summarized above, cellular signaling cascades recognize and distinguish between intact three-dimensional collagen vs two-dimensional "degraded" collagen, and will increase MMP expression upon contact with intact ECM [22]. The initial adhesion to collagen triggers integrin clustering, which in turn activates Src [19]. Activated Src acts downstream to upregulate EGR1, a transcription factor that can directly bind the MT1-MMP promoter and induce expression. This mechanotransduction pathway effectively upregulates MT1-MMP in response to contact with intact collagen via the Src-EGR1 pathway [19]. This hypothesized signaling cascade is supported by findings that demonstrate an increase in proteinase expression upon treatment with anti-integrin antibodies or upon alterations of the actin cytoskeleton [54–57].

Highlighting the distinction between integrin occupancy and aggregation, signals are differentially transduced in response to intact versus proteolytically degraded ECM components [58–62]. For example, cells in the presence of intact 3D collagen, but not proteolyzed collagen, increased their expression of MT1-MMP and subsequently demonstrated increased activation of pro-MMP-2 [63]. Furthermore, MT1-MMP co-localizes with aggregates of collagen-binding integrins such as $\alpha 2\beta 1$ and $\alpha 3\beta 1$ in ovarian cancer, where binding to these integrins upregulates the pathways mentioned previously, such as EGR1 [64–66]. Beyond cell-matrix interactions, there is evidence that supports direct binding between MT1-MMP and collagen components of the ECM via the hemopexin domain [67]. Moreover, integrin localization at focal adhesion sites with the ECM is hypothesized to allow for close proximity of proteinases and their substrates upon contact [68]. Additionally, pro-MMP-2 has been demonstrated to adhere to $\alpha 2\beta 1$ integrin associated with collagen, which can be subsequently activated by co-localized MT1-MMP [69].

As is well established in cancer metastasis, the epithelial-to-mesenchymal transition is a common process undergone by invasive cancer cells. The maintenance of cell-cell junctions via binding of E-cadherin and β-catenin results in reduced expression of MMPs [22]. Loss of E-cadherin is initiated by a variety of mechanisms and leads to dissociation of β-catenin from the cytoplasmic face of the membrane. Soluble β-catenin can then translocate to the nucleus and act as a transcription factor, binding to Tcf/Lef-responsive elements in the MT1-MMP promoter to increase expression [70,71]. Relevant to ovarian cancer metastasis, the engagement of collagen-binding integrins by disseminating ovarian cancer cells results in proteolytic shedding of the E-cadherin ectodomain, catalyzed by MMP-9 [72]. The presence of the soluble ectodomain in ascites can further disrupt cell:cell junctions, releasing additional β-catenin for transcriptional control of MT1-MMP. Multivalent engagement of collagen binding integrins can also induce E-cadherin internalization, leading to elevated levels of nuclear β-catenin with resulting transcriptional activation of Wnt/β-catenin target genes [73]. Via this mechanism, Wnt/β-catenin target gene expression can be regulated in the absence of activating mutations in this pathway. Furthermore, key EMT regulators Snail1 and Slug can promote expression of MT1-MMP. The ability of Snail1 to induce an invasive phenotype is entirely dependent on the expression and upregulation of membrane-type MMPs, where silencing MMP expression abrogates the ability to invade the basement membrane [74]. Moreover, induced expression of Slug through the STAT3/NFκB/EGR axis promotes expression of MT1-MMP, effectively promoting invasion of cells and degradation of ECM components [75].

Beyond mechanical signaling, such as cell-matrix adhesion and disruption of cell-cell junctions, signaling molecules in the environment can trigger changes in the expression and activity of MT1-MMP. As indicated above, regulation of MT1-MMP cell surface activity is an important step in the transition from a matrix-degradative invasive phenotype to a proliferative expansive phenotype. This switch is regulated by an increase in EGF signaling, where prior to EGF stimulation, MT1-MMP is retained at the cell surface, while upon EGF stimulation, MT1-MMP is signaled to internalize into the cell [16,32,50]. Furthermore, EGF treatment increases expression of MT1-MMP at both the protein and mRNA levels to further promote the invasion program of cells [50]. Alternatively, under cellular starvation conditions, MT1-MMP expression can be stimulated via the inactivation of mTOR. The inactivation of mTOR in response to nutrient deficiency begins a signaling cascade that recruits MT1-MMP to the leading edge of invadopodia protrusions, providing a mechanism for cellular escape of nutrient deficient environments [76].

Overall, the collective data support a model wherein MT1-MMP expression, activity, and subcellular localization are finely tuned via cross-talk between cell-cell and cell-matrix mediated signaling mechanisms. MT1-MMP-dependent augmentation of either growth-promoting or growth-inhibitory mechanisms provides an additional layer of regulatory control.

7. "Host" factors influencing MT1-MMP expression and ECM remodeling

The tumor microenvironment of ovarian cancer is composed of heterogeneous cell populations with intercellular communication networks spanning many cell types beyond the cancer cells themselves, including normal host cells such as peritoneal mesothelial and epithelial cells and an array of immune cells. The homeostasis of intercellular communication is impacted by the health status of the individual and is a major determinant of cancer progression and survival. Specifically, these factors can impact the risk of developing ovarian cancer in the context of modulation by MT1-MMP. Of particular note is the impact that aging and obesity have on the metastatic success of ovarian cancer, which is highly dependent on MT1-MMP activity [18,29,77–81]. Here, we will review the current knowledge of the impact of aging and obesity on ovarian cancer metastatic success, highlighting key changes to MT1-MMP activity.

7.1 Aging

Aging is defined as a functional decline that occurs over time [82,83] and is considered an inevitable deterioration of cellular function. The correlative relationship between aging and cancer progression has been long-standing in the oncology research field, yet the molecular mechanisms that contribute to enhanced disease progression in elderly patients remain unclear. In ovarian cancer, epidemiological data from 1993 demonstrated a substantial increase in both incidence and severity in post-menopausal women, and this pattern remains pertinent today [84,85]. Indeed, age is the most significant risk factor for ovarian cancer, with a median age of 63 at diagnosis and 71 for ovarian cancer-related death [2]. The human data are phenocopied by pre-clinical murine models, using multiple syngeneic cell lines and two distinct mouse strains, that show significantly enhanced metastatic tumor burden in aged mice relative to young [29,31]. As scientific research has shifted focus towards consideration of aging as a relevant variable, the relationship between aging and cancer prevalence is revealed to be paradoxical. As individuals increase in age, there is an overall increase in cancer risk, linked consequently to the probability of accumulated somatic mutations throughout the lifespan; however, at the age of 90, there is an observed drastic decline in cancer risk [86]. Moreover, larger animals, with more cells (and therefore presumably more opportunities for acquired mutations), do not exhibit an increase in cancer risk. These clinical findings are coupled by "Peto's Paradox," which highlights the lack of correlation across the animal kingdom between lifespan, body size, and cancer incidence [87,88]. The lack of increased cancer incidence in large animals suggests that evolutionary mechanisms are better adapted at suppressing cancer in other organisms than in humans. These combined observations suggest that cancer incidence increasing with age in humans is a result of microenvironmental changes rather than direct genetic modifications.

Evolution has not left humans defenseless. Our understanding of age-related induction of cellular senescence is a presumed mechanism of tumor suppression, wherein damaged cells are shuttled into a non-proliferative senescent state. However, cellular senescence displays antagonistic pleiotropy, wherein induction into cellular senescence early in life is beneficial for tumor suppression while senescent cell accumulation late in life can promote cancer [89]. Taken together, these observations regarding aging and cancer incidence in humans highlight the molecular complexity of the interrelationship. Here, we discuss changes to the ECM that occur with age and the relationship to other dynamic microenvironmental factors that are modified in the aging host.

The iterations of the Hallmarks of Aging have consistently identified "altered intercellular communication" and the "loss of proteostasis" as prominent characteristics of age-related molecular changes [82,83]. Alterations in intercellular communication occur in many important signaling pathways pertinent to cellular homeostasis. Of particular interest is the senescence-associated secretory phenotype (SASP), a well-documented secretory profile from senescent cells that is comprised of pro-inflammatory cytokines and chemokines, growth factors, lipids, exosomes, and notably, ECM remodeling enzymes [90]. Evidence suggests that the production of SASP plays an important role in the mechanotransduction of pro-fibrotic signals and increased stiffness of aged tissues. Studies have demonstrated that enhanced production of TGF-β by senescent cells can activate mechanosensory molecules YAP/TAZ, which promote the induction of pro-fibrotic gene expression programs [91]. These programs can upregulate the production of MMPs in the microenvironment and promote fibroblast activation and proliferation, all contributing to a positive feedback loop of fibrosis and tissue stiffening [91,92]. The activation of fibroblasts is achieved through the Wnt signaling pathway, which can, in turn, activate other fibrotic pathways such as Notch, RAS, SMAD, and Hedgehog [91]. Furthermore, the loss of proteostatic conditions during aging is a major contributing factor to organ dysfunction. With age, there is an increase in mutant misfolded proteins coupled by insufficient removal of dysfunctional proteins [83]. Additionally, the accumulation of post-translational modifications such as non-enzymatic glycosylation to long-lasting proteins such as collagen, termed advanced glycation endproducts (AGEs), has been shown to accelerate aging. In a study where *D. melangaster* was fed with AGEs, there was an observable reduction in lifespan and healthspan [93]. Overall, the loss of proteostasis and concomitant alteration in intercellular signaling regulators are a pivotal molecular changes that impact the aged microenvironment.

It has been recently demonstrated that the age-related changes that occur within the omentum are critical to the metastatic success of ovarian cancer. This adipocyte-rich immune filter is a preferred location for ovarian cancer metastasis and contains many well-documented age-related tumor-permissive changes. The omentum consists of a collagen-rich extracellular network which displays altered ultrastructure characteristics with age. Harper et al. demonstrated that while there is an overall increase in collagen content with age, there is a corresponding increase in fiber linearity and anisotropic alignment relative to the wavy collagen ultrastructure seen in young hosts [31]. These linear collagen fibers provide a more directed route for ovarian cancer cell migration, allowing for rapid metastatic growth. Omental collagen can also exhibit enhanced crosslinking with age, either via enzymatic means catalyzed by lysyl oxidase, or via non-enzymatic glycation resulting in AGEs. Enhanced degradation of collagen present within tumors of aged hosts is also observed [31]. Through these modifications, enhanced tissue stiffness is achieved and is permissive to tumor growth [94]. Tissue stiffness shifts the microenvironment from simply permissive to promotional, where mechanosensory feedback loops actively assist tumor growth. In response to tissue stiffness in the omentum, TAGLN is activated, which regulates Src activation to drive invasion through the RhoA/Rock pathway [95,96]. To further promote invasion, tissue stiffness activates the YAP/TAZ program, which is a known regulator of the EMT [96]. In addition to promoting invasion, a stiff matrix has been shown to induce the MAPK/MEK pathway to promote proliferation and migration in ovarian cancer cells [97].

In the context of aging, it is important to consider the concept of "truly aged" host collagen versus intratumoral collagen that is actually newly synthesized by cancer- associated fibroblasts (Fig. 3). In contrast, truly aged host collagen has a half-life of 15–117 years [98,99], inhabiting the peritoneal microenvironment for decades [31,96], and this long half-life enables the accumulation of AGE modifications, which are not seen in intratumoral collagen [31]. Furthermore, the remodeling of truly aged collagen to fibrotic structures differs from the localized fibrous stroma that surrounds the tumor [31,96]. Lastly, the ultrastructure of truly aged collagen is distinguished by large fenestrations in contrast to the tumor-specific orientation of collagen fibers in TACS [31]. In women, there is evidence that these age-related changes may be in part due to the hypoestrogenism that occurs during menopause and the perimenopausal transition. One study has demonstrated that with age, there is a significant decline in the collagen type III/type 1

A.

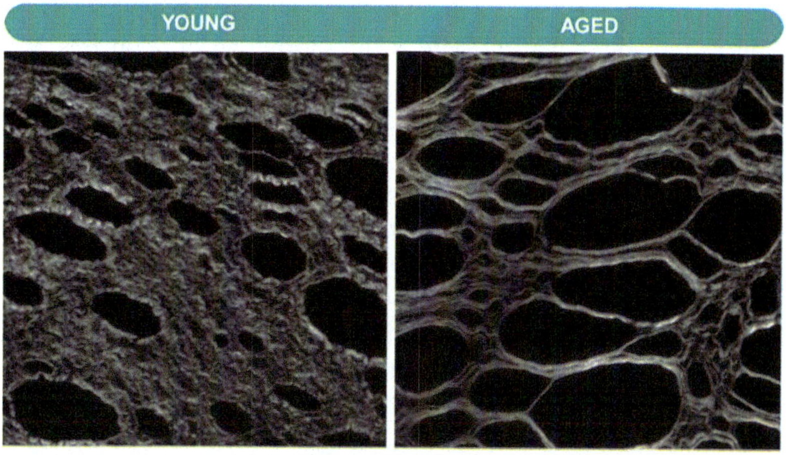

B.

Characteristic	Truly Aged Host Collagen	Tumor-Associated Collagen
Source	Host tissue	Newly synthesized from CAFs
Lifespan/Half-life	Long, Half-life 15-117 years	Recently synthesized, short lifespan
AGE Modifications	Accumulated	No Modifications
Ultrastructural Organizations	Large fenestrations, linear fibers, and anisotropic alignment	TACS signatures with tumor-specific alignment
Remodeling Trends	widespread aged fibrosis	localized tumor-associated fibrotic stroma
Hormonal Influences	Hypoestrogenism drives reduction in Col I:Col III ratio	Tumor secreted factors and mechanosensory loops drive remodeling

Fig. 3 Aged versus Young Collagen. (A) Omental collagen ultrastructure imaged with second harmonic generation microscopy demonstrates altered linearity, fenestration size, and anisotropy between young and aged mice. (B) Comparison of truly aged host collagen and fibrotic collagen deposition around the tumor.

ratio and that this decline corresponds to a drop in estrogen levels [100]. Taken together, these data suggest that age-related modifications to truly aged collagen provide a permissive environment for tumor growth, where rapid tumor invasion and migration is readily achieved in aged compared to young omenta.

To further foster tumor growth, the accumulation of senescent cells, which is a recognized hallmark of aging, substantially contributes to tumor progression [82,83]. Senescent cells are non-proliferating, metabolically active cells that are characterized by distinct deleterious phenotypes such as genomic instability, mitochondrial dysfunction, increased lysosomal activity, and, importantly, the production of the SASP [101]. The most recognized inducer into the senescent state is the shortening of telomeres, however, cells can also enter this state through oxidative stress or the accumulation of DNA damage, including that induced by chemotherapy [102]. The SASP is composed of pro-inflammatory cytokines and chemokines [102] which can dampen the effectiveness of anti-tumor immunity, while growth factors can promote cancer cell migration and metastasis as well as stimulate vascularization [102,103]. In the context of the omental tumor microenvironment, remodeling enzymes present in SASP facilitate ECM remodeling by senescent cells to enable permissive growth of cancer. Senescent cells of many types can produce serine proteases, ADAMTS, hyaluronidases, and MMPs, notably MT1-MMP [104]. These enzymes actively contribute to the remodeling of the omental collagen structure, contributing to the aforementioned properties of increased linearity, anisotropy, and modifications permissive to tumor progression. Importantly, senescent cells are not extensive in the microenvironment, but a single senescent cell can impact surrounding tissues through paracrine signaling and thereby have dramatic impacts [102].

Underlying the aging phenotypes, MT1-MMP acts as a proteolytic driver of structural alterations that enhance this permissive and eventually promotional aged tumor microenvironment. The localization of MT1-MMP and the integrin signaling pathways that regulate it ensure directed precision in proteolytic remodeling of the ECM [22]. With age and the increased production of SASP, TGF-β, and NF-κB can function as transcription factors to directly or indirectly bind the promoter region of the MT1-MMP gene [105–109], further contributing to age-associated collagen remodeling.

7.2 Obesity

Another risk factor that strongly influences collagen ultrastructure and the regulation of MMP expression is obesity. According to the World Health Organization, obesity is an excess of adipose tissue, defined by a body mass index (BMI) greater than 30. In 2024, it was estimated that around 895 million people worldwide were impacted by obesity, and 2.5 billion people were considered overweight [110]. While adipose tissue in the body plays an important role in normal physiological functions such as energy storage, regulating insulin sensitivity, and the immune response, excessive accumulation can disrupt normal tissue homeostasis [111]. This imbalance of tissue signaling contributes to the enhanced risk for obese subjects for many diseases, notably cancer [112]. Specifically, obesity is a known risk factor for ovarian cancer, worsening the incidence rate, metastatic progression, and overall survival in obese patients versus non-obese patients [3–10]. This is recapitulated by studies in pre-clinical models of diet-induced obesity. Mice fed on a high-fat diet (HFD), or "Western" diet, exhibit enhanced peritoneal metastatic burden from ovarian cancer and tumors upregulate expression and nuclear localization of sterol regulatory element binding protein 1 (SREBP1), a master regulator of lipid synthesis and lipid transport [10]. In human ovarian cancer tumors, nuclear SREBP1 is correlated with body mass index [113]. Tumor fibrosis is also upregulated in HFD mice. Moreover, compared to mice on a low-fat diet (LFD), HFD mice respond poorly to standard-of-care chemotherapy (paclitaxel and carboplatin), with increased residual tumor burden following treatment [113].

Multiple additional molecular mechanisms contribute to the poor outcome of obese patients with ovarian cancer. From the adipocytes of obese patients, the hormone leptin is secreted in a fat mass- and adipocyte size-dependent manner [114,115]. Interestingly, estrogen has been shown to increase leptin production, indicating a sexual dimorphic modulation of adipose signaling [111,116]. While leptin plays an important role in energy, lipid, and carbohydrate metabolism, it has also been shown to create a pro-metastatic environment [117–119]. Specifically, leptin has been demonstrated to modulate key pro-cancer pathways such as JAK/STAT, MAPK/ERK, PI3K/AKT, and mTOR [120–122]. While leptin has been shown to promote inflammation, angiogenesis, and fibrosis, it is most interesting to consider its role in the overexpression of MMPs during tumor progression. Leptin increases expression of many MMP family members including MMP2, 7, 9, 13, and notably MT1-MMP. MT1-MMP is specifically upregulated in the presence of leptin to increase cell migration and invasion capacities [123]. Furthermore, leptin has been shown to induce FAK and Src, along with promoting the secretion of MMP-2 and MMP-9, to induce invasion in breast cancer cell lines [124]. The adipokine adiponectin is also produced by white adipose tissue [125]. Adiponectin has normal functions in energy and lipid metabolism, as well as regulating inflammation and insulin sensitivity [126]. Adiponectin exerts anti-inflammatory and cyto-protective effects against cancer [127]; however, adiponectin levels decrease in obesity, mitigating its anti-cancer properties in this context [128].

Another way in which obesity helps sustain tumor growth is through altered metabolism pathways. In the 1920s, it was established that tumor cells utilize aerobic glycolysis even in the presence of oxygen, known as the Warburg effect. This metabolic alteration drives tumor progression even in hypoxic environments that arise in solid tumors [129,130]. However, in the obese tumor microenvironment, adipose tissue is a constant supply of nutrients, and under these conditions, an inversion of the Warburg effect is observed. Lactate, as a waste product, is converted into glucose through gluconeogenesis to provide a constant supply of energy and biological building blocks necessary for tumor growth. Through the inversion of the Warburg effect, a further acceleration of tumor progression is observed in obese patients [129].

Similarly, to the expansion of proliferating cells into surrounding tissue in cancer, adipose tissue expansion in obesity also requires ECM degradation. Adipocytes must remodel the surrounding collagen architecture in order to store the lipid excess accumulated in obesity [81]. MT1-MMP has been shown to be a critical remodeling enzyme in this process, as MT1-MMP deficient mice could not remodel collagen type I, resulting in impaired weight gain [81]. Furthermore, human genetic studies have uncovered particular MT1-MMP gene polymorphisms that predispose individuals to obesity and type 2 diabetes due to aggressive collagen remodeling capabilities [81]. As previously mentioned, caveolae localize at lipid rafts in the cell membrane, and recycling of MT1-MMP is hyperactivated in obese patients. The role of MT1-MMP in obesity-related collagen remodeling is two-fold, in which early stage expansion via MT1-MMP activity breaks down collagen type 1 to support adipocyte expansion without rupturing. However, in an individual with advanced obesity, MT1-MMP excessively remodels, creating collagen fragments such as endotrophin. Endotrophin induces heightened inflammation and activates fibrotic programs, which eventually lead to insulin resistance and dysregulated lipid metabolism [80]. Overall, MT1-MMP is a critical remodeling enzyme in obese states that is required for adipocyte expansion into the extracellular matrix, but through excessive activity can promote obesity-related inflammation and metabolic dysregulation.

The upregulation of MT1-MMP in obese patients predisposes individuals to aggressive cancer phenotypes. First, leptin produced from adipocytes can activate the JAK/STAT pathway and PI3K/AKT pathway to upregulate Snail and Twist, respectively, effectively stimulating the EMT program [131]. Upon stimulation with leptin, epithelial breast cancer cells lost the epithelial marker E-cadherin and exhibited increased cell migration, highlighting the pro-metastatic capacity of leptin [132]. As previously mentioned, activation of the EMT program stimulates the upregulation of MT1-MMP and results in directed collagenolysis of the extracellular matrix, promoting cancer invasion. Leptin can also act as an endothelial cell growth factor to promote angiogenesis directly or indirectly. Leptin can directly bind endothelial cells and stimulate proliferation and migration. Alternatively, leptin can indirectly upregulate VEGF to enhance angiogenic programs. Furthermore, the production of MMPs stimulated by leptin remodels the ECM to help support neovascularization [111]. Additionally, the hypoxic state often induced in obesity and cancer results in the upregulation of hypoxia inducible factor 1α (HIF1α), which can bind to the MT1-MMP promoter region [80]. Hypoxia stimulated MT1-MMP expression and promoted ovarian cancer cell invasion and collagenolysis in a chick chorioallantoic membrane (CAM) invasion assay, and these processes were blocked by treatment with a HIF-1α inhibitor [133]. Together, these data highlight multiple mechanisms whereby obesity may regulate MT1-MMP in both normal (adipocytes) and neoplastic cells.

7.3 Adiposenescence: The convergence of aging and obesity

While aging and obesity create distinct advantages for intra-peritoneal metastatic progression, convergence of the two creates a more growth-promoting microenvironment. Aging is accompanied by a shift in the adipose tissue distribution, with an increase in visceral adipose tissue relative to subcutaneous adipose tissue [11]. Additionally, during middle age, body mass index often increases, resulting in enhanced leptin production [134]. Studies have demonstrated that the pro-inflammatory, hypoxic, and metabolically demanding microenvironment of obese patients can expedite senescence induction [11]. Furthermore, pre-adipocytes are predisposed to senescence, a phenomenon known as adiposenescence [11]. The induction of senescence in preadipocytes not only blocks future differentiation of healthy adipocytes but also further exacerbates the pro-inflammatory microenvironment through the production of SASP [11].

8. Therapeutic strategies targeting MT1-MMP

The integral role of MT1-MMP has been clearly demonstrated in the metastatic success of ovarian cancer, and this pro-metastatic function calls for therapeutic interventions to combat its activity. To date, there is no specific small molecule MT1-MMP inhibitor that is approved for use in any indication. In this section, we will review historic attempts to develop MMP inhibitors and discuss the current challenges that face their implementation. Furthermore, the treatment of ovarian cancer patients has presented unique challenges compared to other cancers. In particular, the localization of tumors in the peritoneal cavity discourages intravenous administration due to poor drug penetration [135]. However, direct intraperitoneal (IP) injection results in higher rates of side effects and localized abdominal pain, causing many patients to cease treatment [136]. Beyond side effects, IP administration fails to penetrate large tumors, and the prevalence of metastatic MCAs may represent a distinct target [135]. Taken together with the current challenges facing MMP inhibitors, these findings call for more targeted therapies such as nanoparticle administration, combination with immunotherapies, and possibly the inclusion of pressurized therapies to enhance IP penetration [137] (Table 1).

8.1 First generation MMP inhibitors

First generation MMP inhibitors were substrate analogs designed through modification of the natural substrate and were termed "peptidomimetic" inhibitors. The first generation of MMPIs primarily utilized high-affinity zinc-binding groups, mainly hydroxamate, which chelate the catalytic zinc atom of MMPs [47]. Introductory molecules in this class were batimastat (BB-94) and marimastat (BB-2516), which are broad-spectrum hydroxamate-based peptidomimetic MMP inhibitors. Both drugs displayed anticancer efficacy *in vitro* and in pre-clinical animal models [47,138,139], however early clinical studies were unsuccessful, albeit with early promising indications in ovarian cancer. In a phase I study, batimastat was administered IP to ovarian cancer patients with malignant ascites who required paracentesis. Following a single IP dose of batimastat, several patients required no further paracentesis for 3 months [140]. Batimastat is poorly soluble in water, resulting in low bioavailability when administered orally. Subsequent modifications to develop marimastat enhanced solubility and oral bioavailability. However, due to the broad-spectrum inhibition of virtually all MMPs, these drugs displayed robust toxicities. The main toxicity concern was musculoskeletal syndrome, which is stiffness and/or reduced mobility in muscles, bones, tendons, joints, ligaments, and connective tissues [47]. Recently, the musculoskeletal side effects of batimastat and marimastat have been linked to the off-target inhibition of the ADAM (a disintegrin and metalloproteinase) family of proteinases that regulate circulating TNF-α and its receptors

Table 1 A representative overview of past, current, and potential MMP inhibition, including, but not limited to, peptidomimetic analogs, nanoparticles, antibodies, and potential combination therapies.

Past, Current, and Potential Approaches to MMP Inhibition

Generation	Examples	Mechanism	Limitations	Clinical Outcomes
First Generation	• Batimastat (BB-94) • Marimastat (BB-2516)	• Peptidomimetic substrate analogs • High-affinity hydroxamate zincbinding groups	• Broad-spectrum inhibition • Musculoskeletal syndrome • Marimastat: Fibrosis via MMP1 inhibition) • Batimastat: Poor water solubility/low oral bioavailability	Failed Phase III trials
Second Generation	• Rebimastat (BMS-275,291) • Tanomastat • Prinomastat	• Peptidomimetic analogs with structural modifications using thiol, carboxylic acid, or alternative hydroxamate groups	• Narrower spectrum than first Gen, but still targeted multiple MMPs • Musculoskeletal syndrome was improved but not eliminated	Failed Phase III trials

Continued

Table 1 A representative overview of past, current, and potential MMP inhibition, including, but not limited to, peptidomimetic analogs, nanoparticles, antibodies, and potential combination therapies.—cont'd

Past, Current, and Potential Approaches to MMP Inhibition

Generation	Examples	Mechanism	Limitations	Clinical Outcomes
Third Generation: Nanomaterials	• Gold Nanoparticles (AuNPs)	• Negatively charged, smaller AuNPs bind the zinc domain of MMP-2 and-9 • Selective targeting based on size and charge	• Conflicting *in vitro* data • Specific MT1-MMP nanoparticles are in very early stages	Preclinical/ Experimental
Third Generation: Antibodies	• Engineered Antibodies • Patient-derived Autoantibodies	• Engineered for high binding affinity and exact specificity for MT1-MMP	• Clinical outcomes, safety, and long-term effects remain unknown	Preclinical
Combination Therapies	• MMP Inhibitors + Cisplatin • MMP Inhibitors + Immunotherapy	• Targeting specific resistance pathways (e.g., MMP3 upregulation) • Modulating the tumor immune response via TIMPs/targeted inhibitors	• Contingent on overcoming the delivery and toxicity challenges MMP inhibition	Preclinical/ Animal models

[138,141]. Overall, these drugs displayed minimal anticancer efficacy with similar survival rates compared to standard of care therapies in most cancers, with both drugs failing phase III clinical trials [47,138,142].

8.2 Second generation MMP inhibitors

Second generation MMPIs aimed to enhance bioavailability, pharmacokinetic properties, and limit toxicities of first generation MMPIs. Rebimastat (BMS-275,291), tanomastat, and prinomastat contain a thiol group, carboxylic acid, and hydroxamate zinc binding groups, respectively, to inhibit MMPs. These second generation MMPIs display narrower specificities than first generation broad-spectrum MMPIs, but still target multiple members of the MMP family [47]. Due to improved specificity, these drugs displayed less severe musculoskeletal syndrome; however, this side effect was not completely eliminated. Ultimately, all three drugs failed phase III clinical trials, resulting from limited efficacy in most cancer types [47,138].

8.3 Third generation and beyond: Nanomaterials, antibodies, and PROTACs

The broad-spectrum effects of early MMPIs call for novel methods of MMP inhibition to enhance specificity and reduce toxicity. One novel approach is utilizing nanoparticles to selectively target specific MMPs at multiple steps, including gene expression, protein expression, and inhibiting enzyme activity [47]. Outside of nanomaterials' potential as drug delivery systems, the particles themselves can have direct action against MMPs. Studies have demonstrated that changes in size and charge of gold nanoparticles, AuNPs, can fine-tune inhibition selectivity against MMPs [143–145]. Not only do the specific effects on cancer growth need further investigation, but further fine-tuning of nanoparticles is required to ensure selective inhibition. Additionally, the development of nanoparticles that specifically target MT1-MMP is still in early stages of development and needs further modification before testing efficacy against cancer [146].

An alternative promising approach to selectively target MT1-MMP is antibody-based inhibitors. Antibodies can be engineered and developed for affinity and specificity preferences. Early preclinical studies have shown anticancer efficacy of MMP antibody inhibitors, but clinical outcomes remain unknown [47]. A few studies designed and developed an antibody with high affinity and inhibitory activity against MT1-MMP [147,148]. Additionally, recent clinical studies have investigated the use of autoantibodies targeting tumor-associated proteins isolated from ovarian cancer patients. This study demonstrated that generated self-antibodies against MT1-MMP may be an effective anti-tumor strategy for patients moving forward [147].

Another option for targeting MMPs is proteolysis-targeting chimeras (PROTACs), which utilize the cell intrinsic recycling pathway to mark disease-promoting proteins for selective degradation [149,150]. By either directly targeting MMPs or by indirectly targeting their upstream activation signaling molecules, PROTAC systems can selectively degrade specific MMPs. Preliminary studies have demonstrated that targeted degradation of BRD2 and BRD4, transcription scaffold proteins, selectively reduced MMP9 gene expression to combat neuroinflammation [151]. Moreover, PROTACs have the potential to be used in combination with other therapies such as photodynamic therapy (PDT) [152]. One study has combined photo-activated nanoparticles containing drug therapy against MMP-2, with BRD4 targeted PROTACs to effectively downregulate PD-L1 upregulation as a response to therapy [152].

8.4 Potential for combination therapies with standard of care therapies for ovarian cancer

While clinical trials with currently available MMP inhibitors have been unsuccessful, their potential to treat cancers, specifically ovarian cancer, remains appealing. The utilization of MMP inhibitors in combination with other standard of care (SoC) treatments for ovarian cancer remains a possibility. One study has demonstrated that MMP3 is significantly upregulated in ovarian cancer patients who have developed cisplatin resistance, implying a potential role for MMP3 in this condition. Moreover, the inhibition of MMP3 in combination with cisplatin-based therapies significantly reduced tumor burden in pre-clinical murine models [153]. Lastly, many MMPs have been shown to modulate the immune response directly or indirectly, and their inhibition by protein inhibitors such as the TIMPs or targeted therapies has resulted in improved anti-tumor immune responses. These studies suggest that MMP inhibitors in combination with immunotherapy represent a promising approach to overcoming immunotherapy inefficacy in ovarian cancer [154,155].

9. Challenges facing the study of MT1-MMP and the development of effective therapeutic inventions

To date, the development of effective and safe therapeutics targeting MT1-MMP has been unsuccessful. The non-specific inhibitors that have been developed remain toxic to patients in large part due to the associated musculoskeletal syndrome. The pursuit of refined therapies requires further investigation into the mechanisms that drive specific MMP expression and regulation. On top of the aforementioned challenges to drug development, numerous additional challenges remain to be addressed regarding the role of specific MMPs in ovarian cancer progression.

First, clinical data on the regulation and expression of MMPs and TIMPs is difficult to monitor in real-time. Available clinical data have largely relied on post-mortem or biopsied sample collection to perform histology. While these molecules can be measured in the plasma levels of patients, their activity cannot be readily quantified to distinguish between zymogen, active enzyme, or enzyme-inhibitor complex. Additionally, plasma level measurements of MMPs and TIMPs do not provide a read-out on their organ of origin. Due to this lack of critical information, the relative contributions of MMPs to normal tissue repair and homeostasis versus upregulation for cancer progression cannot be ascertained using current technologies [153]. In a laboratory setting, many studies identify the failure to fully recapitulate the ECM *in vitro* as a major technical limitation [154,155]. The relative lack of ECM models with appropriate three-dimensional complexity, including stromal cell elements, coupled with the inability to effectively sample MT1-MMP activity in the clinical setting, hinders advancements in our detailed mechanistic understanding of how MT1-MMP drives cancer progression.

In order to overcome this problem, more accurate models that better recapitulate the ovarian cancer tumor microenvironment, including organoid cultures composed of multiple cell types, are under development. Such models can also incorporate stromal elements, including adipocytes, collagen matrices of varying stiffness and fiber alignment, and immune cells. Furthermore, *in vivo* models may be tailored to more accurately represent the spectrum of ovarian cancer patients. For example, murine models of diet-induced obesity can be employed to better model the role of obesity, adipocyte fibrosis, and lipid metabolism in ovarian cancer progression in obese patients. Importantly, the majority of current pre-clinical studies utilize young mice, corresponding to premenopausal women; however, only 10–15 percent of ovarian cancers are diagnosed in premenopausal women. Thus, these models fail to recognize the striking ultrastructural and immune differences in the aged vs young peritoneal tumor microenvironment. Future studies should utilize mice at advanced ages to better understand the ways in which aging and obesity, including adiposenescence, influence tumor progression and to address the specific role of MT1-MMP under these more representative host conditions.

10. Conclusion

The changing demands of normal homeostatic growth result in remodeling of the ECM for tissue development, wound healing, and growth. MMPs are essential for the maintenance of these normal programs, yet in the context of cancer can aberrantly promote the same phenotypes of growth, migration, and invasion. MT1-MMP plays a distinctive role in the progression of ovarian cancer through the directed remodeling of collagen in the primary tumor and at metastatic sites such as the omentum. In this review, we discussed the dynamic regulation of MT1-MMP through direct contact with the ECM, regulation of cancer-promoting signaling programs, and its modulation by host factors such as aging and obesity. In aging, MT1-MMP activity can predispose the microenvironment to permissive tumor growth, driven by signaling from the accumulated senescent cells in the aged host. Likewise, in the context of host obesity, the upregulation of MT1-MMP to support adipocyte expansion, combined with a nutrient-rich environment, supports cancer growth. Though a large body of accumulated evidence suggests that MT1-MMP is a prime therapeutic target to diminish ovarian cancer metastasis, clinical efforts to date have failed to successfully and specifically target MT1-MMP without inactivating other MMP family members, leading to adverse side effects. Novel approaches for specific MT1-MMP inhibition, together with the development of more representative *in vitro* and *in vivo* models, will enhance our mechanistic understanding of the role of MT1-MMP in ovarian cancer progression and metastasis and may highlight novel therapeutic approaches in combination with standard-of-care and/or immunotherapies to improve outcomes for women with ovarian cancer.

Acknowledgments

This work was supported in part by grant number 579937 from the American Institute for Cancer Research (MSS); OC22025 (MSS) from the Department of Defense Ovarian Cancer

Research Program, and UO1CA236979 (MSS) from the National Institutes of Health and the National GEM Consortium Fellowship (BCS). All figures created with BioRender.com

References

[1] American Cancer Society Key statistics for ovarian cancer, American Cancer Society, Atlanta, 2026 (Accessed 29 May 2026) https://www.cancer.org/cancer/types/ovarian-cancer/key-statistics.html .
[2] National Cancer Institute SEER*Explorer application, National Cancer Institute, Bethesda, 2026 (Accessed 29 May 2026) https://seer.cancer.gov/ .
[3] C.M. Olsen, C.M. Nagle, D.C. Whiteman, R. Ness, C.L. Pearce, M.C. Pike, et al., Obesity and risk of ovarian cancer subtypes: evidence from the Ovarian Cancer Association Consortium, Endocr. Relat. Cancer 20 (2) (2013) 251–262, doi:10.1530/erc-12-0395.
[4] H.P. Yang, B. Trabert, M.A. Murphy, M.E. Sherman, J.N. Sampson, L.A. Brinton, et al., Ovarian cancer risk factors by histologic subtypes in the NIH-AARP Diet and Health Study, Int. J. Cancer 131 (4) (2011) 938–948, doi:10.1002/ijc.26469.
[5] M.M. Protani, C.M. Nagle, P.M. Webb, Obesity and ovarian cancer survival: a systematic review and meta-analysis, Cancer Prev. Res. 5 (7) (2012) 901–910, doi:10.1158/1940-6207.capr-12-0048.
[6] M.M. Blank, N. Wentzensen, M.A. Murphy, A. Hollenbeck, Y. Park, Dietary fat intake and risk of ovarian cancer in the NIH-AARP Diet and Health Study, Br. J. Cancer 106 (3) (2012) 596–602, doi:10.1038/bjc.2011.572.
[7] H.S. Yang, C. Yoon, S.J. Myung, S.M. Park, Effect of obesity on survival of Women with epithelial ovarian cancer, Int. J. Gynecol. Cancer 21 (9) (2011) 1525–1532, doi:10.1097/igc.0b013e31822eb5f8.
[8] L. Makowski, C. Zhou, Y. Zhong, P.F. Kuan, C. Fan, B.P. Sampey, et al., Obesity increases tumor aggressiveness in a genetically engineered mouse model of serous ovarian cancer, Gynecol. Oncol. 133 (1) (2014) 90–97, doi:10.1016/j.ygyno.2013.12.026.
[9] K.M. Nieman, H.A. Kenny, C.V. Penicka, A. Ladanyi, R. Buell-Gutbrod, M.R. Zillhardt, et al., Adipocytes promote ovarian cancer metastasis and provide energy for rapid tumor growth, Nat. Med. 17 (11) (2011) 1498–1503, doi:10.1038/nm.2492.
[10] Y. Liu, M.N. Metzinger, K.A. Lewellen, S.N. Cripps, K.D. Carey, E.I. Harper, et al., Obesity contributes to ovarian cancer metastatic success through increased lipogenesis, enhanced vascularity, and decreased infiltration of M1 macrophages, Cancer Res. 75 (23) (2015) 5046–5057, doi:10.1158/0008-5472.can-15-0706.
[11] G. Wang, A. Song, Q.A. Wang, Adipose tissue ageing: implications for metabolic health and lifespan, Nat. Rev. Endocrinol. 21 (10) (2025) 623–637, doi:10.1038/s41574-025-01142-8.
[12] C. Auger, S. Kajimura, Adipose tissue remodeling in pathophysiology, Annu. Rev. Pathol. 18 (1) (2023) 71–93, doi:10.1146/annurev-pathol-042220-023633.
[13] M.Y. Ou, H. Zhang, P.C. Tan, S.B. Zhou, Q.F. Li, Adipose tissue aging: mechanisms and therapeutic implications, Cell Death Dis. 13 (4) (2022) 351, doi:10.1038/s41419-022-04752-6.
[14] Z. Piperigkou, K. Kyriakopoulou, C. Koutsakis, S. Mastronikolis, N.K. Karamanos, Key matrix remodeling enzymes: functions and targeting in cancer, Cancers 13 (6) (2021) 1441, doi:10.3390/cancers13061441.
[15] M.C. Vos, A.A. van der Wurff, T.H. van Kuppevelt, L.F. Massuger, The role of MMP-14 in ovarian cancer: a systematic review, J. Ovarian. Res. 14 (1) (2021) 101, doi:10.1186/s13048-021-00852-7.
[16] J. Yang, W.C. Kasberg, A. Celo, Z. Liang, K. Quispe, M.S. Stack, Post-translational modification of the membrane type 1 matrix metalloproteinase (MT1-MMP) cytoplasmic tail impacts ovarian cancer multicellular aggregate dynamics, J. Biol. Chem. 292 (32) (2017) 13111–13121, doi:10.1074/jbc.m117.800904.

[17] K. Sakata, K. Shigemasa, N. Nagai, K. Ohama, Expression of matrix metalloproteinases (MMP-2, MMP-9, MT1-MMP) and their inhibitors (TIMP-1, TIMP-2) in common epithelial tumors of the ovary, Int. J. Oncol. 17 (4) (2000) 673–681, doi:10.3892/ijo.17.4.673.
[18] P. Carey, E. Low, E. Harper, M.S. Stack, Metalloproteinases in ovarian cancer, Int. J. Mol. Sci. 22 (7) (2021) 3403, doi:10.3390/ijms22073403.
[19] M.V. Barbolina, B.P. Adley, E.V. Ariztia, Y. Liu, M.S. Stack, Microenvironmental regulation of membrane type 1 matrix metalloproteinase activity in ovarian carcinoma cells via collagen-induced EGR1 expression, J. Biol. Chem. 282 (7) (2007) 4924–4931, doi:10.1074/jbc.m608428200.
[20] C. Markosian, A.M. Ille, S.K. Burley, W. Arap, R. Pasqualini, Matrix metalloproteinases as candidate antigenic determinants for anti-tumor autoantibodies in human ovarian cancer: a post hoc analysis, FASEB J. 39 (18) (2025) e23481, doi:10.1096/fj.202500739rr.
[21] M.C. Vos, A. van Tilborg, W.J. Brands, D. Boll, D. van Hamont, H. van der Putten, et al., Polymorphisms in MMP-14 and MMP-2 genes and ovarian cancer survival, Cancer Biomark. 25 (3) (2019) 233–241, doi:10.3233/cbm-181826.
[22] H.G. Munshi, M.S. Stack, Reciprocal interactions between adhesion receptor signaling and MMP Regulation, Cancer Metastasis Rev. 25 (1) (2006) 45–56, doi:10.1007/s10555-006-7888-7.
[23] R.T. Aimes, J.P. Quigley, Matrix metalloproteinase-2 is an interstitial collagenase, J. Biol. Chem. 270 (11) (1995) 5872–5876, doi:10.1074/jbc.270.11.5872.
[24] A.A. Kamat, M. Fletcher, L.M. Gruman, P. Mueller, A. Lopez, C.N. Landen, et al., The clinical relevance of stromal matrix metalloproteinase expression in ovarian cancer, Clin. Cancer Res. 12 (6) (2006) 1707–1714, doi:10.1158/1078-0432.ccr-05-2338.
[25] S. Afzal, E.N. Lalani, R. Poulsom, A. Stubbs, G. Rowlinson, H. Sato, et al., MT1-MMP and MMP-2 mRNA expression in human ovarian tumors: possible implications for the role of desmoplastic fibroblasts, Hum. Pathol. 29 (2) (1998) 155–165, doi:10.1016/s0046-8177(98)90226-x.
[26] B. Davidson, I. Goldberg, W.H. Gotlieb, J. Kopolovic, G. Ben-Baruch, J.M. Nesland, et al., High levels of MMP-2, MMP-9, MT1-MMP and TIMP-2 mRNA correlate with poor survival in ovarian carcinoma, Clin. Exp. Metastasis 17 (10) (1999) 799–808, doi:10.1023/a:1006723011835.
[27] D.A. Fishman, L.M. Bafetti, M.S. Stack, Membrane-type matrix metalloproteinase expression and matrix metalloproteinase-2 activation in primary human ovarian epithelial carcinoma cells, Invasion Metastasis 16 (3) (1996) 150–159.
[28] A.F. Drew, T.J. Blick, M.A. Lafleur, E.L.M. Tim, M.J. Robbie, G.E. Rice, et al., Correlation of tumor- and stromal-derived MT1-MMP expression with progression of human ovarian tumors in SCID mice, Gynecol. Oncol. 95 (3) (2004) 437–448, doi:10.1016/j.ygyno.2004.08.032.
[29] E.A. Loughran, A.K. Leonard, T.S. Hilliard, R.C. Phan, M.G. Yemc, E. Harper, et al., Aging increases susceptibility to ovarian cancer metastasis in murine allograft models and alters immune composition of peritoneal adipose tissue, Neoplasia 20 (6) (2018) 621–631, doi:10.1016/j.neo.2018.03.007.
[30] D. Dhaliwal, T.G. Shepherd, Molecular and cellular mechanisms controlling integrin-mediated cell adhesion and tumor progression in ovarian cancer metastasis: a review, Clin. Exp. Metastasis 39 (2) (2021) 291–301, doi:10.1007/s10585-021-10136-5.
[31] E.I. Harper, T.S. Hilliard, E.F. Sheedy, P. Carey, P. Wilkinson, M.D. Siroky, et al., Another wrinkle with age: aged collagen and intra-peritoneal metastasis of ovarian cancer, Aging Cancer 3 (2) (2022) 116–129, doi:10.1002/aac2.12049.
[32] N.M. Moss, M.V. Barbolina, Y. Liu, L. Sun, H.G. Munshi, M.S. Stack, Ovarian cancer cell detachment and multicellular aggregate formation are regulated by membrane type

1 matrix metalloproteinase: a potential role in i.p. metastatic dissemination, Cancer Res. 69 (17) (2009) 7121–7129, doi:10.1158/0008-5472.can-08-4151.
[33] S.G. Bahar, V.R. Rokkam, Omental tumors, StatPearls Publishing, Treasure Island, 2024 (Accessed 29 May 2026) https://www.ncbi.nlm.nih.gov/books/NBK563148/ .
[34] Q. Cai, J. Yang, H. Shen, W. Xu, Cancer-associated adipocytes in the ovarian cancer microenvironment, Am. J. Cancer Res. 14 (7) (2024) 3259–3279, doi:10.62347/xzri9189.
[35] J. Adur, V.B. Pelegati, A.A. de Thomaz, M.O. Baratti, L.A. Andrade, H.F. Carvalho, et al., Second harmonic generation microscopy as a powerful diagnostic imaging modality for human ovarian cancer, J. Biophotonics 7 (1–2) (2012) 37–48, doi:10.1002/jbio.201200108.
[36] M.W. Conklin, J.C. Eickhoff, K.M. Riching, C.A. Pehlke, K.W. Eliceiri, P.P. Provenzano, et al., Aligned collagen is a prognostic signature for survival in human breast carcinoma, Am. J. Pathol. 178 (3) (2011) 1221–1232, doi:10.1016/j.ajpath.2010.11.076.
[37] E.A. Brett, M.A. Sauter, H.G. Machens, D. Duscher, Tumor-associated collagen signatures: pushing tumor boundaries, Cancer Metabol. 8 (1) (2020) 21, doi:10.1186/s40170-020-00221-w.
[38] T.L Moser, S.V Pizzo, L.M Bafetti, D.A Fishman, M.S. Stack, Evidence for preferential adhesion of ovarian epithelial carcinoma cells to type I collagen mediated by the integrin, Int. J. Cancer 67 (5) (1996) 695–701 10.1002/(SICI)1097-0215(19960904)67:5%3C695::AID-IJC18%3E3.0.CO;2-4.
[39] S.M. Ellerbroek, D.A. Fishman, A.S Kearns, L.M. Bafetti, M.S. Stack, Ovarian carcinoma regulation of matrix metalloproteinase-2 and membrane type 1 matrix metalloproteinase through beta1 integrin, Cancer Res. 59 (7) (1999) 1635–1641.
[40] K.M. Burleson, L.K. Hansen, A.P. Skubitz, Ovarian carcinoma spheroids disaggregate on type I collagen and invade live human mesothelial cell monolayers, Clin. Exp. Metastasis 21 (8) (2005) 685–697, doi:10.1007/s10585-004-5768-5.
[41] H. Sato, T. Takino, Y. Okada, J. Cao, A. Shinagawa, E. Yamamoto, et al., A matrix metalloproteinase expressed on the surface of invasive tumour cells, Nature 370 (6484) (1994) 61–65, doi:10.1038/370061a0.
[42] V. Gifford, Y. Itoh, MT1-MMP-dependent cell migration: proteolytic and non-proteolytic mechanisms, Biochem. Soc. Trans. 47 (3) (2019) 811–826, doi:10.1042/bst20180363.
[43] C.M. Overall, E. Tam, G.A. McQuibban, C. Morrison, U.M. Wallon, H.F. Bigg, et al., Domain interactions in the gelatinase A-TIMP-2-MT1-MMP activation complex, J. Biol. Chem. 275 (50) (2000) 39497–39506, doi:10.1074/jbc.m005932200.
[44] T.N. Young, S.V. Pizzo, M.S. Stack, A plasma membrane-associated component of ovarian adenocarcinoma cells enhances the catalytic efficiency of matrix metalloproteinase-2, J. Biol. Chem. 270 (3) (1995) 999–1002, doi:10.1074/jbc.270.3.999.
[45] Y. Itoh, N. Ito, H. Nagase, R.D. Evans, S.A. Bird, M. Seiki, Cell surface collagenolysis requires homodimerization of the membrane-bound collagenase MT1-MMP, Mol. Biol. Cell 17 (12) (2006) 5390–5399, doi:10.1091/mbc.e06-08-0740.
[46] S. Ingvarsen, D.H. Madsen, T. Hillig, L.R. Lund, K. Holmbeck, N. Behrendt, et al., Dimerization of endogenous MT1-MMP is a regulatory step in the activation of the 72-kDa gelatinase MMP-2 on fibroblasts and fibrosarcoma cells, Biochem. J. 389 (7) (2008) 943–953, doi:10.1515/bc.2008.097.
[47] Y. Shi, X. Ma, G. Fang, X. Tian, C. Ge, Matrix metalloproteinase inhibitors (MMPIs) as attractive therapeutic targets: recent progress and current challenges, Nanoimpact 21 (2021) 100293, doi:10.1016/j.impact.2021.100293.
[48] A. Jiang, K. Lehti, X. Wang, S.J. Weiss, J. Keski-Oja, D. Pei, Regulation of membrane-type matrix metalloproteinase 1 activity by dynamin-mediated endocytosis, Proc. Natl. Acad. Sci. U.S.A. 98 (24) (2001) 13693–13698, doi:10.1073/pnas.241293698.

[49] H. Lee, K.L. Sodek, Q. Hwang, T.J. Brown, M. Ringuette, J. Sodek, Phagocytosis of collagen by fibroblasts and invasive cancer cells is mediated by MT1-MMP, Biochem. Soc. Trans. 35 (4) (2007) 704–706, doi:10.1042/bst0350704.
[50] N.M. Moss, Y. Liu, J.J. Johnson, P. DeBiase, J. Jones, L.G. Hudson, et al., Epidermal growth factor receptor–mediated membrane type 1 matrix metalloproteinase endocytosis regulates the transition between invasive versus expansive growth of ovarian carcinoma cells in three-dimensional collagen, Mol. Cancer Res. 7 (6) (2009) 809–820, doi:10.1158/1541-7786.mcr-08-0571.
[51] W. Xue, L. Yang, C. Chen, M. Ashrafizadeh, Y. Tian, R. Sun, Wnt/-catenin-driven EMT regulation in human cancers, Cell. Mol. Life Sci. 81 (1) (2024) 42, doi:10.1007/s00018-023-05099-7.
[52] C.C. Lynch, L.M. Matrisian, Matrix metalloproteinases in tumor–host cell communication, Differentiation 70 (9–10) (2002) 561–573, doi:10.1046/j.1432-0436.2002.700909.x.
[53] K. Yamada, Integrin transmembrane signaling and cytoskeletal control, Curr. Opin. Cell Biol. 7 (5) (1995) 681–689, doi:10.1016/0955-0674(95)80110-3.
[54] J. Aggeler, S.M. Frisch, Z. Werb, Changes in cell shape correlate with collagenase gene expression in rabbit synovial fibroblasts, J. Cell Biol. 98 (5) (1984) 1662–1671, doi:10.1083/jcb.98.5.1662.
[55] Z. Werb, R.M. Hembry, G. Murphy, J. Aggeler, Commitment to expression of the metalloendopeptidases, collagenase and stromelysin: relationship of inducing events to changes in cytoskeletal architecture, J. Cell Biol. 102 (3) (1986) 697–702, doi:10.1083/jcb.102.3.697.
[56] Z. Werb, P.M. Tremble, O. Behrendtsen, E. Crowley, C.H. Damsky, Signal transduction through the fibronectin receptor induces collagenase and stromelysin gene expression, J. Cell Biol. 109 (2) (1989) 877–889, doi:10.1083/jcb.109.2.877.
[57] R.E. Seftor, E.A. Seftor, K.R. Gehlsen, W.G. Stetler-Stevenson, P.D. Brown, E. Ruoslahti, et al., Role of the alpha V beta 3 integrin in human melanoma cell invasion, Proc. Natl. Acad. Sci. U.S.A. 89 (5) (1992) 1557–1561, doi:10.1073/pnas.89.5.1557.
[58] V.P. Terranova, J.E. Williams, L.A. Liotta, G.R. Martin, Modulation of the metastatic activity of melanoma cells by laminin and fibronectin, Science 226 (4677) (1984) 982–985, doi:10.1126/science.6505678.
[59] T. Turpeenniemi-Hujanen, U.P. Thorgeirsson, C.N. Rao, L.A. Liotta, Laminin increases the release of type IV collagenase from malignant cells, J. Biol. Chem. 261 (4) (1986) 1883–1889, doi:10.1016/s0021-9258(17)36025-8.
[60] T. Kanemoto, R. Reich, L. Royce, D. Greatorex, S.H. Adler, N. Shiraishi, et al., Identification of an amino acid sequence from the laminin A chain that stimulates metastasis and collagenase IV production, Proc. Natl. Acad. Sci. U.S.A. 87 (6) (1990) 2279–2283, doi:10.1073/pnas.87.6.2279.
[61] S. Stack, R.D. Gray, S.V. Pizzo, Modulation of plasminogen activation and type IV collagenase activity by a synthetic peptide derived from the laminin A chain, Biochemistry 30 (8) (1991) 2073–2077, doi:10.1021/bi00222a011.
[62] Y. Jia, Z.Z. Zeng, S.M. Markwart, K.F. Rockwood, K.M. Ignatoski, S.P. Ethier, et al., Integrin fibronectin receptors in matrix metalloproteinase-1–dependent invasion by breast cancer and mammary epithelial cells, Cancer Res. 64 (23) (2004) 8674–8681, doi:10.1158/0008-5472.can-04-0069.
[63] C. Gilles, M. Polette, M. Seiki, P. Birembaut, E.W. Thompson, Implication of collagen type I-induced membrane-type 1-matrix metalloproteinase expression and matrix metalloproteinase-2 activation in the metastatic progression of breast carcinoma, Lab. Invest. 76 (5) (1997) 651–660.
[64] K. Wolf, I. Mazo, H. Leung, K. Engelke, U.H. von Andrian, E.I. Deryugina, et al., Compensation mechanism in tumor cell migration, J. Cell Biol. 160 (2) (2003) 267–277, doi:10.1083/jcb.200209006.

[65] S.M. Ellerbroek, Y.I. Wu, C.M. Overall, M.S. Stack, Functional interplay between type I collagen and cell surface matrix metalloproteinase activity, J. Biol. Chem. 276 (27) (2001) 24833–24842, doi:10.1074/jbc.m005631200.
[66] B.G. Gálvez, S. Matías-Román, M. Yáñez-Mó, F. Sánchez-Madrid, A.G. Arroyo, ECM regulates MT1-MMP localization with or integrins at distinct cell compartments modulating its internalization and activity on human endothelial cells, J. Cell Biol. 159 (3) (2002) 509–521, doi:10.1083/jcb.200205026.
[67] E.M. Tam, Y.I. Wu, G.S. Butler, M.S. Stack, C.M. Overall, Collagen binding properties of the membrane type-1 matrix metalloproteinase (MT1-MMP) hemopexin C domain, J. Biol. Chem. 277 (41) (2002) 39005–39014, doi:10.1074/jbc.m206874200.
[68] J. Pöllänen, K. Hedman, L.S. Nielsen, K. Danø, A. Vaheri, Ultrastructural localization of plasma membrane-associated urokinase-type plasminogen activator at focal contacts, J. Cell Biol. 106 (1) (1988) 87–95, doi:10.1083/jcb.106.1.87.
[69] B. Steffensen, H.F. Bigg, C.M. Overall, The involvement of the fibronectin type II-like modules of human gelatinase A in cell surface localization and activation, J. Biol. Chem. 273 (32) (1998) 20622–20628, doi:10.1074/jbc.273.32.20622.
[70] M. Takahashi, T. Tsunoda, M. Seiki, Y. Nakamura, Y. Furukawa, Identification of membrane-type matrix metalloproteinase-1 as a target of the -catenin/TCF4 complex in human colorectal cancers, Oncogene 21 (38) (2002) 5861–5867, doi:10.1038/sj.onc.1205755.
[71] N.D. Marchenko, G.N. Marchenko, R.N. Weinreb, J.D. Lindsey, A. Kyshtoobayeva, H.C. Crawford, et al., -catenin regulates the gene of MMP-26, a novel matrix metalloproteinase expressed both in carcinomas and normal epithelial cells, Int. J. Biochem. Cell Biol. 36 (5) (2004) 942–956, doi:10.1016/j.biocel.2003.12.007.
[72] J. Symowicz, B.P. Adley, K.J. Gleason, J.J. Johnson, S. Ghosh, D.A. Fishman, et al., Engagement of collagen-binding integrins promotes matrix metalloproteinase-9–dependent E-cadherin ectodomain shedding in ovarian carcinoma cells, Cancer Res. 67 (5) (2007) 2030–2039, doi:10.1158/0008-5472.can-06-2808.
[73] R.J. Burkhalter, J. Symowicz, L.G. Hudson, C.J. Gottardi, M.S. Stack, Integrin regulation of -catenin signaling in ovarian carcinoma, J. Biol. Chem. 286 (26) (2011) 23467–23475, doi:10.1074/jbc.m110.199539.
[74] I. Ota, X.Y. Li, Y. Hu, S.J. Weiss, Induction of a MT1-MMP and MT2-MMP-dependent basement membrane transmigration program in cancer cells by Snail1, Proc. Natl. Acad. Sci. U.S.A. 106 (48) (2009) 20318–20323, doi:10.1073/pnas.0910962106.
[75] S.J. Cho, B.Y. Jeong, Y.S. Song, C.G. Park, D.Y. Cho, H.Y. Lee, STAT3 mediates RCP-induced cancer cell invasion through the NF-κB/Slug/MT1-MMP signaling cascade, Arch. Pharm. Res. 45 (7) (2022) 460–474, doi:10.1007/s12272-022-01396-0.
[76] C.Y. Lin, K.Y. Wu, L.M. Chi, Y.H. Tang, H.J. Huang, C.H Lai, et al., Starvation-inactivated mTOR triggers cell migration via a ULK1-SH3PXD2A/Tks5-MMP14 pathway in ovarian carcinoma, Autophagy 19 (12) (2023) 3151–3168, doi:10.1080/15548627.2023.2239633.
[77] E.I. Harper, E.F. Sheedy, M.S. Stack, With great age comes great metastatic ability: ovarian cancer and the appeal of the aging peritoneal microenvironment, Cancers 10 (7) (2018) 230, doi:10.3390/cancers10070230.
[78] K.L Sodek, M.J Ringuette, T.J. Brown, MT1-MMP is the critical determinant of matrix degradation and invasion by ovarian cancer cells, Br. J. Cancer 97 (3) (2007) 358–367, doi:10.1038/sj.bjc.6603863.
[79] X. Guo, P. Asthana, S. Gurung, S. Zhang, S.K. Wong, S. Fallah, et al., Regulation of age-associated insulin resistance by MT1-MMP-mediated cleavage of insulin receptor, Nat. Commun. 13 (1) (2022) 3740, doi:10.1038/s41467-022-31563-2.
[80] X. Li, Y. Zhao, C. Chen, L. Yang, H. Lee, Z. Wang, et al., Critical role of matrix metalloproteinase 14 in adipose tissue remodeling during obesity, Mol. Cell. Biol. 40 (8) (2020) e00564–e00619, doi:10.1128/mcb.00564-19.

[81] T.H. Chun, M. Inoue, H. Morisaki, I. Yamanaka, Y. Miyamoto, T. Okamura, et al., Genetic link between obesity and MMP14-dependent adipogenic collagen turnover, Diabetes 59 (10) (2010) 2484–2494, doi:10.2337/db10-0073.

[82] C. López-Otín, M.A. Blasco, L. Partridge, M. Serrano, G. Kroemer, The hallmarks of aging, Cell 153 (6) (2013) 1194–1217, doi:10.1016/j.cell.2013.05.039.

[83] C. López-Otín, M.A. Blasco, L. Partridge, M. Serrano, G. Kroemer, Hallmarks of aging: an expanding universe, Cell 186 (2) (2023) 243–278, doi:10.1016/j.cell.2022.11.001.

[84] R. Yancik, Ovarian cancer: age contrasts in incidence, histology, disease stage at diagnosis, and mortality, Cancer 71 (S2) (1993) 517–523, doi:10.1002/cncr.2820710205.

[85] American Cancer Society Causes, risk factors, and prevention of ovarian cancer, American Cancer Society, Atlanta, 2026 (Accessed 29 May 2026) https://www.cancer.org/cancer/types/ovarian-cancer/causes-risks-prevention.html .

[86] N. Pavlidis, G. Stanta, R.A. Audisio, Cancer prevalence and mortality in centenarians: a systematic review, Crit. Rev. Oncol. Hematol. 83 (1) (2012) 145–152, doi:10.1016/j.critrevonc.2011.09.007.

[87] A.F. Caulin, C.C. Maley, Peto's paradox: evolution's prescription for cancer prevention, Trends Ecol. Evol. 26 (4) (2011) 175–182, doi:10.1016/j.tree.2011.01.002.

[88] R. Peto, F.J. Roe, P.N. Lee, L. Levy, J. Clack, Cancer and ageing in mice and men, Br. J. Cancer 32 (4) (1975) 411–426, doi:10.1038/bjc.1975.242.

[89] J. Campisi, The biology of replicative senescence, Eur. J. Cancer 33 (5) (1997) 703–709, doi:10.1016/s0959-8049(96)00058-5.

[90] J. Giroud, I. Bouriez, H. Paulus, A. Pourtier, F. Debacq-Chainiaux, O. Pluquet, Exploring the communication of the SASP: dynamic, interactive, and adaptive effects on the microenvironment, Int. J. Mol. Sci. 24 (13) (2023) 10788, doi:10.3390/ijms241310788.

[91] M. Selman, A. Pardo, Fibroageing: an ageing pathological feature driven by dysregulated extracellular matrix-cell mechanobiology, Ageing Res. Rev. 70 (2021) 101393, doi:10.1016/j.arr.2021.101393.

[92] H.H. Hu, G. Cao, X.Q. Wu, N.D. Vaziri, Y.Y. Zhao, Wnt signaling pathway in aging-related tissue fibrosis and therapies, Ageing Res. Rev. 60 (2020) 101063, doi:10.1016/j.arr.2020.101063.

[93] E.N. Tsakiri, K.K. Iliaki, A. Höhn, S. Grimm, I.S. Papassideri, T. Grune, et al., Diet-derived advanced glycation end products or lipofuscin disrupts proteostasis and reduces life span in Drosophila melanogaster, Free Radic. Biol. Med. 65 (2013) 1155–1163, doi:10.1016/j.freeradbiomed.2013.08.186.

[94] O.M.T. Pearce, R.M. Delaine-Smith, E. Maniati, S. Nichols, J. Wang, S. Böhm, et al., Deconstruction of a metastatic tumor microenvironment reveals a common matrix response in human cancers, Cancer Discov. 8 (3) (2018) 304–319, doi:10.1158/2159-8290.cd-17-0284.

[95] X. Wei, H. Lou, D. Zhou, Y. Jia, H. Li, Q. Huang, et al., TAGLN mediated stiffness-regulated ovarian cancer progression via RhoA/ROCK pathway, J. Exp. Clin. Cancer Res. 40 (1) (2021) 318, doi:10.1186/s13046-021-02091-6.

[96] H. Fujimoto, M. Yoshihara, R. Rodgers, S. Iyoshi, K. Mogi, E. Miyamoto, et al., Tumor-associated fibrosis: a unique mechanism promoting ovarian cancer metastasis and peritoneal dissemination, Cancer Metastasis Rev. 43 (3) (2024) 1037–1053, doi:10.1007/s10555-024-10169-8.

[97] V. Mieulet, C. Garnier, Y. Kieffer, T. Guilbert, F. Nemati, E. Marangoni, et al., Stiffness increases with myofibroblast content and collagen density in mesenchymal high grade serous ovarian cancer, Sci. Rep. 11 (1) (2021) 5254, doi:10.1038/s41598-021-83685-0.

[98] A.J. Bailey, R.G. Paul, L. Knott, Mechanisms of maturation and ageing of collagen, Mech. Ageing Dev. 106 (1–2) (1998) 1–56, doi:10.1016/s0047-6374(98)00119-5.

[99] N. Verzijl, J. DeGroot, S.R. Thorpe, R.A. Bank, J.N. Shaw, T.J. Lyons, et al., Effect of collagen turnover on the accumulation of advanced glycation end products, J. Biol. Chem. 275 (50) (2000) 39027–39031, doi:10.1074/jbc.m006700200.

[100] P. Affinito, S. Palomba, C. Sorrentino, C. Di Carlo, G. Bifulco, M.P. Arienzo, et al., Effects of postmenopausal hypoestrogenism on skin collagen, Maturitas 33 (3) (1999) 239–247, doi:10.1016/s0378-5122(99)00077-8.
[101] A. Hernandez-Segura, J. Nehme, M. Demaria, Hallmarks of cellular senescence, Trends Cell Biol. 28 (6) (2018) 436–453, doi:10.1016/j.tcb.2018.02.001.
[102] C. Hinterleitner, H.V. Goldberg, D. McHugh, V.J.A. Barthet, A. Filliol, S.W. Lowe, Senescence in cancer: hallmarks, paradoxes, and therapeutic promise, Cell 189 (8) (2026) 2357–2378, doi:10.1016/j.cell.2026.03.005.
[103] P. Jin, X. Duan, L. Li, P. Zhou, C. Zou, K. Xie, Cellular senescence in cancer: molecular mechanisms and therapeutic targets, Med. Comm. 5 (5) (2024) e542, doi:10.1002/mco2.542.
[104] D.L. Boaru, D. De Leon-Oliva, P. De Castro-Martinez, C. Garcia-Montero, O. Fraile-Martinez, B. García-González, et al., Extracellular matrix dysregulation in aging, calcification, and cancer diseases: insights into cellular senescence, inflammation, and novel therapeutic strategies, Int J Biol Sci 21 (15) (2025) 6808–6881, doi:10.7150/ijbs.119301.
[105] J. Krstic, J.F. Santibanez, Transforming growth factor-beta and matrix metalloproteinases: functional interactions in tumor stroma-infiltrating myeloid cells, ScientificWorldJournal 2014 (2014) 521754, doi:10.1155/2014/521754.
[106] H.J. Kwak, M.J. Park, H.K. Cho, C.M. Park, S.I. Moon, H.C. Lee, et al., Transforming growth factor-beta induces tissue inhibitor of metalloproteinase-1 expression via activation of extracellular signal-regulated kinase and SP1 in human fibrosarcoma cells, Mol. Cancer Res. 4 (3) (2006) 209–220, doi:10.1158/1541-7786.mcr-05-0140.
[107] S. Djediai, N. Gonzalez Suarez, L. El Cheikh-Hussein, S.R. Torres, L. Gresseau, S. Dhayne, et al., MT1-MMP cooperates with TGF-beta receptor-mediated signaling to trigger Snail and induce epithelial-to-mesenchymal-like transition in U87 glioblastoma cells, Int. J. Mol. Sci. 22 (23) (2021) 13006, doi:10.3390/ijms222313006.
[108] N. Tanaka, T. Sakamoto, MT1-MMP as a key regulator of metastasis, Cells 12 (17) (2023) 2187, doi:10.3390/cells12172187.
[109] Y.P. Han, T.L. Tuan, H. Wu, M. Hughes, W.L. Garner, TNF-alpha stimulates activation of pro-MMP2 in human skin through NF-κB mediated induction of MT1-MMP, J. Cell Sci. 114 (1) (2001) 131–139, doi:10.1242/jcs.114.1.131.
[110] World Health Organization Obesity and overweight, World Health Organization, Geneva, 2026 (Accessed 29 May 2026) https://www.who.int/news-room/fact-sheets/detail/obesity-and-overweight .
[111] A. Bocian-Jastrzębska, A. Malczewska-Herman, B. Kos-Kudła, Role of leptin and adiponectin in carcinogenesis, Cancers 15 (17) (2023) 4250, doi:10.3390/cancers15174250.
[112] C.L. Donohoe, J. Lysaght, J. O'Sullivan, J.V. Reynolds, Emerging concepts linking obesity with the hallmarks of cancer, Trends Endocrinol. Metab. 28 (1) (2017) 46–62, doi:10.1016/j.tem.2016.08.004.
[113] Y. Liu, J. Yang, T.S. Hilliard, Z. Wang, J. Johnson, W. Wang, et al., Host obesity alters the ovarian tumor immune microenvironment and impacts response to standard of care chemotherapy, J. Exp. Clin. Cancer Res. 42 (1) (2023) 274, doi:10.1186/s13046-023-02740-y.
[114] T. Skurk, C. Alberti-Huber, C. Herder, H. Hauner, Relationship between adipocyte size and adipokine expression and secretion, J. Clin. Endocrinol. Metab. 92 (3) (2007) 1023–1033, doi:10.1210/jc.2006-1055.
[115] K. Wåhlen, E. Sjölin, P. Löfgren, Role of fat cell size for plasma leptin in a large population based sample, Exp. Clin. Endocrinol. Diabetes 119 (05) (2011) 291–294, doi:10.1055/s-0031-1273738.

[116] Q. Gao, T.L. Horvath, Cross-talk between estrogen and leptin signaling in the hypothalamus, Am. J. Physiol. Endocrinol. Metab. 294 (5) (2008) E817–E826, doi:10.1152/ajpendo.00733.2007.
[117] C.S. Mantzoros, F. Magkos, M. Brinkoetter, E. Sienkiewicz, T.A. Dardeno, S.Y. Kim, et al., Leptin in human physiology and pathophysiology, Am. J. Physiol. Endocrinol. Metab. 301 (4) (2011) E567–E584, doi:10.1152/ajpendo.00315.2011.
[118] A. Ray, M.P. Cleary, The potential role of leptin in tumor invasion and metastasis, Cytokine Growth Factor Rev. 38 (2017) 80–97, doi:10.1016/j.cytogfr.2017.11.002.
[119] A. Caruso, L. Gelsomino, S. Panza, F.M. Accattatis, G.D. Naimo, I. Barone, et al., Leptin: a heavyweight player in obesity-related cancers, Biomolecules 13 (7) (2023) 1084, doi:10.3390/biom13071084.
[120] A. Booth, A. Magnuson, J. Fouts, M. Foster, Adipose tissue, obesity and adipokines: role in cancer promotion, Horm. Mol. Biol. Clin. Investig. 21 (1) (2015) 57–74, doi:10.1515/hmbci-2014-0037.
[121] C. Jiménez-Cortegana, A. López-Saavedra, F. Sánchez-Jiménez, A. Pérez-Pérez, J. Castiñeiras, J.A. Virizuela-Echaburu, et al., Leptin, both bad and good actor in cancer, Biomolecules 11 (6) (2021) 913, doi:10.3390/biom11060913.
[122] M.N. VanSaun, Molecular pathways: adiponectin and leptin signaling in cancer, Clin. Cancer Res. 19 (8) (2013) 1926–1932, doi:10.1158/1078-0432.ccr-12-0930.
[123] H. Wang, H. Cheng, Q. Shao, Z. Dong, Q. Xie, L. Zhao, et al., Leptin-promoted human extravillous trophoblast invasion is MMP14 dependent and requires the cross talk between Notch1 and PI3K/akt signaling, Biol. Reprod. 90 (4) (2014), doi:10.1095/biolreprod.113.114876.
[124] J.C. Juárez-Cruz, M.D. Zuñiga-Eulogio, M. Olea-Flores, E. Castañeda-Saucedo, M.Á. Mendoza-Catalán, C. Ortuño-Pineda, et al., Leptin induces cell migration and invasion in a FAK-Src-dependent manner in breast cancer cells, Endocr. Connect 8 (11) (2019) 1539–1552, doi:10.1530/ec-19-0442.
[125] E. Murawska-Ciałowicz, Adipose tissue—morphological and biochemical characteristic of different depots, Postępy Hig. Med. Dośw. 71 (1) (2017) 466–484, doi:10.5604/01.3001.0010.3829.
[126] T.M. Nguyen, Adiponectin: role in physiology and pathophysiology, Int. J. Prev. Med. 11 (1) (2020) 1–9, doi:10.4103/ijpvm.ijpvm_193_20.
[127] H.M. Choi, H.M. Doss, K.S. Kim, Multifaceted physiological roles of adiponectin in inflammation and diseases, Int. J. Mol. Sci. 21 (4) (2020) 1219, doi:10.3390/ijms21041219.
[128] H. Cao, Adipocytokines in obesity and metabolic disease, J. Endocrinol. 220 (2) (2014) T47–T59, doi:10.1530/joe-13-0339.
[129] R. Akter, M. Awais, V. Boopathi, J.C. Ahn, D.C. Yang, S.C. Kang, et al., Inversion of the Warburg effect: unraveling the metabolic nexus between obesity and cancer, ACS Pharmacol. Transl. Sci. 7 (3) (2024) 560–569, doi:10.1021/acsptsci.3c00301.
[130] O. Warburg, F. Wind, E. Negelein, The metabolism of tumors in the body, J. Gen. Physiol. 8 (6) (1927) 519–530, doi:10.1085/jgp.8.6.519.
[131] M. Olea-Flores, J.C. Juárez-Cruz, M.D. Zuñiga-Eulogio, E. Acosta, E. García-Rodríguez, A.E. Zacapala-Gomez, et al., New actors driving the epithelial–mesenchymal transition in cancer: the role of leptin, Biomolecules 10 (12) (2020) 1676, doi:10.3390/biom10121676.
[132] L. Wang, C. Tang, H. Cao, K. Li, X. Pang, L. Zhong, et al., Activation of IL-8 via PI3K/AKT-dependent pathway is involved in leptin-mediated epithelial-mesenchymal transition in human breast cancer cells, Cancer Biol. Ther. 16 (8) (2015) 1220–1230, doi:10.1080/15384047.2015.1056409.
[133] L. Sun, P. Lin, Z. Qin, Y. Liu, L.L. Deng, C. Lu, Hypoxia promotes HO-8910PM ovarian cancer cell invasion via snail-mediated MT1-MMP upregulation, Exp. Biol. Med. 240 (11) (2015) 1434–1445, doi:10.1177/1535370215570205.

[134] J.M. Friedman, Leptin and the regulation of body weigh, Keio J. Med. 60 (1) (2011) 1–9, doi:10.2302/kjm.60.1.
[135] L.M. Coussens, B. Fingleton, L.M. Matrisian, Matrix metalloproteinase inhibitors and cancer—trials and tribulations, Science 295 (5564) (2002) 2387–2392, doi:10.1126/science.1067100.
[136] H.S. Rasmussen, P.P. McCann, Matrix metalloproteinase inhibition as a novel anticancer strategy: a review with special focus on Batimastat and Marimastat, Pharmacol. Ther. 75 (1) (1997) 69–75, doi:10.1016/s0163-7258(97)00023-5.
[137] P.D. Brown, R. Giavazzi, Matrix metalloproteinase inhibition: a review of anti-tumour activity, Ann. Oncol. 6 (10) (1995) 967–974, doi:10.1093/oxfordjournals.annonc.a059091.
[138] V.E. de Meijer, D.Y. Sverdlov, Y. Popov, H.D. Le, J.A. Meisel, V. Nosé, et al., Broad-spectrum matrix metalloproteinase inhibition curbs inflammation and liver injury but aggravates experimental liver fibrosis in mice, PLoS One 5 (6) (2010), doi:10.1371/journal.pone.0011256.
[139] S.L. Parsons, S.A. Watson, R.J.C. Steele, Phase I/II trial of batimastat, a matrix metalloproteinase inhibitor, in patients with malignant ascites, Eur. J. Surg. Oncol. 23 (6) (1997) 526–531, doi:10.1016/s0748-7983(97)93077-8.
[140] M. Hashimoto, J.I. Sasaki, S. Yamaguchi, K. Kawai, H. Kawakami, Y. Iwasaki, et al., Gold nanoparticles inhibit matrix metalloproteases without cytotoxicity, J. Dent. Res. 94 (8) (2015) 1085–1091, doi:10.1177/0022034515589282.
[141] S. Kang, G. Zhou, P. Yang, Y. Liu, B. Sun, T. Huynh, et al., Molecular mechanism of pancreatic tumor metastasis inhibition by GD@C82(OH)22 and its implication for de novo design of nanomedicine, Proc. Natl Acad. Sci. U.S.A. 109 (38) (2012) 15431–15436, doi:10.1073/pnas.1204600109.
[142] Y. Shi, K. Pu, H. Yao, Y. Chen, X. Zheng, L. Zhao, et al., Gold nanorods inhibit tumor metastasis by regulating MMP-9 activity: implications for radiotherapy, ACS Appl. Mater. Interfaces 15 (7) (2023) 9034–9043, doi:10.1021/acsami.2c20944.
[143] P. Shahmoradipour, R. Emamzadeh, Advances in the use of MT1-MMP-targeted nanobodies and peptides for therapeutic delivery in triple-negative breast cancer, Biomed. Pharmacother. 194 (2026) 118944, doi:10.1016/j.biopha.2025.118944.
[144] D.H. Nam, C. Rodriguez, A.G. Remacle, A.Y. Strongin, X. Ge, Active-site MMP-selective antibody inhibitors discovered from convex paratope synthetic libraries, Proc. Natl. Acad. Sci. U.S.A. 113 (52) (2016) 14970–14975, doi:10.1073/pnas.1609375114.
[145] A.G. Remacle, P. Cieplak, D.H. Nam, S.A. Shiryaev, X. Ge, A.Y. Strongin, Selective function-blocking monoclonal human antibody highlights the important role of membrane type-1 matrix metalloproteinase (MT1-MMP) in metastasis, Oncotarget 8 (2) (2016) 2781–2799, doi:10.18632/oncotarget.13157.
[146] M. Békés, D.R. Langley, C.M. Crews, PROTAC targeted protein degraders: the past is prologue, Nat. Rev. Drug. Discov. 21 (3) (2022) 181–200, doi:10.1038/s41573-021-00371-6.
[147] M. Ambrose, J. Lee, A. Syed, Z. Ahmed, G. Peng, Non-enzymatic protein targeting agents as a promising strategy for cancer treatment, Front. Drug. Discov. 5 (2025) 1520734, doi:10.3389/fddsv.2025.1520734.
[148] K.M. DeMars, C. Yang, C.I. Castro-Rivera, E. Candelario-Jalil, Selective degradation of BET proteins with DBET1, a proteolysis-targeting chimera, potently reduces pro-inflammatory responses in lipopolysaccharide-activated microglia, Biochem. Biophys. Res. Commun. 497 (1) (2018) 410–415, doi:10.1016/j.bbrc.2018.02.096.
[149] F. Tong, Y. Wang, Y. Xu, Y. Zhou, S. He, Y. Du, et al., MMP-2-triggered, mitochondria-targeted PROTAC-PDT therapy of breast cancer and brain metastases inhibition, Nat. Commun. 15 (1) (2024), doi:10.1038/s41467-024-54854-2.

[150] M. Rivera-Serrano, M. Flores-Colón, F. Valiyeva, L.M. Meléndez, P.E. Vivas-Mejía, Upregulation of MMP3 promotes cisplatin resistance in ovarian cancer, Int. J. Mol. Sci. 26 (9) (2025) 4012, doi:10.3390/ijms26094012.
[151] N. Cai, K. Cheng, Y. Ma, S. Liu, R. Tao, Y. Li, et al., Targeting MMP9 in CTNNB1 mutant hepatocellular carcinoma restores CD8+ T cell-mediated antitumour immunity and improves anti-PD-1 efficacy, Gut 73 (6) (2023) 985–999, doi:10.1136/gutjnl-2023-331342.
[152] N. Priego, A. de Pablos-Aragoneses, M. Perea-García, V. Pieri, C. Hernández-Oliver, L. Álvaro-Espinosa, et al., TIMP1 mediates astrocyte-dependent local immunosuppression in brain metastasis acting on infiltrating CD8+ T cells, Cancer Discov. 15 (1) (2024) 179–201, doi:10.1158/2159-8290.cd-24-0134.
[153] S. Boumiza, K. Chahed, Z. Tabka, M.P. Jacob, X. Norel, G. Ozen, MMPs and timps levels are correlated with anthropometric parameters, blood pressure, and endothelial function in obesity, Sci. Rep. 11 (1) (2021) 20052, doi:10.1038/s41598-021-99577-2.
[154] F. Sabeh, R. Shimizu-Hirota, S.J. Weiss, Protease-dependent versus -independent cancer cell invasion programs: three-dimensional amoeboid movement revisited, J. Cell Biol. 185 (1) (2009) 11–19, doi:10.1083/jcb.200807195.
[155] K. Hotary, X.Y. Li, E. Allen, S.L. Stevens, S.J. Weiss, A cancer cell metalloprotease triad regulates the basement membrane transmigration program, Genes Dev. 20 (19) (2006) 2673–2686, doi:10.1101/gad.1451806.

CHAPTER TEN

Challenges and opportunities in the drug design and pharmacological applications of MMP inhibitors

Claudiu T. Supuran [a,*], **Antoine Dufour** [b]
[a] Neurofarba Department, Pharmaceutical and Nutraceutical Section, University of Florence, Sesto Fiorentino, FI, Italy
[b] McCaig Institute for Bone and Joint Health, Department of Physiology and Pharmacology, University of Calgary, Calgary, Canada
*Corresponding author. e-mail address: claudiu.supuran@unifi.it.

Contents

Abstract

The 23 members of the matrix metalloproteinase (MMP) family are involved in a multitude of physiological processes, such as cell migration, invasion, cell signaling, immune response, fibrosis, extracellular matrix remodeling, tissue homeostasis, but also in a broad range of diseases, among which cancer, inflammation (sepsis, arthritis, asthma, chronic obstructive pulmonary disorder), periodontitis, etc. Their biochemical characterization, expression patterns, role as targets, offtargets or antitargets in many physiologic or pathologic conditions made relevant progress over the last decades. MMP inhibitors (MMPIs) belonging to a variety of classes (hydrozamates, carboxylates, thiols, phosphonates, etc.) are well known. The first/second generations such derivatives, all zinc binders, have been investigated in clinical trials as anticancer agents in the late '90s, with poor

Fungal and Protozoan Carbonic Anhydrases, Volume 60
ISSN 1874-6047
https://doi.org/10.1016/bs.enz.2026.03.005

results, probably due to their metabolic instability, toxicity and lack of selectivity for the target isoform(s). Recent developments in the field saw the design of highly selective inhibitors for several MMPs, belonging to the modified tetracyclines class, as well as mechanism based, suicide inhibitors belonging to the substituted thiiranes class, exosite and allosteric inhibitors which spare the zinc ion from the enzyme active site. Two compounds are in clinical use for the management of periodontitis (doxycycline, at submicrobial dosages) and abametapir for the treatment of lice infestation. Clinical trials in various types of cancer, in different inflammatory conditions, and microbial infections (tuberculosis) with some of the new generation MMPIs are ongoing, and the expected results will demonstrate their efficacy, probably in combination therapy with other drugs. More selective new inhibitors, better planned clinical trials and a better understanding of the biology/pharmacology of these enzymes may lead to other breakthroughs in the field.

Keywords: Abametapir, Cancer, Carboxylate, Doxycycline, Hydroxamate, Inflammation, Matrix metalloproteinase (MMP), MMP inhibitors (MMPIs), Modified tetracycline, Periodontitis, Zinc enzymes

1. Introduction

There are 23 human matrix metalloproteinases (MMPs) (labelled as MMP1-MMP28, with some members of the family, such as MMP4, MMP5, and MMP6, being reidentified with proteases described earlier and thus no longer valid), all belonging to the metzincin class of zinc proteases [1–3]. MMPs are multidomain proteins with a rather complex structural organization which has been described in detail in Chapter 1 of the book and also recently reviewed [1–5]. Thus, such information will not be repeated here. The MMPs possess a multitude of substrates among which collagen and other extracellular matrix (ECM) proteins, but also chemokines, cytokines, growth factors, cell surface receptors, endogenous protease inhibitors, and several intracellular proteins (see Chapter 1 and refs. [1,3–9]). Thus, the MMPs are involved through their catalytic activity (or via protein-protein interactions) in a multitude of physiological processes, such as cell migration, invasion, cell signaling, immune response, fibrosis, ECM remodeling, tissue homeostasis, but also in a broad range of diseases (see later in the text), which raised substantial interest in their possible targeting for obtaining modulators of their activity with pharmacological applications, already in the early 1990s [4–10]. Indeed, effective MMP inhibitors (MMPIs) have been available for at least three decades, and most of them belong to the zinc binders [3–10]. Similar to other metalloenzymes studied in detail, such as the carbonic anhydrase [11,12], the metal ion from the MMP active site is coordinated by three His residues (in a tetrahedral

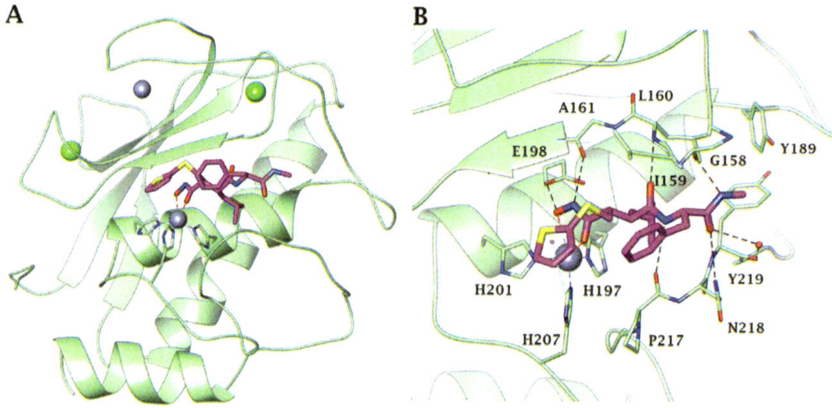

Fig. 1 MMP8 with Batimastat **1** (see Fig. 3 for its chemical structure) bound. **A**. General view of the catalytic domain of the protease, with its two Zn(II) ions in grey and the calcium ions in green. **B**. Detailed view of the inhibitor (in magenta) coordinated in trigonal bipyramidal geometry of the zinc. H-bonds of the inhibitor with various amino acid residues are shown as dotted lines. The catalytic zinc is also coordinated by residues His197, 201 and 207.

geometry of zinc), with the fourth ligand being a water molecule which is nucleophilically activated to attack the scissile peptide bond of the substrate [1,2,7,9]. Most MMPIs belonging to the first and second generations of such compounds basically substituted the water molecule coordinated to zinc, binding usually bidentately, or in some cases also in a monodentate fashion to Zn(II) within the active site, also participating in many favorable interaction with various subsites of the protease (both in the primed and non-primed pockets, e.g., S1, S2 and/or S1', S2,' etc.) [1–3]. As for many other zinc enzymes, the zinc-binding groups (ZBGs) present in MMPIs include the hydroxamate [1–3,6–9,13,14], carboxylate [14], thiol [14,15], phosphonate [16–18], and many other moieties with affinity to coordinate this metal ion [19].

As shown in Figs. 1 and 2 for two such inhibitors [20,21], the hydroxamate batimastat **1** and the barbiturate inhibitor **2**, bind deep within the enzyme active site, making a large number of favorable interactions with the catalytic metal ion and amino acid residues in the neighboring pockets.

Extensive structural biology data afforded the design of effective inhibitors for many MMPs of the 23 human enzymes, in some cases also with selectivity for some members of interest, such as collagenases (MMP1, -8

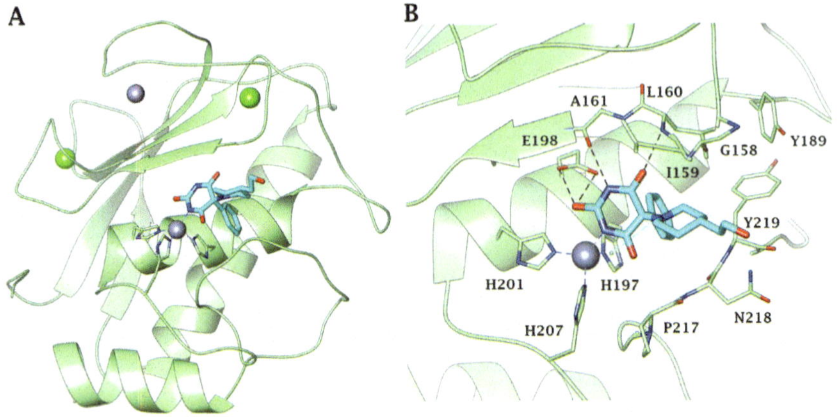

Fig. 2 MMP8 with the barbiturate inhibitor **2** (see Fig. 3 for its chemical structure) bound to it. A. General view of the catalytic domain of the protease, with its two Zn(II) ions in grey and the calcium ions in green. B. Detailed view of the inhibitor (in sky blue) coordinated to the catalytic Zn ion in tetrahedral geometry. H-bonds of the inhibitor with various amino acid residues are shown as dotted lines. The catalytic zinc is also coordinated by residues His197, 201 and 207.

and -13) or gelatinases (MMP2- and -9), though to be involved in cancer in the 1990s, with a large number of hydroxamates (first generation MMPIs) and carboxylates (second generation inhibitors) [1–3,9] investigated in great detail. A large number of chemical structures/scaffolds of such MMPIs is known at the moment, not only hydroxamates/carboxylates, but also thiols, phosphorus-based inhibitors, mechanism-based inhibitors, chemically modified tetracyclines, as well as allosteric and exosite inhibitors, which do not bind the catalytic metal ion [1,3,6,9,14,22].

Unfortunately, most of them have failed in multiple clinical trials, mainly as antitumor agents (more than 50 MMPIs were investigated in such trials for the management of various forms of cancer), raising thus serious concerns regarding this entire class [2]. In this chapter, we will try to dissect why are there so many challenges to develop therapeutics based on MMPIs, also highlighting some recent, more successful approaches, which led to clinically used compounds, such as doxycycline hydrate (Periostat®) for the management of periodontal disease and abatemapir for the treatment of lice infestation [3,14]. However, the field of MMPIs may open many new opportunities for the treatment of a variety of other conditions, as it will be discussed later in the chapter.

2. Challenges: Selectivity/toxicity issues

Several factors are probably responsible for the lack of efficacy observed in clinical trials with the first and second generation MMPIs, which will be discussed in this section.

2.1 Lack of selectivity for the target MMP over offtarget MMPs and other enzymes

As mentioned above, the MMPs are a family of 23 enzymes described so far in humans [1–4]. This situation is rather unusual among enzymes, with most of the crucial such biocatalysts involved in relevant physiological functions possessing a much more limited number of isoforms, usually one or two. For example, acetylcholinesterase (AchE), which hydrolyses the most important neurotransmitter, acetylcholine, has just one isoform [23,24]; monoamine oxidases, which metabolize the three monoamine neurotransmitters, noradrenaline, dopamine and serotonin, have two isoforms, MAO-A and MAO-B [25]; cyclooxygenases, involved in the prostagladin/leukotriene biosynthesis, possess two main isoforms COX-1 and COX-2 (COX-3 is a splicing variant of COX-1, with a dubious physiological functions) [26]. The only other reported case of a crucially important enzyme possessing a multitude of isoforms (in vertebrates) is the carbonic anhydrase (CA) superfamily, with 16 isoforms described so far, CA I–XV (there are two V-type isoforms, CA VA and VB) [11,12,27,28]. Similar to the MMPs, the CAs are also zinc enzymes and they possess a similar metal ion coordination within the active site, with three His residues and a water molecule acting as nucleophile in the reaction of CO_2 hydration to bicarbonate and protons [11,12,27,28]. However, the active site architectures of the two families of enzymes are highly different and the CAs do not possess protease activity (although they act as esterases on a multitude of substrates [29,30]). Of the 16 different mammalian CAs, 13 possess catalytic activity and they are generally widely expressed in many cells and tissues [27,28]. Developing potent and isoform-selective inhibitors for the CAs was rather challenging, but significant progress has been achieved in the last decade, using crystallographic, kinetics and other biochemical/biophysical techniques, which rendered this goal achievable [27,28]. In fact, nowadays there are many isoform-selective CA inhibitors for all the human isoforms with pharmacological applications [31].

The case of the MMPs is however more challenging due to the higher number of isoforms compared to the CAs, their diverse and multiple substrates [1,4,5], participation in a protease web, interconnecting their activity with that of other such enzymes [5], and last but not least, due to the fact that only a limited number of MMPs were considered as drug targets (MMP1-3, -8, -9, -12, and -14 [3]), and as thus, the detailed information on selectivity for different classes of compounds is scarce. To make things even more complex, there are other families of metalloproteases, such as that of "a disintegrin and metalloproteinases" (ADAMs), of which around 30 different members were described [32], with the tumor necrosis factor-alpha converting enzyme (TACE, ADAM17) being the most investigated one [33], which possess the same active site metal ion, in the same coordination environment, possessing proteolytic activity/inhibition profiles which might overlap with those of some MMPs. The ADAMTS (a disintegrin and metalloproteinase with thrombospondin motif) is another family of such enzymes to be considered [34]. An off-targeting of MMPIs belonging to the hydroxamate, carboxylate and sulfonamide/N–hydroxy-sulfonamide classes to various CAs has also been reported [35], with some of these compounds showing significant inhibition of both families of enzymes (from the micro- to the nanomolar range) [35]. All these factors may diminish the efficacy of a MMPI on the target isoform, "diluting" its pharmacological effects due to such a cross-reactivity with other MMPs (offtargets/anti-targets) or other zinc enzymes, such as the ADAMs, ADAMTSs, the CAs, etc.

2.2 Expression patterns of target/offtarget/antitarget MMPs

More than 600 (but there are estimates that the real number may be much higher) different substrates are known to date for the MMPs, of which only around 31 percent are ECM proteins, with the remaining 69 percent being non-ECM proteins, such as other proteases, chemokines, cytokines, growth factors, cell–cell adhesion molecules as well as cytosolic and even nuclear proteins (although the ways by which MMPs can be localized in such districts of the cells are poorly understood at the moment) [1,5,36]. Connected with the fact that unlike most other proteases, the MMPs do not have strict or obvious consensus amino acid recognition sequences that are cleaved by the proteolytic action of the enzyme, and considering the huge number of their substrates, MMPs thus play numerous additional biological roles except ECM remodeling, such as in cell signaling, cell migration, invasion, immune regulation, and transcriptional control [1–5,36].

Enhanced MMP expression of many individual/multiple isoforms has been reported in a variety of diseases, such as cancer (MMP2, -3, -7, -8, -9, -11, -12, -14, -19, -26); osteoarthritis (MMP1, -8, -13, etc.); periodontitis (MMP8, -13); asthma and lung emphysema (MMP12); vascular diseases (MMP1, -2, -9, -12); as well as a range of inflammatory conditions, such as sepsis and septic shock (MMP1, -2, -7, -8, -9 and -13); intestinal inflammation, including inflammatory bowel disease (IBD), necrotizing enterocolitis (NEC), Crohn's disease (MMP1, -3, -9 and -13)—but MMP10 seems to have a protective role in these conditions [2]; brain inflammation, leading to stroke, traumatic brain injury, multiple sclerosis, hypoxic ischemia, Alzheimer's disease, meningitis, etc., provoked due to blood brain barrier (BBB) damage (MMP9), but it should be mentioned that MMP2 and -9 seem to also have a protective role in the repair after nervous system injury [1,2]). Thus, intervening with a specific MMPI for the protease(s) showing such a dysregulated activity in all these diseases seems to be in principle beneficial for the management of such conditions. However, as shown by Dufour and Overall, the reality is much more complex: 10 of the 24 murine MMPs were shown to be anti-targets and to possess beneficial properties, their inhibition possibly being detrimental for the organism [5]. They include MMP2, -3, -7, -8, -9, -11, -12, 13, -14, and -19, and their beneficial exerted effects are, among others: inhibition of angiogenesis and metastasis, promotion of spinal cord injury recovery (MMP2); migration of CD_{4+} T lymphocytes and stimulation of apoptosis (MMP3); mucosal defense against bacterial infection (MMP7); apoptosis, inhibition of tumor growth and metastasis, regulation of inflammation in arthritis (MMP8); bacterial clearance, dampening of systemic autoimmunity, host response to bacterial peritonitis, inhibition of angiogenesis and metastasis, reduction of hemorrhage and brain edema (MMP9); reduction of metastasis formation (MMP11); inhibition of angiogenesis and metastasis, abrogation of neutrophils and macrophage recruitment, antibacterial activity, effector of T-cell polarization (MMP12); bone formation and remodeling (MMP13); macrophages regulation, osteocyte processing, upregulation of anti-inflammatory cytokines, nuclear trafficking, (MMP14); inhibition of angiogenesis and tumor invasion (MMP-19). Since the first and second generation MMPIs were in most cases pan-inhibitors, they interfered both with the targets and the anti-target MMPs, which may explain some of their serious side effects and lack of efficacy (as antitumor agents) [1–5]. As noted by Dufour and Overall [5], long-term broad-spectrum MMPI drug administration should be avoided to reduce severe side effects and

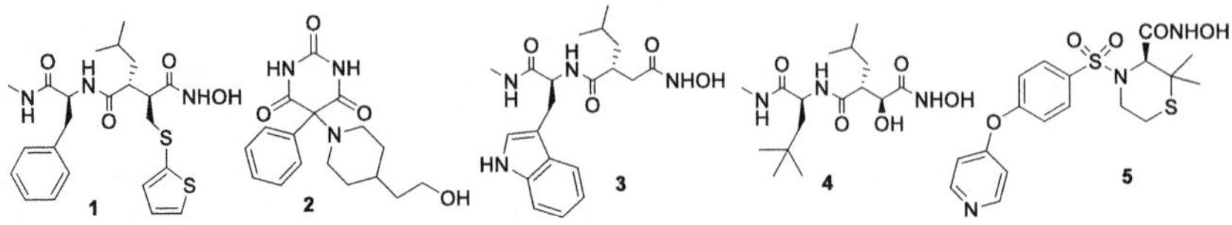

Fig. 3 MMPIs of the hydroxamate/barbiturate type, **1–5**. Batimastat **1**, RO200-1770 **2**, ilomastat **3**, marimastat **4** and prinimastat **5**.

inhibition of the anti-target enzymes, whereas ultraselective inhibitors for a certain target of the 23 human MMPs should be developed and used in the various pathologies where the exact isoform(s) involved are well-known. Considering the lessons of the past, with the more than 50 drugs of the class having failed in clinical trials, the future MMPIs to enter new trials should possess a precise, well-tailored inhibition profile, such as to spare the beneficial, anti-target enzyme, and be ultraselective and effective for the target one(s).

2.3 Poor pharmacokinetics, stability issues and side effects/toxicity

The first/second generation MMPIs, such as compounds shown in Fig. 3, mainly belonged to the hydroxamate class of inhibitors. Batimastat **1** and ilomastat **3** were not bioavailable for oral administration and were given parenterally; the newer derivatives such as marimastat **4** or prinomastat **5** (and many other such peptidomimetic MMPIs) were orally active and they entered clinical trials for the treatment of various advanced cancers (e.g., pancreas, brain, lung, prostate, tumors affecting the gastrointestinal tract, etc.) but they showed a lack of efficacy and serious side effects, such as stiffened joints and musculoskeletal pain, among others [1–6]. All of them failed in Phase III clinical trials, and nowadays only ilomastat shows some cosmetic applications in skincare products for its antiaging properties [37] or for the management of ocular scars [38].

It has been disclosed at the end of the 1990s, when it became apparent that all these more than 50 clinical trials of MMPIs as anticancer agents were a failure, that MMPIs might be more effective in earlier stages of cancers [39], whereas all clinical trials have been performed with patients in advanced stages of the disease with a focus on treating metastasis [1–3]. Apart the lack of selectivity for the target MMP isoforms (only prinomastat shows some selective inhibitory effects on 4 isoforms involved in cancer, e.g., MMP2, -9,

-13 and -14 [40]) and inhibition of anti-target such enzymes, all hydroxamates are generally poorly bioavailable, due to the highly polar nature of the hydroxamate moiety, although some of the more hydrophobic inhibitors of Fig. 3 (compounds **4** and **5**) were possible to be administered orally [41]. Furthermore, hydroxamates are not metabolically stable, as they are metabolized in the liver by the uridine diphosphate-glucuronosyltransferases (UGTs) and cytochrome P450s (CYPs) pathways, which has as a consequence their short half-lives after oral administration [41]. Hydroxamates are also quite toxic compounds due to at least three different mechanisms: (i) they can chelate other metal ions than the Zn(II) in MMPs, such as the zinc ion in other metalloenzymes (CAs, histone deacetylases (HDACs), etc.), but also iron in hemoglobin/myoglobin, and other heme and non-heme-containing iron enzymes [41–44]; (ii) they can undergo hydrolysis with generation of hydroxylamine, a toxic and metabolically highly reactive compound [45]; (iii) they can lead to the formation of isocyanates via the Lossen rearrangement, generating thus highly reactive intermediates which thereafter react with many nucleophiles found in biological fluids, leading to genotoxicity and mutagenicity [42].

All these data presented here point out to the many critical issues related to the use of hydroxamate MMPIs in the clinical trials as anticancer agents and should be attentively considered for future clinical trials.

2.4 Differences between animal models and humans

The failure of the anticancer clinical trials with MMPIs was also considered to be due to the inappropriate mice models used, as these models could not be easily applicable to the patients enrolled in the trials due to several reasons: (i) most mice models harbored tumors in initial phases whereas the patients in the clinical trials were in advanced stages of the disease and patients with metastases [46,47]; (ii) the nude mice were used in many such models, although it is well known that these animals possess a highly compromised immune system, and as thus, they do not realistically mimic the situation of the cancer patients [48,49]; (iii) the pharmacokinetics of mice and humans may be quite different, as shown in some studies using MMPIs [50]. Furthermore, mice possess 24 MMP (due do a duplication in MMP1: MMP1a and MMP1b) isoforms whereas humans just 23 [1,5]. These differences may have thus contributed to the lack of efficacy observed in the mentioned clinical trials with MMPIs as antitumor agents [2].

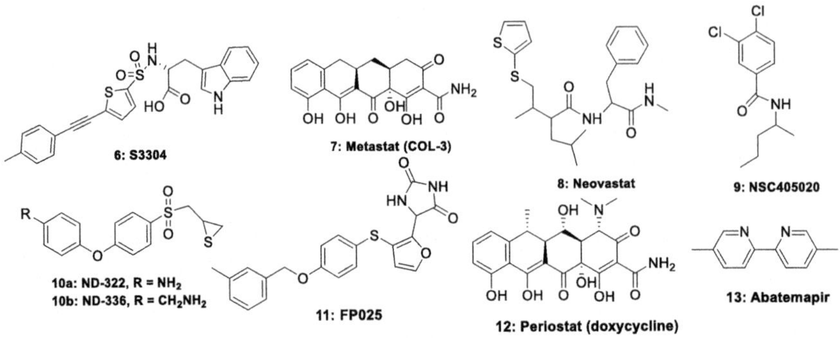

Fig. 4 MMPIs inhibitors **6–13** investigated in various clinical studies.

3. Opportunities

Considering the knowledge accumulated in the last two decades on the roles, expression patterns, functional aspects and regulation of MMP activities, as well as in the design of much more selective and less toxic inhibitors, these enzymes continue to attract attention for their targeting in a variety of diseases in need of novel pharmacological treatments and innovative therapeutic approaches.

3.1 Cancer

Several compounds belonging to various chemical classes, different from the hydroxamates, of types **6–9** (Fig. 4), entered clinical trials as anticancer agents, more recently, with alternate results which will be mentioned in the next paragraphs.

S-3304, compound **6**, is a d-tryptofan derivative possessing a long hydrophobic chain (*p*-tolylethynyl-thiophenylsulfonamide) which accommodates in the S1' pocket of the enzyme, conferring high MMP9 selectivity to this inhibitor [51,52]. The compound showed a good pharmacokinetic and safety profile [51] and entered Phase I clinical trials for the treatment of patients with advanced solid tumors [52]. However, its efficacy was quite low and the clinical trial was terminated, presumably, because the efficacy of this type of compound, as mentioned earlier, might be seen in patients at initial stages of the disease.

Metastat, or COL-3 (compound **7** in Fig. 4) is a chemically modified tetracycline derivative which also acts as MMPI, presumably binding to the zinc ion from the enzyme active site by coordinating with the neighboring

$CONH_2$ and OH functionalities [1–4]. Several clinical trials were done with this orally active compound [53] for various types of tumors, such as high-grade gliomas [54], soft tissue sarcomas [55], and Kaposi's sarcoma [56]. The efficacy was not excellent, except for the last type of tumor, for which COL-3, administered at 50 mg/d, was active and well tolerated in the patients with AIDS-related Kaposi's sarcoma [56]. Metastat is thus considered a promising agent for the treatment of this opportunistic neoplasm [56]. More recently, a synergistic effect of COL-3 and paclitaxel has been observed, leading to enhanced anti–proliferative and pro–apoptotic effects in breast cancer cells [57]. COL-3 was also shown to exert strong anti-proliferative and pro-apoptotic effects in concentrations achievable in human plasma for the management of acute myeloid leukemia [58].

Neovastat, compound **8** in Fig. 4, was isolated from shark cartilage, being shown to possess broad activity as an MMP inhibitor and antiangiogenic properties [59,60]. The drug, alone or in combination with other agents, was shown to possess potent antiangiogenic properties, decreasing lung metastases formation in the Lewis lung carcinoma model in mice [59], also possessing bone antimetastatic effects [61]. It showed some efficacy also in patients with non-small cell lung cancer at stages IIIA/B [62,63] and in a Phase II clinical trials in patients with early relapse or refractory multiple myeloma [62]. The Food and Drug Administration (FDA) granted the status of orphan drug to neovastat in 2002 [64].

An interesting compound is NDC405020 (**9** in Fig. 4), a potent MMP14 inhibitor which does not bind to the metal ion from the enzyme active site [65]. Being an exosite inhibitor, which binds to the hemopexin-like domain of MMP14, it does not inhibit other isoforms but by blocking the collagenolytic activity of MMP14 shows potent inhibitory effects on the MCF7-β3/MT xenograft tumor model in mice [66]. More recently, the compound was also shown to be effective in suppressing thyroid carcinoma cell migration and invasion [67]. MMP14 was demonstrated to act in concert with another anticancer drug target, carbonic anhydrase IX (CAIX), being demonstrated that CA IX enhances MMP14-mediated collagen degradation by directly contributing H^+ ions required for the catalytic activity of MMP14 [68]. Thus, compounds that would inhibit CA IX and MMP14 might have interesting anticancer activity. At the moment, only dual inhibitors of CA II and MMP9 have been reported [69].

ND-322, compound **10a** (Fig. 4) is a mechanism based, suicide inhibitor that initially slowly coordinates to the zinc ion from the active site through its thiirane sulfur, followed by a nucleophilic attack from the carboxylate of a neighbouting glutamic acid residue to the 3-membrede ring, which generates a thiolate, strongly coordinating to the zinc ion and potently inhibiting thus the enzyme [70,71]. ND-322, a potent MMP2 and -14 inhibitor, was shown to reduce melanoma cell growth, migration, and invasion, and to delay metastatic dissemination of this skin tumor [72].

Although none of these compounds were approved as anticancer agents, compared to the first generation inhibitors they showed better efficacy and much fewer side effects. This is probably due to the fact that they incorporate non-hydroxamate ZBGs (the carboxylate moiety in **6**, the chelating carboxamido-enol moiety of the tetracycline **7**, whereas **8** and **9** do not possess obvious such moieties and presumably bind in allosteric sites or exosites. Thus, significant but not enough progress has been achieved with MMPIs of non-hydroxamate type as potential anticancer agents, and presumably in the future more effective and less toxic such derivatives may lead to a breakthrough in the field.

3.2 Inflammatory diseases, sepsis, arthritis, asthma, chronic obstructive pulmonary disorder and other conditions

As outlined earlier in the chapter, there is an important interplay between many MMPs activity and chemokines, cytokines and various other signaling molecules involved in inflammation [1,2,7]. MMPs are connected with both detrimental and beneficial roles in inflammation, including a range of conditions such as cancer, asthma, endotoxin shock and arthritis [7]. In sepsis and endotoxemia, the following isoforms are involved: MMP1, -3, -7, -8, -9, -10, -12, -13 and -14 [2,7]. Their role is variable, but for example MMP1 disrupts the endothelial barrier, induces intravascular coagulation, lung vascular permeability and enhances the cytokine storms by activation of the protease-activated receptor 1 (PAR1) in a murine model of sepsis [7]. MMP3 contributes to the disruption of the blood–brain barrier, whereas high MMP9 levels correlate with decreased survival in humans under septic shock [7]. In arthritis, the following isoforms are involved: MMP1-3, -8-10 and -13 [7]. MMP1, -3, -8 and -9 mediate irreversible joint destruction and their levels were higher in the synovial fluid of rheumatoid arthritis patients compared to controls [7,73–75]. However, first-generation MMPIs of the hydroxamate type, such as cipemastat were ineffective in several clinical trials for the management of rheumatoid arthritis and osteoarthritis [76,77]. This hydroxamate, also known as Trocade, was, on the other hand, shown to lower mortality and brain injury in an animal model of experimental pneumococcal meningitis [78]. In asthma, MMP2, -8 and 9 seem to be involved in the regulation of inflammatory cell apoptosis [7].

FP-025, compound **11** (Fig. 4), also known as aderamastat, is an MMP12 inhibitor under development for respiratory conditions, such as asthma, chronic inflammatory airway diseases, pulmonary fibrosis and chronic obstructive pulmonary disorder (COPD) [79–81]. It has a hydanthoine moiety as ZBG and a long, lipophilic benzyloxyphenyl-thioether moiety which presumably binds the S1' pocket, explaining its high selectivity for inhibiting MMP12 [3]. Two clinical trials with this compound are ongoing (NCT02238834 (Study I); NCT03304964 (Study II) for patients with asthma, COPD, and pulmonary fibrosis [79].

Recently, the use of MMPIs in the management of tuberculosis, in combination with antitubercular drugs, has been proposed, considering the role of these proteases in inflammatory tissue damage within the lungs, which helps the *Mycobacterium tuberculosis* to promote and disseminate the infection [82]. A recent clinical trial showed the efficacy of doxycycline **12**, in combination with antitubercular drugs, in reducing the severity of the disease, without significant side effects [83].

Brain inflammation, including brain/hypoxic ischemia, traumatic brain injury, pneumococcal meningitis, and middle cerebral artery occlusion are other conditions in which different MMPs are involved, such as MMP2 and -9 [1–3]. The use of various inhibitors (of the first generation, such as batimastat **1** but also newer compounds) has thus been proposed and in some cases also showed efficacy [84,85]. The broad-spectrum tetracycline inhibitors COL-3 (**7**) and doxycycline (compound **12**), an approved, clinically used agent for the management of periodontitis (see next section), seem to be efficacious in various such conditions, e.g., acute intracerebral hemorrhage [53,86], focal/global cerebral ischemia [87–89]. Doxycycline was also shown to be effective in suppressing MMP9 levels in brain arteriovenous malformations [90].

Other studies demonstrated that doxycycline **12** improves the proteolytic balance in abdominal aortic aneurysm, affecting the aortic wall neutrophil content, providing a rationale for the use of this drug in other conditions characterized by a massive neutrophil influx, such as Kawasaki disease and Behçet disease [91]. Doxycycline was also shown to prevent or reduce the risk of secondary myocardial infarctions due to a systemic anti-proteolytic and anti-inflammatory shield as a consequence of MMP inhibition [92] or in patients undergoing cardiopulmonary bypass [93].

Inflammatory bowel diseases (IBD) are among others characterized by increased levels of MMPs, such as MMP2 and -9 [1,94], and the use of selective MMPI for these isoforms has been proposed as a therapeutic option for IBD, as well as Crohn's disease [94–97]. Doxycycline **12** was the most used agent to date also for the management of IBD [95–97].

ND-366, compound **10b** (Fig. 4) is a mechanism-based, MMP9 inhibitor (of the same class as **10a** discussed earlier in the chapter) in clinical development for the treatment of diabetic foot ulcers (DFUs) [98–100]. Selective inhibition of MMP-9 with ND-366 lowered inflammation associated with DFU, reduced reactive oxygen species (ROS), and increased angiogenesis, without affecting MMP8, an isoform involved in the natural repair mechanisms [101].

3.3 Periodontal diseases

Periostat, doxycycline **12** was the first MMPI to be approved by FDA (in 1998, and by EMEA in Europe, few years later), as a drug for the management of periodontal disease, acting through inhibition of MMPs (gelatinases MMP2 and -9 and enamelysin, MMP20) found in high levels in patients with dental diseases, their inhibition being associated with periodontal wound repair [102–106]. It should be noted that sub-antimicrobial dosages of **12** are used in the management of periodontitis, and for this reason the long-term usage is possible without negative side-effects associated with the drug, e.g., gastrointestinal disturbances or development of tetracycline-resistant microorganisms [103].

3.4 Pediculocides

Abametapir **13** (Fig. 4) is a bipyridyl derivative possessing potent metal chelating properties and acting as pan-inhibitor of many zinc enzymes, including MMPs involved in lice eggs development, hatching, and adult survival [107,108]. The drug was approved by the FDA in 2020 as a topical pediculocide, as it possesses a direct ovicidal activity inhibiting 100 percent of treated louse eggs from hatching and needs only a single application [107]. Although possessing such a simple chemical structure, the drug is effective and resolves a stringent issue, since significant drug resistance to other pediculocides has been registered [107–109]. Abametapir, in combination with flavesone, seems to be highly effective also for scabies treatment [110].

4. Conclusions

The involvement of MMPs in a huge number of physiologic processes connected not only with ECM remodeling but with the processing of many chemokines, cytokines, growth/transcription factors, other proteases, etc., made the translation from bench to bedside quite challenging for the inhibitors of these enzymes. The main challenges were connected with the large number of isoforms (23 MMPs are present in the human genome and 24 in the mouse genome due to a duplication in MMP1) and their rather similar active site architecture around the catalytic zinc ion, cross reactivity of the first generations MMPIs with other metalloenzymes, such as the ADAMs, ADAMTs, CAs, HDACs, etc., toxicity and metabolic instability of these inhibitors, mainly those based on the hydroxamate ZBG. However, a deeper understanding of the genetics, biochemistry, expression patterns and regulation of these enzymes over the last decade has allowed for significant progress in the field, correlated also with the design of novel inhibitors, with less toxicity, and a much higher selectivity for the target over the anti-target such enzymes. Although not enough progress has been registered in anticancer trials of MMPIs, several promising clinical trials related to inflammatory conditions, such as sepsis, arthritis, asthma, COPD and other similar conditions, or in some bacterial infections (e.g., *Mycobacterium tuberculosis*) allows some optimism for the field. Correlated with the success of the modified tetracyclines in the management of periodontitis (doxycycline, the first MMPI to be approved for clinical use, but hopefully, some of its derivatives may have the same fate in the future) and the recent approval of abametapir for the management of lice infestation, the future of the MMP and their inhibitors research looks bright. We estimate that more selective drugs, better planned clinical trials and a better understanding of the biology/pharmacology of these enzymes may lead to breakthroughs also in other fields, such as cancer, inflammation and other pathologies.

References

[1] L.G.N. de Almeida, H. Thode, Y. Eslambolchi, S. Chopra, D. Young, S. Gill, L. Devel, A. Dufour, Matrix metalloproteinases: from molecular mechanisms to physiology, pathophysiology, and pharmacology, Pharmacol. Rev. 74 (3) (2022) 712–768.

[2] R.E. Vandenbroucke, C. Libert, Is there new hope for therapeutic matrix metalloproteinase inhibition? Nat. Rev. Drug Discov. 13 (12) (2014) 904–927.

[3] A. Trabocchi, E. Lenci, Matrix metalloproteases, in: C.T. Supuran, W.A. Donald (Eds.), Metalloenzymes—From Bench to Bedside, Elsevier—Academic Press, London, UK, 2024, pp. 197–206.

[4] N. Das, C. Benko, S.E Gill, A. Dufour, The pharmacological TAILS of matrix metalloproteinases and their inhibitors, Pharmaceuticals (Basel) 14 (1) (2020) 31.
[5] A. Dufour, C.M. Overall, Missing the target: matrix metalloproteinase antitargets in inflammation and cancer, Trends Pharmacol. Sci. 34 (4) (2013) 233–242.
[6] E. Lenci, L. Cosottini, A. Trabocchi, Novel matrix metalloproteinase inhibitors: an updated patent review (2014–2020), Expert Opin. Ther. Pat. 31 (6) (2021) 509–523.
[7] A. Dufour, Degradomics of matrix metalloproteinases in inflammatory diseases, Front. Biosci. 7 (1) (2015) 150–167.
[8] S. Banerjee, S.K. Baidya, N. Adhikari, T. Jha, An updated patent review of matrix metalloproteinase (MMP) inhibitors (2021-present), Expert Opin. Ther. Pat. 33 (10) (2023) 631–649.
[9] M. Whittaker, C.D. Floyd, P. Brown, A.J. Gearing, Design and therapeutic application of matrix metalloproteinase inhibitors, Chem. Rev. 99 (9) (1999) 2735–2776.
[10] A. Dufour, N.S. Sampson, J. Li, C. Kuscu, R.C. Rizzo, J.L. Deleon, J. Zhi, N. Jaber, E. Liu, S. Zucker, J. Cao, Small-molecule anticancer compounds selectively target the hemopexin domain of matrix metalloproteinase-9, Cancer Res. 71 (14) (2011) 4977–4988.
[11] C.T. Supuran, Carbonic anhydrases: novel therapeutic applications for inhibitors and activators, Nat. Rev. Drug Discov. 7 (2008) 168–181.
[12] C.T. Supuran, A simple yet multifaceted 90 years old, evergreen enzyme: carbonic anhydrase, its inhibition and activation, Bioorg. Med. Chem. Lett. 93 (2023) 129411.
[13] C.T. Supuran, Hydroxamates as carbonic anhydrase inhibitors, in: S.P. Gupta (Ed.), Hydroxamic Acids: A Unique Family of Chemicals with Multiple Biological Activities, Springer Verlag, Heidelberg, 2013, pp. 55–69.
[14] H. Laronha, I. Carpinteiro, J. Portugal, A. Azul, M. Polido, K.T. Petrova, M. Salema-Oom, J. Caldeira, Challenges in matrix metalloproteinases inhibition, Biomolecules 10 (5) (2020) 717.
[15] Y. Jin, M.D. Roycik, D.B. Bosco, Q. Cao, M.H. Constantino, M.A. Schwartz, Q.X. Sang, Matrix metalloproteinase inhibitors based on the 3-mercaptopyrrolidine core, J. Med. Chem. 56 (11) (2013) 4357–4373.
[16] K. Voos, E. Schönauer, A. Alhayek, J. Haupenthal, A. Andreas, R. Müller, R.W. Hartmann, H. Brandstetter, A.K.H. Hirsch, C. Ducho, Phosphonate as a Stable Zinc-Binding Group for "pathoblocker" inhibitors of clostridial collagenase H (ColH), ChemMedChem 16 (8) (2021) 1257–1267.
[17] R. Reich, A. Hoffman, A. Veerendhar, A. Maresca, A. Innocenti, C.T. Supuran, E. Breuer, Carbamoylphosphonates control tumor cell proliferation and dissemination by simultaneously inhibiting carbonic anhydrase IX and matrix metalloproteinase-2. Toward nontoxic chemotherapy targeting tumor microenvironment, J. Med. Chem. 55 (17) (2012) 7875–7882.
[18] G. Luisi, G. Angelini, C. Gasbarri, A. Laghezza, M. Agamennone, F. Loiodice, C.T. Supuran, C. Campestre, P. Tortorella, Dual targeting of cancer-related human matrix metalloproteinases and carbonic anhydrases by chiral N-(biarylsulfonyl)-phosphonic acids, J. Enzyme Inhib. Med. Chem. 32 (1) (2017) 1260–1264.
[19] , in: C.T. Supuran, W.A. Donald (Eds.), Metalloenzymes—From Bench to Bedside, Elsevier—Academic Press, London, UK, 2024, pp. 1–615.
[20] F. Grams, M. Crimmin, L. Hinnes, P. Huxley, M. Pieper, H. Tschesche, W. Bode, Structure determination and analysis of human neutrophil collagenase complexed with a hydroxamate inhibitor, Biochemistry 34 (43) (1995) 14012–14020.
[21] H. Brandstetter, F. Grams, D. Glitz, A. Lang, R. Huber, W. Bode, H.W. Krell, R.A. Engh, The 1.8-A crystal structure of a matrix metalloproteinase 8-barbiturate inhibitor complex reveals a previously unobserved mechanism for collagenase substrate recognition, J. Biol. Chem. 276 (20) (2001) 17405–17412.

[22] S. Santamaria, E. Nuti, G. Cercignani, G. La Regina, R. Silvestri, C.T. Supuran, A. Rossello, Kinetic characterization of 4,4′-biphenylsulfonamides as selective non-zinc binding MMP inhibitors, J. Enzyme Inhib. Med. Chem. 30 (6) (2015) 947–954.
[23] P. Begines, J.G. Fernández-Bolaños, Ó. López, An updated patent review of acetylcholinesterase inhibitors for the treatment of Alzheimer's disease (2021-present), Expert Opin. Ther. Pat. 36 (2) (2026) 159–190.
[24] A. Chatonnet, N. Lenfant, P. Marchot, M.E. Selkirk, Natural genomic amplification of cholinesterase genes in animals, J. Neurochem. 142 (Suppl 2) (2017) 73–81.
[25] M.A. Shaldam, S. Carradori, M. Balaha, P. Guglielmi, F. Diomede, I. D'Agostino, H.O. Tawfik, Patents involving monoamine oxidase (MAO): a comprehensive update (2022-2025) on its inhibitors and applications, Expert Opin. Ther. Pat. 36 (4) (2026) 363–387.
[26] M.N. Romanelli, Cyclooxygenase, in: C.T. Supuran, W.A. Donald (Eds.), Metalloenzymes from Bench to Bedside, Elsevier/Academic Press, London, 2024, pp. 431–447.
[27] C.T. Supuran, Multi- and polypharmacology of carbonic anhydrase inhibitors, Pharmacol. Rev. 77 (1) (2025) 100004.
[28] S. Giovannuzzi, C.T. Supuran, Human carbonic anhydrase modulators: the past, present, and future, Trends Pharmacol. Sci. 46 (9) (2025) 836–847.
[29] A. Innocenti, C.T. Supuran, Paraoxon, 4-nitrophenyl phosphate and acetate are substrates of α- but not of β-, γ- and ζ-carbonic anhydrases, Bioorg. Med. Chem. Lett. 20 (21) (2010) 6208–6212.
[30] A. Innocenti, A. Scozzafava, S. Parkkila, L. Puccetti, G. De Simone, C.T. Supuran, Investigations of the esterase, phosphatase, and sulfatase activities of the cytosolic mammalian carbonic anhydrase isoforms I, II, and XIII with 4-nitrophenyl esters as substrates, Bioorg. Med. Chem. Lett. 18 (7) (2008) 2267–2271.
[31] K. D'Ambrosio, A. Di Fiore, V. Alterio, E. Langella, S.M. Monti, C.T Supuran, G. De Simone, Multiple binding modes of inhibitors to human carbonic anhydrases: an update on the design of isoform-specific modulators of activity, Chem. Rev. 125 (1) (2025) 150–222.
[32] D. Cuffaro, S.D. Scilabra, D.P. Spanò, M. Calligaris, E. Nuti, A. Rossello, A disintegrin and metalloproteinases (ADAMs) and tumor necrosis factor-alpha converting enzyme (TACE), in: C.T. Supuran, W.A. Donald (Eds.), Metalloenzymes—From Bench to Bedside, Elsevier—Academic Press, London, UK, 2024, pp. 207–237.
[33] F. Lu, M. Fournier, M. Freeman, The cytoplasmic domain of the pseudoprotease iRhom2 mediates distinct signaling mechanisms to control activation of the cell surface protease ADAM17, J. Biol. Chem. 301 (10) (2025) 110643.
[34] Y. Yuan, P. Guo, Y. Song, Z. Yu, B. Song, The ADAMTS family: from extracellular matrix proteases to orchestrators of fibrosis, Cell Commun. Signal. 24 (1) (2026) 187, doi:10.1186/s12964-026-02743-0.
[35] A. Scozzafava, C.T. Supuran, Carbonic anhydrase and matrix metalloproteinase inhibitors: sulfonylated amino acid hydroxamates with MMP inhibitory properties act as efficient inhibitors of CA isozymes I, II, and IV, and N-hydroxysulfonamides inhibit both these zinc enzymes, J. Med. Chem. 43 (20) (2000) 3677–3687.
[36] A. Dufour, C.M. Overall, Subtracting matrix out of the equation: new key roles of matrix metalloproteinases in innate immunity and disease, in: I. Sagi, J.P. Gaffney (Eds.), Matrix Metalloproteinase Biology, John Wiley & Sons, Inc, Hoboken NJ, 2015, pp. 131–152.
[37] F. Li, J. Zhi, R. Zhao, Y. Sun, H. Wen, H. Cai, W. Chen, X. Jiang, R. Bai, Discovery of matrix metalloproteinase inhibitors as anti-skin photoaging agents, Eur. J. Med. Chem. 267 (2024) 116152.

[38] A.H.A. Mohamed-Ahmed, A. Lockwood, H. Li, M. Bailly, P.T. Khaw, S. Brocchini, An ilomastat-CD eye drop formulation to treat ocular scarring, Invest. Ophthalmol. Vis. Sci. 58 (9) (2017) 3425–3431.

[39] G. Bergers, K. Javaherian, K.M. Lo, J. Folkman, D. Hanahan, Effects of angiogenesis inhibitors on multistage carcinogenesis in mice, Science 284 (5415) (1999) 808–812.

[40] D.R. Shalinsky, J. Brekken, H. Zou, C.D. McDermott, P. Forsyth, D. Edwards, S. Margosiak, S. Bender, G. Truitt, A. Wood, N.M. Varki, K. Appelt, Broad antitumor and antiangiogenic activities of AG3340, a potent and selective MMP inhibitor undergoing advanced oncology clinical trials, Ann. N. Y. Acad. Sci. 878 (1999) 236–270.

[41] Z. Xu, J. Liu, Current scenario of hydroxamic acid derivatives with in vivo antitumor therapeutic potential, Arch. Pharm. (Weinheim) 358 (12) (2025) e70178.

[42] S. Shen, A.P. Kozikowski, Why hydroxamates may not be the best histone deacetylase inhibitors—what some may have forgotten or would rather forget? ChemMedChem 11 (1) (2016) 15–21.

[43] A. Akdemir, Cytochrome P450 (inhibitors for the metabolism of drugs), in: C.T. Supuran, W.A. Donald (Eds.), Metalloenzymes—From Bench to Bedside, Elsevier—Academic Press, London, UK, 2024, pp. 449–458.

[44] M. Ferraroni, Iron enzymes, nonheme containing, in: C.T. Supuran, W.A. Donald (Eds.), Metalloenzymes—From Bench to Bedside, Elsevier—Academic Press, London, UK, 2024, pp. 467–484.

[45] J.H. Weisburger, E.K. Weisburger, Biochemical formation and pharmacological, toxicological, and pathological properties of hydroxylamines and hydroxamic acids, Pharmacol. Rev. 25 (1) (1973) 1–66.

[46] W.C. Powell, J.D. Knox, M. Navre, T.M. Grogan, J. Kittelson, R.B. Nagle, G.T. Bowden, Expression of the metalloproteinase matrilysin in DU-145 cells increases their invasive potential in severe combined immunodeficient mice, Cancer Res. 53 (2) (1993) 417–422.

[47] X. Wang, X. Fu, P.D. Brown, M.J. Crimmin, R.M. Hoffman, Matrix metalloproteinase inhibitor BB-94 (batimastat) inhibits human colon tumor growth and spread in a patient-like orthotopic model in nude mice, Cancer Res. 54 (17) (1994) 4726–4728.

[48] G.W. Sledge Jr., M. Qulali, R. Goulet, E.A. Bone, R. Fife, Effect of matrix metalloproteinase inhibitor batimastat on breast cancer regrowth and metastasis in athymic mice, J. Natl. Cancer Inst. 87 (20) (1995) 1546–1550.

[49] J.A. Low, M.D. Johnson, E.A. Bone, R.B. Dickson, The matrix metalloproteinase inhibitor batimastat (BB-94) retards human breast cancer solid tumor growth but not ascites formation in nude mice, Clin. Cancer Res. 2 (7) (1996) 1207–1214.

[50] O. Santos, C.D. McDermott, R.G. Daniels, K. Appelt, Rodent pharmacokinetic and anti-tumor efficacy studies with a series of synthetic inhibitors of matrix metalloproteinases, Clin. Exp. Metastasis 15 (5) (1997) 499–508.

[51] T.G. Mant, D. Bradford, D.M. Amin, J. Pisupati, Y. Kambayashi, Y. Yano, K. Tanaka, T. Yamada-Sawada, Pharmacokinetics and safety assessments of high-dose and 4-week treatment with S-3304, a novel matrix metalloproteinase inhibitor, in healthy volunteers, Br. J. Clin. Pharmacol. 63 (5) (2007) 512–526.

[52] A.A. Chiappori, S.G. Eckhardt, R. Bukowski, D.M. Sullivan, M. Ikeda, Y. Yano, T. Yamada-Sawada, Y. Kambayashi, K. Tanaka, M.M. Javle, T. Mekhail, C.L. O'bryant, P.J. Creaven, A phase I pharmacokinetic and pharmacodynamic study of s-3304, a novel matrix metalloproteinase inhibitor, in patients with advanced and refractory solid tumors, Clin. Cancer Res. 13 (7) (2007) 2091–2099.

[53] S. Syed, C. Takimoto, M. Hidalgo, J. Rizzo, J.G. Kuhn, L.A. Hammond, G. Schwartz, A. Tolcher, A. Patnaik, S.G. Eckhardt, E.K. Rowinsky, A phase I and pharmacokinetic study of Col-3 (Metastat), an oral tetracycline derivative with potent matrix metalloproteinase and antitumor properties, Clin. Cancer Res. 10 (19) (2004) 6512–6521.

[54] M.A. Rudek, P. New, T. Mikkelsen, S. Phuphanich, J.B. Alavi, L.B. Nabors, S. Piantadosi, J.D. Fisher, S.A. Grossman, Phase I and pharmacokinetic study of COL-3 in patients with recurrent high-grade gliomas, J. Neurooncol. 105 (2) (2011) 375–381.
[55] Q.S. Chu, B. Forouzesh, S. Syed, M. Mita, G. Schwartz, J. Cooper, J. Curtright, E.K. Rowinsky, A phase II and pharmacological study of the matrix metalloproteinase inhibitor (MMPI) COL-3 in patients with advanced soft tissue sarcomas, Invest. New Drugs 25 (4) (2007) 359–367.
[56] B.J. Dezube, S.E. Krown, J.Y. Lee, K.S. Bauer, D.M. Aboulafia, Randomized phase II trial of matrix metalloproteinase inhibitor COL-3 in AIDS-related Kaposi's sarcoma: an AIDS Malignancy Consortium Study, J. Clin. Oncol. 24 (9) (2006) 1389–1394.
[57] R.M. El-Tabba', P. Mathew, W. Masocha, M.A. Khajah, COL–3 enhances the anti–proliferative and pro–apoptotic effects of paclitaxel in breast cancer cells, Oncol. Rep. 41 (1) (2019) 630–642.
[58] H. Song, M. Fares, K.R. Maguire, A. Sidén, Z. Potácová, Cytotoxic effects of tetracycline analogues (doxycycline, minocycline and COL-3) in acute myeloid leukemia HL-60 cells, PLoS One 9 (12) (2014) e114457.
[59] E. Dupont, P. Falardeau, S.A. Mousa, V. Dimitriadou, M.C. Pepin, T. Wang, A.-J. Ma, Antiangiogenic and antimetastatic properties of Neovastat (AE-941), an orally active extract derived from cartilage tissue, Clin. Exp. Metastasis 19 (2) (2002) 145–153.
[60] S. Liekens, E. De Clercq, J. Neyts, Angiogenesis: regulators and clinical applications, Biochem. Pharmacol. 61 (3) (2001) 253–270.
[61] M.H. Weber, J. Lee, F.W. Orr, The effect of Neovastat (AE-941) on an experimental metastatic bone tumor model, Int. J. Oncol. 20 (2) (2002) 299–303.
[62] P. Falardeau, P. Champagne, P. Poyet, C. Hariton, E. Dupont, Neovastat, a naturally occurring multifunctional antiangiogenic drug, in phase III clinical trials, Semin Oncol. 28 (6) (2001) 620–625.
[63] C. Lu, J.J. Lee, R. Komaki, R.S. Herbst, L. Feng, W.K. Evans, H. Choy, P. Desjardins, B.T. Esparaz, M.T. Truong, S. Saxman, J. Kelaghan, A. Bleyer, M.J. Fisch, Chemoradiotherapy with or without AE-941 in stage III non-small cell lung cancer: a randomized phase III trial, J. Natl. Cancer Inst. 102 (12) (2010) 859–865.
[64] No authors listed, FDA grants orphan-drug status to Aeterna's Neovastat for kidney cancer, Expert Rev. Anticancer Ther. 2 (6) (2002) 618.
[65] A.G. Remacle, V.S. Golubkov, S.A. Shiryaev, R. Dahl, J.L. Stebbins, A.V. Chernov, A.V. Cheltsov, M. Pellecchia, A.Y. Strongin, Novel MT1-MMP small-molecule inhibitors based on insights into hemopexin domain function in tumor growth, Cancer Res. 72 (9) (2012) 2339–2349.
[66] K. Zarrabi, A. Dufour, J. Li, C. Kuscu, A. Pulkoski-Gross, J. Zhi, Y. Hu, N.S. Sampson, S. Zucker, J. Cao, Inhibition of matrix metalloproteinase 14 (MMP-14)-mediated cancer cell migration, J. Biol. Chem. 286 (38) (2011) 33167–33177.
[67] D. Rocco, V. Marotta, D. Palumbo, M. Vitale, Inhibition of metalloproteinases-2, -9, and -14 suppresses papillary thyroid carcinoma cell migration and invasion, Int. J. Mol. Sci. 26 (16) (2025) 7956.
[68] M. Swayampakula, P.C. McDonald, M. Vallejo, E. Coyaud, S.C. Chafe, A. Westerback, G. Venkateswaran, J. Shankar, G. Gao, E.M.N. Laurent, Y. Lou, K.L. Bennewith, C.T. Supuran, I.R. Nabi, B. Raught, S. Dedhar, The interactome of metabolic enzyme carbonic anhydrase IX reveals novel roles in tumor cell migration and invadopodia/MMP14-mediated invasion, Oncogene 36 (45) (2017) 6244–6261.
[69] E. Lenci, A. Angeli, L. Calugi, R. Innocenti, F. Carta, C.T. Supuran, A. Trabocchi, Multitargeting application of proline-derived peptidomimetics addressing cancer-related human matrix metalloproteinase 9 and carbonic anhydrase II, Eur. J. Med. Chem. 214 (2021) 113260.

[70] M. Ikejiri, M.M. Bernardo, R.D. Bonfil, M. Toth, M. Chang, R. Fridman, S. Mobashery, Potent mechanism-based inhibitors for matrix metalloproteinases, J. Biol. Chem. 280 (40) (2005) 33992–34002.
[71] M.M. Bernardo, S. Brown, Z.H. Li, R. Fridman, S. Mobashery, Design, synthesis, and characterization of potent, slow-binding inhibitors that are selective for gelatinases, J. Biol. Chem. 277 (13) (2002) 11201–11207.
[72] E. Zhang, V. Thakur, B. Bedogni, Matrix metalloproteinase inhibition in melanoma, Exp. Dermatol. 35 (1) (2026) e70203.
[73] I. Tchetverikov, H.K. Ronday, B. Van El, G.H. Kiers, N. Verzijl, J.M. TeKoppele, T.W. Huizinga, J. DeGroot, R. Hanemaaijer, MMP profile in paired serum and synovial fluid samples of patients with rheumatoid arthritis, Ann. Rheum. Dis. 63 (7) (2004) 881–883.
[74] P.S. Burrage, K.S. Mix, C.E. Brinckerhoff, Matrix metalloproteinases: role in arthritis, Front. Biosci. 11 (2006) 529–543.
[75] B. Grillet, R.V.S. Pereira, J. Van Damme, A. Abu El-Asrar, P. Proost, G. Opdenakker, Matrix metalloproteinases in arthritis: towards precision medicine, Nat. Rev. Rheumatol. 19 (6) (2023) 363–377.
[76] E.J. Lewis, J. Bishop, K.M. Bottomley, D. Bradshaw, M. Brewster, M.J. Broadhurst, P.A. Brown, J.M. Budd, L. Elliott, A.K. Greenham, W.H. Johnson, J.S. Nixon, F. Rose, B. Sutton, K. Wilson, Ro 32-3555, an orally active collagenase inhibitor, prevents cartilage breakdown in vitro and in vivo, Br. J. Pharmacol. 121 (3) (1997) 540–546.
[77] F.J. Hemmings, M. Farhan, J. Rowland, L. Banken, R. Jain, Tolerability and pharmacokinetics of the collagenase-selective inhibitor Trocade™ in patients with rheumatoid arthritis, Rheumatology (Oxford) 40 (5) (2001) 537–543.
[78] F.D. Liechti, D. Grandgirard, D. Leppert, S.L. Leib, Matrix metalloproteinase inhibition lowers mortality and brain injury in experimental pneumococcal meningitis, Infect. Immun. 82 (4) (2014) 1710–1718.
[79] K. Abd-Elaziz, C. Voors-Pette, K.L. Wang, S. Pan, Y. Lee, J. Mao, Y. Li, B. Chien, D. Lau, Z. Diamant, First-in-man safety, tolerability, and pharmacokinetics of a novel and highly selective inhibitor of matrix metalloproteinase-12, FP-025: results from two randomized studies in healthy subjects, Clin. Drug Investig. 41 (1) (2021) 65–76.
[80] L. Ravanetti, T. Dekker, L. Guo, A. Dijkhuis, B.S. Dierdorp, Z. Diamant, S. Florquin, R. Lutter, Efficacy of FP-025: a novel matrix metalloproteinase-12 (MMP-12) inhibitor in murine allergic asthma, Allergy 78 (2) (2023) 559–562.
[81] K.S. Abd-Elaziz, R. Cheng, J. Chen, H. Maarse, Y. Lee, W. Yang, B. Chien, Z. Diamant, J. Kosterink, D.J. Touw, Validation of a method for the determination of Aderamastat (FP-025) in K2EDTA human plasma by LC-MS/MS, J. Chromatogr. B Analyt. Technol. Biomed. Life Sci. 1245 (2024) 124244.
[82] A. Jhilta, K. Jadhav, R. Singh, E. Ray, A. Kumar, A.K. Singh, R.K. Verma, Breaking the cycle: matrix metalloproteinase inhibitors as an alternative approach in managing tuberculosis pathogenesis and progression, ACS Infect. Dis. 10 (8) (2024) 2567–2583.
[83] Q.H. Miow, A.F. Vallejo, Y. Wang, J.M. Hong, C. Bai, F.S. Teo, A.D. Wang, H.R. Loh, T.Z. Tan, Y. Ding, H.W. She, S.H. Gan, N.I. Paton, J. Lum, A. Tay, C.B. Chee, P.A. Tambyah, M.E. Polak, Y.T. Wang, A. Singhal, P.T. Elkington, J.S. Friedland, C.W. Ong, Doxycycline host-directed therapy in human pulmonary tuberculosis, J. Clin. Invest. 131 (15) (2021) e141895.
[84] T. Pfefferkorn, G.A. Rosenberg, Closure of the blood-brain barrier by matrix metalloproteinase inhibition reduces rtPA-mediated mortality in cerebral ischemia with delayed reperfusion, Stroke 34 (8) (2003) 2025–2030.
[85] B.Q. Zhao, S. Wang, H.Y. Kim, H. Storrie, B.R. Rosen, D.J. Mooney, X. Wang, E.H. Lo, Role of matrix metalloproteinases in delayed cortical responses after stroke, Nat. Med. 12 (4) (2006) 441–445.

[86] J.C. Copin, P. Merlani, T. Sugawara, P.H. Chan, Y. Gasche, Delayed matrix metalloproteinase inhibition reduces intracerebral hemorrhage after embolic stroke in rats, Exp. Neurol. 213 (1) (2008) 196–201.
[87] D. Burggraf, A. Trinkl, M. Dichgans, G.F. Hamann, Doxycycline inhibits MMPs via modulation of plasminogen activators in focal cerebral ischemia, Neurobiol. Dis. 25 (3) (2007) 506–513.
[88] H. Lee, J.W. Park, S.P. Kim, E.H. Lo, S.R. Lee, Doxycycline inhibits matrix metalloproteinase-9 and laminin degradation after transient global cerebral ischemia, Neurobiol. Dis. 34 (2) (2009) 189–198.
[89] P.W. Pires, C.T. Rogers, J.L. McClain, H.S. Garver, G.D. Fink, A.M. Dorrance, Doxycycline, a matrix metalloprotease inhibitor, reduces vascular remodeling and damage after cerebral ischemia in stroke-prone spontaneously hypertensive rats, Am. J. Physiol. Heart Circ. Physiol. 301 (1) (2011) H87–H97.
[90] T. Hashimoto, M.M. Matsumoto, J.F. Li, M.T. Lawton, W.L. Young, University of California, San Francisco, BAVM Study Group, Suppression of MMP-9 by doxycycline in brain arteriovenous malformations, BMC Neurol. 5 (1) (2005) 1.
[91] H. Abdul-Hussien, R. Hanemaaijer, J.H. Verheijen, J.H. van Bockel, R.H. Geelkerken, J.H. Lindeman, Doxycycline therapy for abdominal aneurysm: improved proteolytic balance through reduced neutrophil content, J. Vasc. Surg. 49 (3) (2009) 741–749.
[92] I. Kormi, H. Alfakry, T. Tervahartiala, P.J. Pussinen, J. Sinisalo, T. Sorsa, The effect of prolonged systemic doxycycline therapy on serum tissue degrading proteinases in coronary bypass patients: a randomized, double-masked, placebo-controlled clinical trial, Inflamm. Res. 63 (5) (2014) 329–334.
[93] W. Gao, F. Fang, T.J. Xia, Y. Zhang, J. Sun, Q. Wu, W. Wang, Doxycycline can reduce glycocalyx shedding by inhibiting matrix metalloproteinases in patients undergoing cardiopulmonary bypass: a randomized controlled trial, Microvasc. Res. 142 (2022) 104381.
[94] M.M. Heimesaat, I.R. Dunay, D. Fuchs, D. Trautmann, A. Fischer, A.A. Kühl, C. Loddenkemper, A. Batra, B. Siegmund, H.W. Krell, S. Bereswill, O. Liesenfeld, Selective gelatinase blockage ameliorates acute DSS colitis, Eur. J. Microbiol. Immunol. (Bp) 1 (3) (2011) 228–236.
[95] T.W. Lee, L. Russell, M. Deng, P.R. Gibson, Association of doxycycline use with the development of gastroenteritis, irritable bowel syndrome and inflammatory bowel disease in Australians deployed abroad, Intern. Med. J. 43 (8) (2013) 919–926.
[96] J. Breton, A. Kastl, N. Hoffmann, R. Rogers, A.B. Grossman, P. Mamula, J.R. Kelsen, R.N. Baldassano, L. Albenberg, Efficacy of combination antibiotic therapy for refractory pediatric inflammatory bowel disease, Inflamm. Bowel Dis. 25 (9) (2019) 1586–1593.
[97] J.M. Rhodes, S. Subramanian, P.K. Flanagan, G.W. Horgan, K. Martin, J. Mansfield, M. Parkes, A. Hart, H. Dallal, T. Iqbal, J. Butterworth, K. Culshaw, C. Probert, Randomized trial of ciprofloxacin doxycycline and hydroxychloroquine versus budesonide in active Crohn's disease, Dig. Dis. Sci. 66 (8) (2021) 2700–2711.
[98] Z. Peng, T.T. Nguyen, W. Song, B. Anderson, W.R. Wolter, V.A. Schroeder, D. Hesek, M. Lee, S. Mobashery, M. Chang, The selective MMP-9 inhibitor (R)-ND-336 alone or in combination with linezolid accelerates wound healing in infected diabetic mice, ACS Pharmacol. Transl. Sci. 4 (1) (2020) 107–117.
[99] T.T. Nguyen, D. Ding, W.R. Wolter, R.L. Pérez, M.M. Champion, K.V. Mahasenan, D. Hesek, M. Lee, V.A. Schroeder, J.I. Jones, E. Lastochkin, M.K. Rose, C.E. Peterson, M.A. Suckow, S. Mobashery, M. Chang, Validation of matrix metalloproteinase-9 (MMP-9) as a novel target for treatment of diabetic foot ulcers in humans and discovery of a potent and selective small-molecule MMP-9 inhibitor that accelerates healing, J. Med. Chem. 61 (2018) 8825–8837.

[100] C.E. Gabriel, T.T. Nguyen, E.M. Gargano, J.F. Fisher, M. Chang, S. Mobashery, Metabolism of the selective matrix metalloproteinase-9 inhibitor (R)-ND-336, ACS Pharmacol. Transl. Sci. 4 (3) (2021) 1204–1213.
[101] M. Chang, Targeting matrix metalloproteinase-9 for therapeutic intervention in diabetic foot ulcers, ACS Pharmacol. Transl. Sci. 7 (10) (2024) 2901–2911.
[102] R. Gapski, J.L. Barr, D.P. Sarment, M.G. Layher, S.S. Socransky, W.V. Giannobile, Effect of systemic matrix metalloproteinase inhibition on periodontal wound repair: a proof of concept trial, J. Periodontol. 75 (3) (2004) 441–452.
[103] G.J. Boelen, L. Boute, J. d'Hoop, M. EzEldeen, I. Lambrichts, G. Opdenakker, Matrix metalloproteinases and inhibitors in dentistry, Clin. Oral Investig. 23 (7) (2019) 2823–2835.
[104] T. Sorsa, L. Tjäderhane, Y.T. Konttinen, A. Lauhio, T. Salo, H.M. Lee, L.M. Golub, D.L. Brown, P. Mäntylä, Matrix metalloproteinases: contribution to pathogenesis, diagnosis and treatment of periodontal inflammation, Ann. Med. 38 (5) (2006) 306–321.
[105] R. Gapski, H. Hasturk, T.E. Van Dyke, R.J. Oringer, S. Wang, T.M. Braun, W.V. Giannobile, Systemic MMP inhibition for periodontal wound repair: results of a multi-centre randomized-controlled clinical trial, J. Clin. Periodontol. 36 (2) (2009) 149–156.
[106] R. Cosgarea, C.A. Ramseier, S. Jepsen, N.B. Arweiler, P.M. Jervøe-Storm, I. Batori-Andronescu, R. Rößler, T. Conrad, S. Eick, A. Sculean, One-year clinical, microbiological and immunological results of local doxycycline or antimicrobial photodynamic therapy for recurrent/persisting periodontal pockets: a randomized clinical trial, Antibiotics (Basel) 11 (6) (2022) 738.
[107] A.D. Woods, C.L. Porter, S.R. Feldman, Abametapir for the treatment of head lice: a drug review, Ann. Pharmacother. 56 (3) (2022) 352–357.
[108] A. Chakraborty, Abametapir: a new solution to an age old problem, Int. J. Trichol. 16 (1–6) (2024) 55–57.
[109] J.M. Clark, New chemistries for the control of human head lice, *Pediculus humanus capitis*: a mini-review, Pestic. Biochem. Physiol. 181 (2022) 105013.
[110] D.D. Fernando, G.R. Samarawickrama, N.A. Nammunige, G. Hartel, V. Bowles, K. Fischer, In vitro assessment of a combination therapy of abametapir and flavesone as a novel topical single-dose scabies treatment, J. Am. Acad. Dermatol. (2026) S0190-9622(26)00058-7, doi:10.1016/j.jaad.2026.01.019.

Printed and bound by CPI Group (UK) Ltd, Croydon, CR0 4YY
24/09/2026
02222378-0003